Statistical Distributions in Scientific Work

Volume 6 – Applications in Physical, Social, and Life Sciences

NATO ADVANCED STUDY INSTITUTES SERIES

Proceedings of the Advanced Study Institute Programme, which aims at the dissemination of advanced knowledge and the formation of contacts among scientists from different countries

The series is published by an international board of publishers in conjunction with NATO Scientific Affairs Division

A	Life Sciences	Plenum Publishing Corporation
B	Physics	London and New York
C	Mathematical and Physical Sciences	D. Reidel Publishing Company Dordrecht, Boston and London
D	Behavioural and Social Sciences	Sijthoff & Noordhoff International Publishers
E	Applied Sciences	Alphen aan den Rijn and Germantown U.S.A.

Series C – Mathematical and Physical Sciences

Volume 79 – Statistical Distributions in Scientific Work

Volume 6 – Applications in Physical, Social, and Life Sciences

Statistical Distributions in Scientific Work

Volume 6 – Applications in Physical, Social, and Life Sciences

Proceedings of the NATO Advanced Study Institute held at the Università degli Studi di Trieste, Trieste, Italy, July 10 – August 1, 1980

edited by

CHARLES TAILLIE
Department of Statistics, The Pennsylvania State University, University Park, Pennsylvania, U.S.A.

GANAPATI P. PATIL
Department of Statistics, The Pennsylvania State University, University Park, Pennsylvania, U.S.A.

and

BRUNO A. BALDESSARI
Instituto di Calcolo delle Probabilità, Facoltà di Scienze Statistische Demografiche e Attuariali, Università degli Studi di Roma, Italy

D. Reidel Publishing Company

Dordrecht : Holland / Boston : U.S.A. / London : England

Published in cooperation with NATO Scientific Affairs Division

Library of Congress Cataloging in Publication Data

NATO Advanced Study Institute (1980: Trieste, Italy)
Statistical distributions in scientific work.

(NATO advanced study institutes series. Series C, Mathematical and physical sciences; v. 79)
Includes bibliographical references and indexes.
Contents: v. 4. Models, structures, and characterizations – v. 5. Inferential problems and properties – v. 6. Applications in physical, social, and life sciences.
1. Distribution (Probability theory)–Congresses. 2. Mathematical statistics–Congresses. I. Taillie, C. II. Patil, Ganapati P. III. Baldessari, Bruno. IV. North Atlantic Treaty Organization. Division of Scientific Affairs. V. Title. VI. Series.
QA273.6.N37 1980 519.5 81-12043
ISBN 90-277-1334-0 (v. 6) AACR2

Published by D. Reidel Publishing Company
P.O. Box 17, 3300 AA Dordrecht, Holland

Sold and distributed in the U.S.A. and Canada
by Kluwer Boston Inc.,
190 Old Derby Street, Hingham, MA 02043, U.S.A.

In all other countries, sold and distributed
by Kluwer Academic Publishers Group,
P.O. Box 322, 3300 AH Dordrecht, Holland

D. Reidel Publishing Company is a member of the Kluwer Group

Printed in The Netherlands

TABLE OF CONTENTS

Statistical Distributions in Scientific Work

COMMITTEE

(Chairman: G. P. Patil)

Arnold, B.
Baldessari, B. A.
Cacoullos, T.
Douglas, J. B.
Engen, S.
Folks, J. L.
Gross, A. J.
Kemp, C. D.
Kotz, S.
Mardia, K. V.
Mosimann, J. E.
Ord, J. K.
Patil, G. P.
Ratnaparkhi, M. V.
Rao, C. R.
Shapiro, S. S.
Stene, J.
Taillie, C.
Tiago de Oliveira, J.
Warren, W. G.
Wise, M. E.

INTERNATIONAL SUMMER SCHOOL ON MODERN STATISTICAL DISTRIBUTION THEORY AND ITS APPLICATIONS (Trieste, Italy, 1980)

Director: B. A. Baldessari
Co-Director: C. Taillie
Host: L. Rondini
Secretary: B. Alles
Scientific Directors: S. Kotz, J. E. Mosimann, J. K. Ord, and G. P. Patil
Local Arrangements: E. Feoli, A. Kostoris (Secretary), S. Orviati, L. Rondini (Chairman), M. Strassoldo, and M. Umani

SPONSORS

NATO Advanced Study Institutes Program
NATO Scientific Affairs Division
International Transfer of Science and Technology, Belgium
Department of Statistics, The Pennsylvania State University
Instituto di Calcolo delle Probabilita, Universita di Roma
Instituto di Statistica, Universita di Trieste
International Statistical Ecology Program
Office of Naval Research, USA
Consiglio Nazionale delle Ricerche, Italy
Regione Autonoma Friuli, Venezia Giulia, Italy
National Institutes of Health, USA
- National Cancer Institute
- National Heart, Lung, and Blood Institute
- Fogarty International Center
- Division of Computer Research and Technology

The Participants and Their Home Institutions and Organizations

PARTICIPANTS

Ahmad, M., Saudi Arabia
Aitchison, J., Hong Kong
Akdeniz, F., Turkey
Al-Ani, S., Iraq
Alles, B., Pennsylvania
Al-Zaid, A. H., England
Ammassari, P., Italy
Arnold, B. C., California
Bajusz, B. A., Pennsylvania
Baldessari, B. A., Italy
Bargmann, R. E., Georgia
Basu, A. P., Missouri
Battaglia, F., Italy
Beirlant, J., Belgium
Blaesild, P., Denmark
Block, H. W., Pennsylvania
Blumenthal, S., Illinois
Bochicchio, A., Italy
Bologna, S., Italy
Boswell, M. T., Pennsylvania
Braumann, C. A., Portugal
Brown, B. E., Massachusetts
Cacoullos, T., Greece
Canfield, R. V., Utah
Capobianco, M., New York
Chanda, K., Texas
Charalambides, C. A., Greece
Chieppa, M., Italy
Chung, C. J. F., Canada
Cobb, L., South Carolina
Cocchi, D., Italy
Cook, W. L., Idaho
Csörgo, M., Canada
Damiani, S., Italy
De Lucia, L., Italy
Dennis, B., Pennsylvania
D'Esposito, M. R., Italy
Diana, G., Italy
Do Couto, H. T. Z., Brazil
Dussauchoy, A., France
Dyer, D. D., Texas
Eicker, F., Germany
Enns, E. G., Canada
Feoli, E., Italy
Ferrari, P., Italy
Ferreri, C., Italy
Finocchiaro, M., Italy
Folks, J. L., Oklahoma
Friday, D. S., Colorado
Frishman, F., Maryland
Galambos, J., Pennsylvania
Gallo, F., Italy
George, E. O., Nigeria
Giavelli, G., Italy
Gili, A., Italy
Gomes, M. I. L., Portugal
Gross, A. J., South Carolina
Gupta, P. L., Maine
Gupta, R. C., Maine
Gyires, B., Hungary
Hengeveld, R., Netherlands
Hennemuth, R. C., Massachusetts
Inal, H. C., Turkey
Janardan, K. G., Illinois
Jancey, R. C., Canada
Kemp, A. W., England
Kemp, C. D., England
Kostoris, A., Italy
Kotz, S., Maryland
Landenna, G., Italy
Langsaeter, T., Norway
Laud, P. W., Illinois
Lee, J. C., New Jersey
Linder, E., Pennsylvania
Lindsay, B., Pennsylvania
Lukacs, E., Washington, D.C.
Marasini, D., Italy
Mardia, K. V., England
Marvulli, R., Italy
McDonald, J. B., Utah
Mineo, Italy
Miserocchi, M., Italy
Morgan, P. R., Maryland
Mosimann, J. E., Maryland
Mudholkar, G. S., New York
Norton, R. M., South Carolina
Oksoy, D., Turkey
Orviati, S., Italy
Palmer, J. E., Massachusetts
Panaretos, J., Ireland
Papageorgiou, H., Greece
Parrish, R. S., Georgia
Patil, G. P., Pennsylvania
Pesarin, F., Italy
Plachky, D., West Germany
Policello, G. E., Ohio
Pollastri, A., Italy
Poterasu, V., Romania
Prato, G., Italy
Provasi, C., Italy
Ramachandran, G., England
Ratnaparkhi, M. V., Maryland
Reid, D. D., Australia
Rigatti-Luchini, S., Italy
Rondini, L., Italy
Rossi, O., Italy
Roux, J. J. J., South Africa
Saunders, R., Illinois
Sclove, S., Illinois
Seshadri, V., Canada
Shantaram, R., Michigan
Shapiro, S. S., Florida
Shimizu, R., Japan
Singh, J., Pennsylvania
Sobel, M., California
Srivastava, M. S., Canada
Srivastava, R. C., Ohio
Stene, J., Denmark
Stepanek, V., Czechoslovakia
Strassoldo, M., Italy
Taillie, C., Pennsylvania
Tatlidil, H., Turkey
Tiago de Oliveira, J., Portugal
Tiku, M. L., Canada
Tranquilli, G. B., Italy
Tuncer, Y., Turkey
Umani, M., Italy
Uppuluri, V. R., Tennessee
Vandemaele, M., Belgium
Vedaldi, R., Italy
Vik, G., Norway
Villasenor, J. A., Mexico
Weber, J. E., Arizona
Wise, M. E., Netherlands
Xekalaki, E., Ireland
Zanni, R., Italy

AUTHORS NOT LISTED ABOVE

Amato, P., Italy
Barndorff-Nielsen, O., Denmark
Becker, P. J., South Africa
Bowman, K. O., Tennessee
Brain, C., Florida
Chen, T. L., California
Crain, B. R., Oregon
Davies, L., West Germany
Davis, A. S., Wisconsin
DeAngelis, R. J., Pennsylvania
DuFrain, R. J., Tennessee
Ehlers, P. F., Canada
Ghosh, J. K., India
Goodman, I. R., Washington, D.C.
Hansen, E., California
Hernandez, F., Wisconsin
James, I., Australia
Jensen, J. L., Denmark
Johnson, R. A., Wisconsin
Loukas, S., England
Malley, J. D., Maryland
Mathai, A. M., Canada
Mazumder, B. S., India
McNolty, F., California
Midlarsky, M. I., Colorado
Miller, M. C., III, South Carolina
Mumme, D. C., Idaho
Olsen, D. R., Texas
Ord, J. K., England
Rukhin, A. L., Indiana
Savits, T. H., Pennsylvania
Schaeffer, D. J., Illinois
Shanmugam, R., Alabama
Shenton, L. R., Georgia
Singh, M., Canada
Stulır, S., Canada
Subbaiah, P., California
Thomsen, W., West Germany
Trivedi, M. C., New York
Wenocur, R. S., Pennsylvania
Weier, D. R., South Carolina

Foreword

The International Summer School on Statistical Distributions in Scientific Work was held in Trieste during July 1980 for a period of three weeks. The emphasis was on research, review, and exposition concerned with the interface between modern statistical distribution theory and real world problems and issues involving science, technology, and management. Both theory and applications received full attention at the School. The program consisted of a Short Intensive Preparation Course, a NATO Advanced Study Institute, and a Research Conference. While the relative composition of these activities varied somewhat in terms of instruction, exposition, research-review, research, and consultation, the basic spirit of each was essentially the same. Every participant was both a professor and a student.

The summer school was sponsored by the NATO Advanced Study Institutes Program; Consiglio Nazionale delle Ricerche, Italy; Regione Autonoma Friuli Venezia Giulia, Italy; National Institutes of Health, USA; Office of Naval Research, USA; The Pennsylvania State University; Universita di Roma; Universita di Trieste; International Statistical Ecology Program; International Transfer of Science and Technology, Belgium; and the participants and their home institutions and organizations.

Research papers, research-review expositions and instructional lectures were specially prepared for the program. These materials have been refereed and revised, and are now available in a series of several edited volumes and monographs.

BACKGROUND

It is now close to two decades since the International Symposium on Classical and Contagious Distributions was held in Montreal in 1963. It was the first attempt to identify the area of discrete distributions as a subject area by itself. The symposium was a great success in that it stimulated growth in the field and more importantly provided a certain direction to it. Next came the Biometric Society Symposium on Random Counts in Scientific Work at the annual meetings of the American Association for the Advancement of Science held in 1968. The first symposium had emphasized models and structures, the second one focused its attention on the useful role of discrete distributions in applied work.

Seven years ago, a Modern Course on Statistical Distributions in Scientific Work was held at the University of Calgary in 1974 under sponsorship of the NATO Scientific Affairs Division. The Program consisted of an Advanced Study Institute (ASI) followed by a Research Conference on Characterizations of Statistical Distributions. The purpose of the ASI was to provide an open forum with focus on different aspects of statistical distributions arising in scientific or statistical work. The purpose of the characterizations conference was to bring together research workers investigating characterization problems that have motivation in scientific concepts and formulations or that have application or potential use for statistical theory. The program was a great success. Participants still remember it very fondly for its scientific impact and its social and professional contact.

C. Taillie et al. (eds.), Statistical Distributions in Scientific Work, Vol. 6, xi–xiii.

CALGARY PROGRAM

The edited Proceedings of the Calgary Program consist of three substantive volumes. They have been acknowledged to include a wealth of material ranging over a broad spectrum of the theory and applications of distributions and families of distributions. Most papers have been acknowledged for their content by reviewers in professional journals. The reviews have on the whole stressed the importance of these Proceedings as a successful effort to unify the field and to focus on main achievements in the area. Moreover, many of the papers which appeared in the Proceedings have been, and continue to be, quoted extensively in recent research publications. The Calgary Program of 1974 has had a definite and positive impact on stimulating further developments in the field of statistical distributions and their applications.

At the same time, essentially for economic reasons, the sciences, technology, and society are recognizing ever-expanding needs for quantification. The random quantities arising in conceptualization and modeling, in simulation, in data analysis, and in decision-making lead increasingly to various kinds of distributional problems and requests for solution. Statistical distributions remain an important and focal area of study. It is no surprise that the subject area of statistical distributions in scientific work is still advancing steadily.

Interestingly, the Calgary participants perceived this future need and concern. In anticipation, several prominent participants formed a Committee on Statistical Distributions in Scientific Work to discuss future plans and activities that would help consolidate and strengthen the subject area of statistical distributions and its applications on a continuing basis. The Committee identified the following needs and activities: (i) Preparation of a Comprehensive Dictionary and Bibliography of Statistical Distributions in Scientific Work, (ii) Preparation of Monographs and Modules on Important Distributions, Concepts, and Methods with Applications, and (iii) Planning and Organization of a Sequel to the Calgary Program.

DISTRIBUTIONAL ACTIVITIES

A well sustained seven year effort has produced a comprehensive three-volume set entitled *A Modern Dictionary and Bibliography of Statistical Distributions in Scientific Work*. The three volumes are: Volume 1, Discrete Models; Volume 2, Continuous Univariate Models; and Volume 3, Multivariate Models. The Dictionary covers several hundred distributional models and gives wherever possible their genesis, structural properties and parameters, random number generations, tabulations, graphs, and inter-relations through verbal statements as well as schematic diagrams. The Bibliography covers over ten thousand publications. Besides the usual reference information, each entry provides users listing (citation index), reviews, classification by distribution, inference and application, plus any special notes. The massive effort by the dictionary bibliography team consisting of M. T. Boswell, S. W. Joshi, G. P. Patil, M. V. Ratnaparkhi, and J. J. J. Roux needs to be specially acknowledged. So also the continuing interest and response of the professional community. It is hoped that the dictionary and bibliography effort will be a continuing activity serving the community with updated information from time to time.

On the monographs front, a lucid volume by J. B. Douglas, entitled *Analysis with Standard Contagious Distributions,* has been published. It should be of value to all those who are working with contagious distributions in one context or the other. More monographs are under preparation as follows:

Aitchison, J.: *Distributions on the Simplex of Their Applications*

Arnold, B. C.: *Pareto Distributions and Applications*

Cobb, L.: *Catastrophe Theory and Distributional Problems*

Folks, J. L. and Chhikara, R. S.: *Inverse Gaussian Distribution and Applications*

Mosimann, J. E.: *Analysis Using Size and Shape Variables*

Ord, J. K. and Patil, G. P.: *Introduction to Probabilty and Statistical Modeling*

Regarding the planning and organization of a sequel to the Calgary Program, the NATO Advanced Study Institutes Program encouraged part of the Committee to meet and assisted the Committee to have indepth discussions at Parma, Italy, in 1978. The following members were in attendance: B. A. Baldessari, T. Cacoullos, S. Engen, S. Kotz, J. E. Mosimann, J. K. Ord, G. P. Patil, C. Taillie, J. Tiago de Oliveira, W. G. Warren, and M. E. Wise. The intensive and open deliberations proved to be very constructive. The Committee felt unanimously that a follow-up to the Calgary ASI was very much needed, and that it should be held in 1980. Several institutions offered to host such an ASI. It was decided that the program be held in Italy. Bruno Baldessari and Livia Rondini assured the necessary support in this connection.

TRIESTE PROGRAM

A major purpose of the program was to give a unified and integrated view of different classes of distributions and to describe novel methodologies related to statistical distributions and/or their applications. Also, contributions on the description and characterization of distributions which are useful in a variety of fields of application were welcomed.

An application was prepared for the NATO ASI Program with G. P. Patil as the Chairman of the Organizing Committee, with B. Baldessari as the Director and C. Taillie as the Co-Director, with S. Kotz, J. E. Mosimann, J. K. Ord, and G. P. Patil as the Scientific Directors, and with L. Rondini as the Host. The NATO ASI program provided a positive response. Requests for the additional support needed were granted from within Italy and the USA. Participants and their institutions also extended a helping hand.

Spread over the three week period, the School had over 140 scientific participants and 50 accompanying persons from various countries around the world. The scientific program was more than full, and yet the overall program had a relaxing touch. Everything that the hosts, L. Rondini, A. Kostoris, S. Orviati, M. Strassoldo, M. Umani, and E. Feoli, did has been simply sweet and gratifying.

The Trieste program was a great success. Many have wondered as to when it would be again that they would meet and participate in another timely activity on statistical distributions in scientific work. If you have any thoughts or suggestions, please do not hesitate to let us know. I look forward to hearing from you.

April 30, 1981 G. P. Patil

Program Acknowledgments

For any program to be successful, mutual understanding and support among all participants are essential in directions ranging from critical to constructive and from cautious to constructive. The present program is grateful to the members of the Committee, and to the referees, advisors, sponsors and the participants for their timely advice and support.

Trieste is a beautiful place and so is the surrounding region. The Mediterranean around, the mountains nearby, and the campus on the top of a mountain provide a very scenic mosaic conducive for scholarship and communication. Italy has had a long tradition of research on distributional problems and related issues arising from uncertainty. It was only natural that the International Summer School on Statistical Distributions in Scientific Work met at Trieste.

The success of the program was due, in no small measure, to the endeavors of the Local Arrangements Committee. We thank L. Rondini, A. Kostoris, S. Orviati, M. Strassoldo, M. Umani, and E. Feoli for their hospitality and support.

And finally those who have assisted with the arduous task of preparing the materials for publication. Barbara Alles has been an ever cheerful and industrious secretary in the face of every adversity. Bonnie Burris, Bonnie Henninger, and Sandy Rothrock prepared the final versions of the manuscripts. Rani Venkataramani helped with the subject and author indexes. George Otto did the figures and artwork.

All of these nice people have done a fine job indeed. To all of them, our sincere thanks.

April 30, 1981

B. A. Baldessari
G. P. Patil
C. Taillie

Reviewers of Manuscripts

With appreciation and gratitude, the program acknowledges the valuable services of the following referees who have served as reviewers of manuscripts submitted to the program for possible publication. The editors thank the reviewers for their critical and constructive reviews.

M. Ahmad
University of Petroleum and Minerals

B. C. Arnold
University of California

A. C. Atkinson
Imperial College, London

L. J. Bain
University of Missouri

R. E. Bargmann
University of Georgia

O. Barndorff-Nielsen
Aarhus University

V. Barnett
University of Sheffield

A. K. Basu
Laurentian University

A. P. Basu
University of Missouri

L. V. Bellavista
University of Palermo

H. W. Block
University of Pittsburgh

S. Blumenthal
University of Illinois

M. T. Boswell
The Pennsylvania State University

L. A. Bruckner
Los Alamos Scientific Laboratory

R. V. Canfield
Utah State University

Ch. A. Charalambides
University of Athens

R. M. Cormack
University of St. Andrews

B. R. Crain
Portland State University

R. C. Dahiya
Old Dominion University

J. Darroch
Flinders University

A. P. Dawid
City University of London

B. Dennis
The Pennsylvania State University

P. J. Diggle
University of Newcastle upon Tyne

I. R. Dunsmore
University of Sheffield

A. Dussauchoy
Université Claude-Bernard

D. S. Friday
National Bureau of Standards

O. Frank
University of Lund

J. Galambos
Temple University

J. L. Gastwirth
George Washington University

D. V. Gokhale
University of California

M. I. Gomes
Faculty of Sciences of Lisbon

D. R. Grey
University of Sheffield

A. J. Gross
Medical University of South Carolina

A. M. Gross
Bell Laboratories

R. C. Gupta
University of Maine

D. M. Hawkins
CSIR, South Africa

R. Hengeveld
Catholic University, Nijmegen

T. Hettmansperger
The Pennsylvania State University

J. J. Higgens
University of South Florida

P. Holgate
Birkbeck College, London

H. K. Hsieh
University of Massachusetts

I. James
University of Western Australia

K. G. Janardan
Sangamon State University

N. L. Johnson
University of North Carolina

R. A. Johnson
University of Wisconsin

A. W. Kemp
University of Bradford

R. A. Kempton
Plant Breeding Institute, Cambridge

J. R. Kettenring
Bell Laboratories

C. G. Khatri
Gujarat University

S. Kotz
University of Maryland

P. R. Krishnaiah
University of Pittsburgh

I. J. Lauder
University of Hong Kong

J. C. Lee
Wright State University

H. J. Malik
University of Guleph

R. M. Marcus
Equitable Life Assurance Society

A. M. Mathai
McGill University

D. G. Morrison
Columbia University

N. E. Morton
University of Hawaii

J. E. Mosimann
National Institutes of Health

G. S. Mudholkar
University of Rochester

H. Nagao
University of Tsukuba

R. M. Norton
College of Charleston

J. K. Ord
University of Warwick

H. Papageorgiou
University of Athens

J. K. Patel
University of Missouri

P. K. Pathak
University of New Mexico

G. Policello
Ohio State University

B. Ramachandran
Indian Statistical Institute

G. Ramachandran
Building Research Establishment

M. V. Ratnaparkhi
National Institutes of Health

D. S. Robson
Cornell University

H. J. Rossberg
Karl Marx University

J. J. J. Roux
University of South Africa

D. B. Rubin
Educational Testing Service

R. M. Schrader
University of New Mexico

A. K. Sen
University of Illinois

S. S. Shapiro
Florida International University

R. Shimizu
Institute of Statistical Mathematics

S. Shirahata
Osaka University

R. W. Shorrock
Bell Canada

R. Simon
National Institutes of Health

J. Singh
Temple University

M. D. Springer
University of Arkansas

M. S. Srivastava
University of Toronto

P. R. Tadikamalla
University of Pittsburgh

C. Taillie
The Pennsylvania State University

J. Taigo de Oliveira
Faculty of Sciences of Lisbon

G. L. Tietjen
Los Alamos Scientific Laboratory

R. C. Tripathi
University of Texas

A. A. Tsiatis
St. Jude Children's Research Hospital

G. G. Walter
University of Wisconsin

J. K. Wani
University of Calgary

W. G. Warren
Forintek Canada Corp.

J. E. Weber
University of Arizona

M. E. Wise
Leiden University

M. C. K. Yang
University of Florida

S. Zacks
SUNY at Binghamton

Contents of Edited Volumes

Volume 4
MODELS, STRUCTURES, AND CHARACTERIZATIONS **455 pp.**
Continuous Models: J. AITCHISON, Statistical Predictive Distributions. O. BARNDORFF-NIELSEN and P. BLAESILD, Hyperbolic Distributions and Ramifications: Contributions to Theory and Application. P. BLAESILD and J. L. JENSEN, Multivariate Distributions of Hyperbolic Type. L. COBB, The Multimodal Exponential Families of Statistical Catastrophe Theory. J. L. FOLKS and A. S. DAVIS, Regression Models for the Inverse Gaussian Distribution. V. SESHADRI, A Note on the Inverse Gaussian Distribution. V. R. R. UPPULURI, Some Properties of the Log-Laplace Distribution. J. J. J. ROUX and P. J. BECKER, Compound Distributions Relevant to Life Testing. I. JAMES, Distributions Associated with Neutrality Properties for Random Proportions. J. E. MOSIMANN and J. D. MALLEY, The Independence of Size and Shape Before and After Scale Change. J. AITCHISON, Distributions on the Simplex for the Analysis of Neutrality.

Discrete Models: E. XEKALAKI, Chance Mechanisms for the Univariate Generalized Waring Distribution and Related Characterizations. C. FERRERI, On a New Family of Discrete Distributions. R. SHANMUGAM and J. SINGH, On the Stirling Distribution of the First Kind. P. L. GUPTA and J. SINGH, On the Moments and Factorial Moments of a MPSD. T. CACOULLOS and H. PAPAGEORGIOU, On Bivariate Discrete Distributions Generated by Compounding. Ch. A. CHARALAMBIDES, Bivariate Generalized Discrete Distributions and Bipartitional Polynomials. M. AHMAD, A Bivariate Hyper-Poisson Distribution. B. GYIRES, On the Multinomial Distributions Generated by Stochastic Matrices and Applications.

Structural Properties: A. L. RUKHIN, Distributions with Sufficient Statistics for Multivariate Location Parameter and Transformation Parameter. E. LUKACS, Analytic Distribution Functions. S. L. SCLOVE, Some Recent Statistical Results for Infinitely Divisible Distributions. A. M. MATHAI, An Alternate Simpler Method of Evaluating the Multivariate Beta Function and an Inverse Laplace Transform Connected with Wishart Distribution. D. PLACHKY and W. THOMSEN, On a Theorem of Polya. J. C. LEE, Asymptotic Distributions of Functions of Eigenvalues.

Computer Generation: M. T. BOSWELL and R. J. DeANGELIS, A Rejection Technique for the Generation of Random Variables with the Beta Distribution. C. D. KEMP and S. LOUKAS, Fast Methods for Generating Bivariate Discrete Random Variables. A. W. KEMP, Frugal Methods of Generating Bivariate Discrete Random Variables.

Characterizations: J. PANARETOS, A Characterization of the Negative Multinomial Distribution. R. C. GUPTA, On the Rao-Rubin Characterization of the Poisson Distribution. R. C. SRIVASTAVA, On Some Characterizations of the Geometric Distribution. M. V. RATNAPARKHI, On Splitting Model and Related Characterizations of Some Statistical Distributions. C. TAILLIE and G. P. PATIL, Rao-Rubin Condition for a Certain Class of Continuous Damage Models. J. J. J. ROUX and M. V. RATNAPARKHI, On Matrix-Variate Beta Type I Distribution and Related Characterization of Wishart Distribution. J. PANARETOS, On the Relationship Between the Conditional and Unconditional Distribution of a Random Variable. M. V. RATNAPARKHI, Some Bivariate Distributions of (X,Y) where the Conditional Distribution of Y, Given X, is Either Beta or Unit-Gamma. E. O. GEORGE and G. S. MUDHOLKAR, Some Relationships Between the Logistic and the Exponential Distributions. R. C. SRIVASTAVA, Some Characterizations of the Exponential Distribution Based on Record Values. C. TAILLIE, A Note on Srivastava's Characterization of the Exponential Distribution Based on Record Values. R. SHANTARAM, On the Stochastic Equation $X+Y=XY$. R. SHIMIZU and L. DAVIES, On the Stability of Characterizations of Non-Normal Stable Distributions.

Volume 5
INFERENTIAL PROBLEMS AND PROPERTIES **439 pp.**
Distributional Testing and Goodness-of-Fit: S. S. SHAPIRO and C. BRAIN, A Review of Distributional Testing Procedures and Development of a Censored Sample Distributional Test. A. J. GROSS and S. S. SHAPIRO, A Goodness-of-Fit Procedure for Testing Whether a Reliability Growth Model Fits Data that Show Improvement. K. C. CHANDRA, Chi-Square Goodness-of-Fit Tests Based on Dependent Observations. F. PESARIN, An Asymptotically Distribution-Free Goodness-of-Fit Test for Families of Statistical Distributions Depending on Two Parameters. A. W. KEMP, Conditionality Properties for the Bivariate Logarithmic Distribution with an Application to Goodness of Fit.

Parameter Estimation: S. BLUMENTHAL, A Survey of Estimating Distributional Parameters and Sample Sizes from Truncated Samples. B. R. CRAIN and L. COBB, Parameter Estimation for Truncated Exponential Families. B. G. LINDSAY, Properties of the Maximum Likelihood Estimator of a Mixing Distribution. G. E. POLICELLO II, Conditional Maximum Likelihood Estimation in Gaussian Mixtures. W. L. COOK and D. C. MUMME, Estimation of Pareto Parameters by Numerical Methods. M. CHIEPPA and P. AMATO, A New Estimation Procedure for the Three-Parameter Lognormal Distribution.

Hypothesis Testing: M. CSORGO, On the Asymptotic Distribution of the Multivariate Cramer-von Mises and Hoeffding-Blum-Kiefer-Rosenblatt Independence Criteria. G. S. MUDHOLKAR and P. SUBBAIAH, Complete Independence in the Multivariate Normal Distribution. D. R. WEIER and A. P. BASU, On Tests of Independence Under Bivariate Exponential Models. M. S. SRIVASTAVA, On Tests for Detecting Change in the Multivariate Mean. G. LANDENNA and D. MARASINI, A Two-Dimensional t-Distribution and a New Test with Flexible Type I Error Control. M. L. TIKU and M. SINGH, Testing Outliers in Multivariate Data.

Approximations: G. S. MUDHOLKAR and M. C. TRIVEDI, A Normal Approximation for the Multivariate Likelihood Ratio Statistics. K. O. BOWMAN and L. R. SHENTON, Explicit Accurate Approximations for Fitting the Parameters of LU. R. S. PARRISH and R. E. BARGMANN, A Method for the Evaluation of Cumulative Probabilities of Bivariate Distributions Using the Pearson Family. F. HERNANDEZ and R. A. JOHNSON, Transformation of a Discrete Distribution to Near Normality.

Reliability and Life Testing: H. W. BLOCK and T. H. SAVITS, Multivariate Distributions in Reliability Theory and Life Testing. I. R. GOODMAN and S. KOTZ, Hazard Rates Based on Isoprobability Contours. J. GALAMBOS, Failure Time Distributions: Estimates and Asymptotic Results. P. LAUD and R. SAUNDERS, A Note on Shock Model Justification for IFR Distributions. R. C. GUPTA, On the Mean Residual Life Function in Survival Studies. A. P. BASU, Identifiability Problems in the Theory of Competing and Complementary Risks — A Survey. D. S. FRIDAY, Dependence Concepts for Stochastic Processes.

Miscellaneous: J. AITCHISON, Some Distribution Theory Related to the Analysis of Subjective Performance in Inferential Tasks. E. G. ENNS, P. F. EHLERS, and S. STUHR, Every Body Has Its Moments. M. CAPOBIANCO, Some Distributions in the Theory of Graphs. A. GILI, Cograduation Between Statistical Distributions and Its Applications — A General Review.

Volume 6
APPLICATIONS IN PHYSICAL, SOCIAL, AND LIFE SCIENCES 445 pp.

Applications in the Physical Sciences: K. V. MARDIA, Recent Directional Distributions with Applications. J. K. GHOSH and B. S. MAZUMDER, Size Distribution of Suspended Particles-Unimodality, Symmetry and Lognormality. D. DYER, Offshore Oil/Gas Lease Bidding and the Weibull Distribution. F. McNOLTY and E. HANSEN, Statistical Distributions Occurring in Photoelectron Phenomena, Radar and Infrared Applications. K. G. JANARDAN and D. J. SCHAEFFER, Application of Discrete Distributions for Estimating the Number of Organic Compounds in Water. R. SHANMUGAM and J. SINGH, Some Bivariate Probability Models Applicable to Traffic Accidents and Fatalities. V. STEPANEK, Role and Use of Statistical Distributions in Information Theory as Applied to Chemical Analysis.

Applications in the Social Sciences: S. L. SCLOVE, Modeling the Distribution of Fingerprint Characteristics. M. I. MIDLARSKY, Stochastic Modeling in Political Science Research. P. R. MORGAN, Statistical Distribution Models in the Behavioral Sciences: A Review of Theory and Applications. J. B. McDONALD, Some Issues Associated with the Measurement of Income Inequality. C. TAILLIE, Lorenz Ordering Within the Generalized Gamma Family of Income Distributions. J. K. ORD, G. P. PATIL and C. TAILLIE, The Choice of a Distribution to Describe Personal Incomes. J. K. ORD, G. P. PATIL and C. TAILLIE, Relationships Between Income Distributions for Individuals and for Households.

Applications in the Life Sciences: M. E. WISE, Spike Interval Distributions for Neurons and Random Walks with Drift to a Fluctuating Threshold. J. STENE, Probability Distributions Arising from the Ascertainment and the Analysis of Data on Human Families and Other Groups. K. G. JANARDAN, D. J. SCHAEFFER and R. J. DuFRAIN, A Stochastic Model for the Study of the Distribution of Chromosome

Aberrations in Human and Animal Cells Exposed to Radiation or Chemicals. D. COCCHI, A Model for the Analysis of Platelet Survival. B. DENNIS, Extinction and Waiting Times in Birth-Death Processes: Applications to Endangered Species and Insect Pest Control. D. D. REID, The Poisson Lognormal Distribution and Its Use as a Model of Plankton Aggregation. A. J. GROSS and M. C. MILLER III, Some Applications of Statistical Distribution Theory to Biology and Medicine.

Extreme Values and Order Statistics: R. V. CANFIELD, D. R. OLSEN, and T. L. CHEN, Extreme Value Theory with Application to Hydrology. G. RAMACHANDRAN, Properties of Extreme Order Statistics and Their Application to Fire Losses and Earthquake Magnitudes. J. TIAGO DE OLIVEIRA, Statistical Choice of Univariate Extreme Models. M. I. GOMES, An i-Dimensional Limiting Distribution Function of Largest Values and Its Relevance to the Statistical Theory of Extremes. R. S. WENOCUR, Waiting Times and Return Periods to Exceed the Maximum of a Previous Sample. R. S. WENOCUR, Waiting Times and Return Periods Related to Order Statistics: An Application of Urn Models.

Essentially because of the present economic conditions, the sciences, technology, and society are recognizing ever-expanding needs for quantification. The random quantities arising in conceptualization and modeling, in simulation, in data analysis, and in decision making lead increasingly to various kinds of distributional problems and requests for solution. Statistical distributions remain an important and focal area of study.

Preface

These three volumes constitute the edited Proceedings of the NATO Advanced Study Institute on Statistical Distribution Theory and its Applications held at the University of Trieste from July 10-August 1, 1980. The general title of the volume is *Statistical Distributions in Scientific Work,* a continuation from the Proceedings of an earlier program held at the University of Calgary during the summer of 1974, which brought out volumes 1, 2, and 3. The present volumes are: Volume 4 — Models, Structures, and Characterizations; Volume 5 — Inferential Problems and Properties; and Volume 6 — Applications in Physical, Social, and Life Sciences. These are based on the research-review expositions, instructional lectures, and research papers specially prepared for the program by the invited researchers and expositors.

The planned activities of the Institute consisted of lucid perceptive lectures and expositions, seminar lectures, study group discussions, tutorials, and individual study. The activities included meetings of editorial committees to discuss editorial matters for these proceedings which consist of the contributions that have gone through the usual refereeing process. The overall perspective of the program is provided by the Chairman of the Organizing Committee, Professor G. P. Patil, in his Foreword to the Volumes as summarized from his inaugural address to the Institute.

The Proceedings are being published in three volumes. All together, they consist of 15 topical sections of 100 contributions of 1260 pages of research, review, and exposition. Subject and author indexes also appear at the end of each volume. Effort has been made to keep the title and the content of each volume mutually consistent. However, it is quite possible that a different composition would have looked equally natural!

We view this program as a continuation of the tradition established by the pioneering 1963 Montreal Symposium which identified and consolidated statistical distributions as a separate field of statistical inquiry. The tradition was further carried on and amplified by the 1974 Calgary program. It was reassuring to see several participants at Trieste that were present at Montreal and/or Calgary. A number of new and young faces were also visible at Trieste. The papers in these Proceedings should reflect the recent and current developments and mirror the growth and maturity of the discipline and its integration within the general framework of applied statistics and related quantitative studies.

While working in the field of statistical distributions in general, it is often tempting to tackle isolated problems involving formal generalizations. One at times loses sight of the underlying probabilistic model even in this process. While this generalization approach may be quite acceptable from the mathematical point of view, it does however result, on occasion, in statistically unjustified theoretical exercises. There has been some justified criticism voiced by practitioners that we are losing touch with reality. A purpose of the Trieste program was to help generate a constructive dialogue between theory and application.

The program covered a broad spectrum of topics. Models and structures theme touched base with continuous models, discrete models, properties, computer generation, and characterizations. Inferential problems and properties included distribu-

C. Taillie et al. (eds.), Statistical Distributions in Scientific Work, Vol. 6, xxi–xxii.

tional testing and goodness-of-fit, parameter estimation, hypothesis testing, approximations, reliability and life testing. Real world problems were drawn from the physical sciences, social sciences and life sciences, and also included work on extreme values and order statistics. Thus, the formal and informal dialogues provided a panorama of the distributional field both in theory and in application. These published volumes constitute an effort to share those Proceedings with the interested reader. The spark and the spontaneity of a lively dialogue do not necessarily transmit themselves through written proceedings. We hope and trust, however, that the reader will instead reap the benefit from the careful preparation and editing through which each paper has gone.

In any collaborative effort of this magnitude and nature, the enthusiastic support of a large number of individuals and institutions is a prerequisite for success. We are extremely grateful to all of our sponsors, participants, and the hosts. Also to our ever-cheerful program secretary, Barbara Alles, who has managed to keep the program moving in every sense of the word.

These three volumes have been included in the ongoing NATO Advanced Study Institutes Series. They are published by the D. Reidel Publishing Company, a member of the Board of Publishers of the NATO ASI Series. It is only proper that we conclude here with our sincere thanks to both the Publisher and the NATO Scientific Affairs Division for these co-operative arrangements.

April 30, 1981

Charles Taillie
Ganapati P. Patil
Bruno A. Baldessari

RECENT DIRECTIONAL DISTRIBUTIONS WITH APPLICATIONS

K. V. MARDIA

Department of Statistics
University of Leeds
Leeds, LS2 9JT, UK

SUMMARY. The paper reviews new directional distributions motivated by applications since Mardia (1975a,b). The major new developments have taken place in three broad topics. Firstly, in modelling data which is clustered around a small circle on a sphere, and secondly in defining appropriate multivariate directional distributions and thirdly, weighted distributions on a rotating sphere. Various other families of distributions have also appeared. In addition, various distributions on generalized spaces such as cylindrical and shape distributions have been constructed. These distributions will be discussed in the context of their practical relevance to various areas of scientific application.

KEY WORDS. distributions on cylinder, distributions on triangle, families of distributions, multivariate direction distributions, offset normal distribution, rotated spheres, small circle distribution, truncated sphere, weighted distribution.

1. INTRODUCTION

Let θ be a circular random variable. The most important distribution on a circle is von Mises whose p.d.f. is

$$f(\theta) = \{2\pi I_0(\kappa)\}^{-1}\exp\{\kappa \cos(\theta-\mu_0)\},\ 0 < \theta \leqslant 2\pi,\ 0 < \mu_0 \leqslant 2\pi,\ \kappa > 0.$$

We will say θ is $M(\mu_0,\kappa)$. Let θ and ϕ be colatitude and

C. Taillie et al. (eds.), Statistical Distributions in Scientific Work, Vol. 6, 1–19.

longitude respectively. The Fisher distribution is important on the sphere and it has p.d.f. given by

$$f(\theta,\phi) = \{\kappa/4\pi \sinh \kappa\}\exp[\kappa\{\cos \mu_0 \cos\theta + \sin \mu_0 \sin \theta$$

$$\times \cos(\phi-\nu_0)\}]\sin \theta,\ 0 < \theta < \pi,\ 0 < \phi \leqslant 2\pi,\ \kappa>0.$$

These distributions and other important distributions such as Bingham's axial distribution and Down's distribution on a Stiefel manifold are defined and reviewed in Mardia (1972) and Mardia (1975a,b). Since these papers, Bingham (1974), Khatri and Mardia (1977), Mardia and Khatri (1977), Mardia and Zemroch (1977), Jupp and Mardia (1979) give their further properties and results. However, we will concentrate on the new distributions which have since then appeared from practical considerations in directional statistics. Here, directional statistics stands for any distribution on non-Euclidean space.

2. SMALL CIRCLE DISTRIBUTIONS

Often there is the need to model a small circle distribution on the sphere, that is, observations are concentrated near a parallel of latitude relative to $\underset{\sim}{\mu}$, the axis of symmetry. Interest in the problem arose from some data in the field of plate tectonics (see Mardia and Gadsden, 1977).

2.1 Fitting a Small Circle by Minimization. Let $\underset{\sim}{x}_i' = (x_i, y_i, z_i)$, $i=1,\ldots,n$, be n observations on the unit sphere, and let the small circle be defined by

$$\lambda x + \mu y + \nu z = \cos \alpha,\ \underset{\sim}{x}'\underset{\sim}{x} = 1,\ \underset{\sim}{\mu}'\underset{\sim}{\mu} = 1,\ \underset{\sim}{\mu}' = (\lambda,\mu,\nu).$$

By minimizing the sum of the squared deviations

$$V = 1 - \frac{\cos\alpha}{n} \sum_{i=1}^{n} \underset{\sim}{\mu}'\underset{\sim}{x}_i - \frac{\sin\alpha}{n} \sum_{i=1}^{n} \{1 - (\underset{\sim}{\mu}'\underset{\sim}{x}_i)^2\}^{\frac{1}{2}}, \tag{1}$$

Mardia and Gadsden (1977) show that $\hat{\alpha}$ and $\hat{\underset{\sim}{\mu}}$ are solutions of the equations

$$\sum_{i=1}^{n} \underset{\sim}{\mu}'\underset{\sim}{x}_i = \frac{\cos\alpha}{\sin\alpha} \sum_{i=1}^{n} \{1 - (\underset{\sim}{\mu}'\underset{\sim}{x}_i)^2\}^{\frac{1}{2}}, \tag{2}$$

$$\cos\alpha \sum_{i=1}^{n} \underset{\sim}{x}_i = \sin\alpha \sum_{i=1}^{n} \underset{\sim}{x}_i{}'\underset{\sim}{\mu}'\underset{\sim}{x}_i \{1 - (\underset{\sim}{\mu}'\underset{\sim}{x}_i)^2\}^{\frac{1}{2}} - \rho\underset{\sim}{\mu}, \tag{3}$$

where ρ is a Lagrange multiplier introduced for the constraint $\underset{\sim}{\mu}'\underset{\sim}{\mu}=1$. Equations (2) and (3) have to be solved numerically. An iterative method is given in Mardia and Gadsden (1977).

Let $F(\underset{\sim}{\mu},\kappa)$ denote the Fisher distribution with mean directional vector $\underset{\sim}{\mu}$ and concentration parameter κ. Consider the model

$$\underset{\sim}{x}_i \sim F\{\underset{\sim}{\mu}(\phi_i), \kappa\}, \qquad i=1,\ldots,n,$$

where $\underset{\sim}{\mu}(\phi_i) = \underset{\sim}{\Gamma}(\cos\alpha, \sin\alpha\sin\phi_i, \sin\alpha\cos\phi_i)$, with $\underset{\sim}{\Gamma}$ as an orthogonal matrix with the first column of $\underset{\sim}{\Gamma}$ as $\underset{\sim}{\mu}$. Now the least squares estimators of $\underset{\sim}{\mu}$ and α are the same as the maximum likelihood estimators for the model. Note that the m.l.e. of ϕ_i is ϕ_i^* which is the longitude for $\underset{\sim}{z}_i = \hat{\underset{\sim}{\Gamma}}\underset{\sim}{x}_i$, $i=1,\ldots,n$. The model is still under investigation.

2.2 Maximum Entropy Small Circle Distribution. Let θ and ϕ denote colatitude and longitude. From maximum entropy considerations, Mardia and Gadsden (1977) proposed the following small circle distribution with probability element (p.e.) of θ,ϕ as

$$dF(\theta,\phi) = C(\kappa,\alpha)\exp\{\kappa\cos(\theta-\alpha)\}\sin\theta\, d\theta d\phi, \quad 0 \leqslant \theta \leqslant \pi,\ 0 \leqslant \phi \leqslant 2\pi, \tag{4}$$

when the pole is given by the z-axis. In terms of direction cosines $\underset{\sim}{x}$, the p.e. is

$$C(\kappa,\alpha)\exp(\kappa[\beta(\underset{\sim}{\mu}'\underset{\sim}{x}) + \gamma\{1 - (\underset{\sim}{\mu}'\underset{\sim}{x})^2\}^{\frac{1}{2}}])dS, \tag{5}$$

where $\beta = \cos\alpha$, $\gamma = \sin\alpha$ and dS is the uniform measure. The normalizing constant is given by

$$C(\kappa,\alpha)^{-1} = 2\pi \sum_{i=0}^{\infty} \sum_{j=0}^{\infty} (\kappa\beta)^{2i} (\kappa\gamma)^{j} \beta(i+\tfrac{1}{2}, \tfrac{1}{2}j+1)\{(2i)!\, j!\}^{-1}.$$

For $\alpha=0$, (4) reduces to the Fisher distribution and for $\alpha = \pi/2$ we have

$$dF(\theta,\phi) = C \exp(\kappa\sin\theta) \sin\theta \, d\theta d\phi,$$

a girdle distribution which has been investigated by Selby (1964). For $\alpha\neq0$, the distribution has a dimple at the pole.

Maximum likelihood estimation in general is complicated, but simplifies for large κ. For given $\underset{\sim}{\mu}$ and α we have for large κ,

$$\hat{\kappa}^{-1} = 2V + 3V^2,$$

where V is defined in equation (1).

Mardia and Gadsden (1977) analyze two sets of small-circle data: (a) 66 hot spots and areas of vulcanism in the North Pacific, known as the Hawaiian Trend and (b) 15 data points as to the local directions of the earth's magnetic field in Australia about 100 million years ago.

2.3 Generalized Dimroth-Watson Small Circle Distribution. Because of the intractibility of (5), Bingham and Mardia (1978) proposed the distribution with p.e.

$$f(\underset{\sim}{x};\, \tau,\nu,\underset{\sim}{\mu}) \frac{dS}{4\pi} = C(\tau,\nu)^{-1} \exp\{-\tau(\underset{\sim}{\mu}'\underset{\sim}{x}-\nu)^2\} \frac{dS}{4\pi}\,, \quad (6)$$

where $0 \leqslant \nu < \infty$, $-\infty < \tau < \infty$, $\underset{\sim}{\mu}'\underset{\sim}{\mu}=1$, $\underset{\sim}{x}'\underset{\sim}{x}=1$. For $\nu=0$, (6) becomes the Dimroth-Watson distribution, while for $\tau \to 0$ such that $2\tau\nu \to \kappa$, it becomes the Fisher distribution. Let $P(\theta) = -\tau(\cos\theta-\nu)^2$. When $\nu \geqslant 1$ and $\tau > 0$, $P(\theta)$ has a maximum at $\theta=0$ and a minimum at $\theta=\pi$ and no other extrema. When $\nu \geqslant 1$ and $\tau < 0$, the extrema are reversed. Thus when $\nu \geqslant 1$ the distribution is of polar type concentrated about $\underset{\sim}{\mu}(\tau>0)$, $-\underset{\sim}{\mu}(\tau<0)$. When $0 < \nu < 1$, $\tau > 0$, we have a small circle distribution because $P(\theta)$ has a maximum at $\theta = \alpha \equiv \text{arc} \cos \nu$ and minima at $\theta=0$ and $\theta=\pi$. When $0 < \nu < 1$, $\tau < 0$, there is a minimum at $\theta=\alpha$ and maxima at $\theta=0$ and $\theta=\pi$.

The normalizing constant in (6) is given by

$$C(\tau,\nu) = \tfrac{1}{2}(1-\nu)\,{}_1F_1\{\tfrac{1}{2};\ \tfrac{3}{2};\ -\tau(1-\nu)^2\} + \tfrac{1}{2}(1+\nu)\,{}_1F_1\{\tfrac{1}{2};\ \tfrac{3}{2};\ -\tau(1+\nu)^2\},$$

where ${}_1F_1$ is a confluent hypergeometric function.

Bingham and Mardia (1978) give an iterative method of finding the maximum likelihood estimators for $0 < \nu < 1$ and $\tau > 0$, the case of interest. Define S_1, S_2, and $\underset{\sim}{S}$ by

$$S_1 = \underset{\sim}{\mu}'\underset{\sim}{\bar{x}},\quad S_2 = n^{-1}\sum_{i=1}^{n}(\underset{\sim}{\mu}'\underset{\sim}{x}_i - S_1)^2,\quad \underset{\sim}{S} = n^{-1}\sum_{j=1}^{n}\underset{\sim}{x}_j\underset{\sim}{x}_j' - \underset{\sim}{\bar{x}}\underset{\sim}{\bar{x}}',$$

and let $t_1 \geqslant t_2 \geqslant t_3$ be the eigenvalues of $\underset{\sim}{S}$ with corresponding eigenvectors $\underset{\sim}{u}_1, \underset{\sim}{u}_2, \underset{\sim}{u}_3$. When $6t_3 < (1-\underset{\sim}{u}_3'\underset{\sim}{\bar{x}})^2$ or $S_2/(1-S_1)^2 < 1/6$, approximate maximum likelihood estimates are given by

$$\hat{\underset{\sim}{\mu}} = \underset{\sim}{u}_3,\quad \hat{\nu} = \underset{\sim}{u}_3'\underset{\sim}{\bar{x}},\quad \hat{\tau} = (2t_3)^{-1}.$$

Bingham and Mardia (1978) give various tests of hypothesis and confidence regions when τ is large.

2.4 Rotated Fisher Small Circle. In studying the arrival directions of cosmic rays it is necessary to model a small-circle distribution (Edwards and Mardia, 1980). A suitable model for the z-axis as the pole is given by

$$f(\theta,\phi) = \frac{\kappa}{4\pi\sinh\kappa}\, e^{\kappa\cos\alpha\cos\theta}\, I_0(\kappa\sin\alpha\sin\theta)\sin\theta. \qquad (7)$$

The model is derived by rotating a Fisher distribution around a small circle of constant colatitude, $\theta=\alpha$. For general pole, $\underset{\sim}{\mu}$,

$$f(\underset{\sim}{\ell}) = \frac{\kappa}{4\pi\sinh\kappa}\, e^{\kappa\cos\alpha\,(\underset{\sim}{\ell}'\underset{\sim}{\mu})}\, I_0\left[\kappa\sin\alpha\{1-(\underset{\sim}{\ell}'\underset{\sim}{\mu})^2\}^{\frac{1}{2}}\right]. \qquad (8)$$

The distribution (7), unlike (6) and (4), is not in the exponential family, however the normalizing constant is simple and does not depend on α. Note the close comparison between (5) and (8).

For known $\underset{\sim}{\mu}$, maximum likelihood estimates of α and κ can be easily found.

3. MULTIVARIATE DIRECTIONAL DISTRIBUTIONS

From maximum entropy considerations, Mardia (1975a) constructed a suitable distribution when two unit random vectors $\underset{\sim}{\ell}_1$ and $\underset{\sim}{\ell}_2$ are correlated, of the form,

$$\text{const} \times \exp\{\underset{\sim}{a}_1'\underset{\sim}{\ell}_1 + \underset{\sim}{a}_2'\underset{\sim}{\ell}_2 + \text{tr}\, \underset{\sim}{A}\underset{\sim}{\ell}_1\underset{\sim}{\ell}_2'\},\quad \underset{\sim}{\ell}_1, \underset{\sim}{\ell}_2 \in S_p, \tag{9}$$

i.e., a family of bivariate von Mises-Fisher distributions. Kent (1979) explores a particular case which has ellipse-shaped probability contours about the pole. Mardia (1979) discusses general problems of testing dependence. The marginal distributions of $\underset{\sim}{\ell}_1$, $\underset{\sim}{\ell}_2$ however, are not of the von Mises-Fisher form except for trivial cases.

Mardia (1975) in the author's reply also introduced a family of bivariate matrix Bingham-von Mises-Fisher distributions. For $\underset{\sim}{X}$ and $\underset{\sim}{Y}$ on the Stiefel Manifolds $V_m(R^p)$ and $V_n(R^q)$, the density of $(\underset{\sim}{X},\underset{\sim}{Y})$ has the form

$$C \exp\{\text{tr}(\underset{\sim}{F}\underset{\sim}{X}) + \text{tr}(\underset{\sim}{G}\underset{\sim}{Y}) + \text{tr}(\underset{\sim}{X}'\underset{\sim}{A}\underset{\sim}{X}\underset{\sim}{B}) + \text{tr}(\underset{\sim}{Y}'\underset{\sim}{S}\underset{\sim}{Y}\underset{\sim}{T}) + \text{tr}(\underset{\sim}{Y}'\underset{\sim}{C}\underset{\sim}{Y}\underset{\sim}{D})$$
$$+ \text{tr}(\underset{\sim}{X}'\underset{\sim}{S}\underset{\sim}{Y}\underset{\sim}{T}) + \text{tr}(\underset{\sim}{Y}'\underset{\sim}{U}\underset{\sim}{X}\underset{\sim}{V})\}, \tag{10}$$

where the matrices are appropriately constrained. Jupp and Mardia (1980) consider the generalized exponential family of densities given by

$$\exp\{\alpha(\underset{\sim}{A},\underset{\sim}{t}_1) + \underset{\sim}{t}_1'\underset{\sim}{A}\underset{\sim}{t}_2 + \beta(\underset{\sim}{A},\underset{\sim}{t}_2) - \kappa(\underset{\sim}{A})\}, \tag{11}$$

where $\underset{\sim}{t}_1 \equiv \underset{\sim}{t}_1(X)$, $\underset{\sim}{t}_2 \equiv \underset{\sim}{t}_2(Y)$, and $\underset{\sim}{X}$ and $\underset{\sim}{Y}$ take values on general Riemannian manifolds M and N. The interest lies in a correlation coefficient ρ^2, for bidirectional distributions and it is shown that if $\underset{\sim}{X}$ and $\underset{\sim}{Y}$ are independent on M and N, then under suitable regularity conditions,

$$\rho^2 = \text{tr}(\underset{\sim}{\Sigma}_{11} \underset{\sim}{A} \underset{\sim}{\Sigma}_{22} \underset{\sim}{A}') + O(||\underset{\sim}{A}||^3),$$

where $\underset{\sim}{\Sigma}_{11}$ and $\underset{\sim}{\Sigma}_{22}$ are the covariance matrices of $\underset{\sim}{t}_1$ and $\underset{\sim}{t}_2$ for the distribution with $\underset{\sim}{A} = \underset{\sim}{0}$, and where $||\underset{\sim}{A}||^2 = \text{tr}(\underset{\sim}{A}\,\underset{\sim}{A}')$.

Both (9) and a sub-class of (10) are contained in the family defined by (11). By this approach, Jupp and Mardia (1980) obtain various important correlation coefficients including that of Mardia and Puri (1978). Further, it is shown that the likelihood ratio for the test of independence is asymptotically equivalent to the sample counterpart of ρ^2. Stephens (1979) gives spherical correlation coefficients following ideas of Mackenzie (1957) and Downs (1974).

4. MISCELLANEOUS CONTRIBUTIONS

4.1 Off-Set Normal Distribution. Cairns (1975) and Fraser (1979) propose a model for directional data which allows for skewness and is called the projected normal distribution but it is really the off-set normal distribution with the parameters identified.

Consider the bivariate normal distribution with mean $(\lambda,0)'$ and covariance matrix $\underset{\sim}{I}$, the p.d.f. of which we denote $\phi_2\{\underset{\sim}{x}; (\lambda,0)', \underset{\sim}{I}\}$. The p.d.f. of $\underset{\sim}{\ell}$ is given by

$$f(\underset{\sim}{\ell}) = \int_{r=0}^{\infty} \phi_2\{r\ell_1, r\ell_2; (\lambda,0)', \underset{\sim}{I}\}\, r dr, \tag{12}$$

where $\underset{\sim}{\ell}' = (\ell_1, \ell_2)$, $\underset{\sim}{\ell}'\underset{\sim}{\ell} = 1$. On performing the integration, (12) becomes

$$f(\underset{\sim}{\ell}) = \frac{1}{\sqrt{2\pi}}\, \phi(\lambda) + \lambda\ell_1 \Phi(\lambda\ell_1)\phi(\lambda\ell_2), \tag{13}$$

where $\Phi(x)$ is the distribution function and $\phi(x)$ the p.d.f. of a $N(0,1)$ variable.

From the standard form given by (13), the vector $\underset{\sim}{\ell}$ is subjected to the matrix $\underset{\sim}{A}$ defined by

$$\underset{\sim}{A} \equiv \underset{\sim}{A}_1 \underset{\sim}{A}_2 \underset{\sim}{A}_3 = \begin{pmatrix} \cos\alpha & -\sin\alpha \\ \sin\alpha & \cos\alpha \end{pmatrix} \begin{pmatrix} 1 & \tau \\ 0 & 1 \end{pmatrix} \begin{pmatrix} \sigma & 0 \\ 0 & \sigma^{-1} \end{pmatrix},$$

to give the generalized variable on the unit circle, $\underset{\sim}{A}\underset{\sim}{\ell}/||\underset{\sim}{A}\underset{\sim}{\ell}||$. Note that $\underset{\sim}{A}_1$ is a rotation matrix, $\underset{\sim}{A}_2$ skews the plane parallel to the first axis and A_3 scales the first axis by σ and the second axis by σ^{-1}. Thus it is the distribution of $\theta|r=1$ from $N_2\left[\underset{\sim}{A}\begin{pmatrix}\lambda\\0\end{pmatrix}, \underset{\sim}{A}\,\underset{\sim}{A}'\right]$. Fraser (1979) fits the projected normal distribution to the famous Turtle data, obtaining maximum likelihood estimates numerically.

This work is related to the offset Normal distribution of Mardia (1972) and Saw (1978).

4.2 Weighted Distributions. A problem in Cosmic Rays had led us to explore weighted distributions on the sphere (Edwards and Mardia, 1980). Using the terminology of Rao (1965), a random variable x has a probability of being observed proportional to $u(x)$, where x has true underlying p.d.f. $g(x)$, the observed p.d.f. of x being defined by $f(x) = u(x)g(x)/\int u(x)g(x)dx$.

We consider the case where the weights depend only on one variable, i.e., $f(\theta,\phi) = C\, u(\theta)g(\theta,\phi)$. Assuming $u(\theta)$ can be approximately written in the form of a step function, closed forms for the maximum likelihood estimators can be found for small concentration, as well as approximations to the likelihood ratio tests of uniformity against a Fisher or Dimroth-Watson alternative. The case of truncation is a special case of equal weights on part of the sphere and zero weights elsewhere. Unweighted variables simply have equal weights on the complete sphere. Let the weighted Fisher distribution be defined by

$$f_\kappa(\theta,\phi) = \frac{1}{2\pi C(\kappa,\mu_0)}\, u_j\, e^{\kappa\{\cos\mu_0\cos\theta + \sin\mu_0\sin\theta\cos(\phi-\nu_0)\}} \sin\theta,$$

$$\theta_{j-1}<\theta<\theta_j, \quad 0<\phi<2\pi, \quad j=1,\dots,m, \tag{14}$$

where u_j is constant over the range (θ_{j-1}, θ_j). Asymptotically the likelihood ratio test of uniformity,

$$f_0(\theta,\phi) = \frac{u_j \sin\theta}{2\pi \sum_{j=1}^{m} u_j\{\cos\theta_{j-1} - \cos\theta_j\}},$$

against a Fisher alternative reduces to

$$n\bar{R}^2 = n\left[\frac{\bar{R}_x^2}{\operatorname{var}\bar{R}_x} + \frac{\bar{R}_y^2}{\operatorname{var}\bar{R}_y} + \frac{\{\bar{R}_z - E(\bar{R}_z)\}^2}{\operatorname{var}\bar{R}_z}\right] \sim \chi_3^2,$$

where $\bar{R}_x = \frac{1}{n}\sum_{i=1}^{n} \sin\theta_i \cos\phi_i$, $\bar{R}_y = \frac{1}{n}\sum_{i=1}^{n} \sin\theta_i \sin\phi_i$,

$$\bar{R}_z = \frac{1}{n}\sum_{i=1}^{n} \cos\theta_i, \quad E(\bar{R}_z) = \frac{D_2(u)}{4D_1(u)},$$

$$\operatorname{var}(\bar{R}_x) = \operatorname{var}(\bar{R}_y) = \frac{1}{24n}\left\{9 - \frac{D_3(u)}{D_1(u)}\right\},$$

$$\operatorname{var}(\bar{R}_z) = \frac{1}{48n}\left\{12 + \frac{4D_3(u)}{D_1(u)} - \frac{3D_2^2(u)}{D_1^2(u)}\right\},$$

with $D_k(u) = \sum_{j=1}^{m} u_j\{\cos k\theta_{j-1} - \cos k\theta_j\}$, $k=1,2,\ldots$.

Similar results can be found for a Dimroth-Watson distribution.

4.3 Some Conditional Forms. The inference for weighted distributions can be based on conditional distributions of $\phi|\theta$ if weights are known. Downs (1966) investigated the distribution on the sphere of the sub-vector (ℓ,m) constrained to lie on the unit circle. For the Fisher distribution this is equivalent to the conditional distribution of $\phi|\theta$, namely

$$f(\phi|\theta) = \frac{1}{2\pi I_0(\kappa\sin\mu_0\sin\theta)} \exp\{\kappa\sin\mu_0\sin\theta\cos(\theta-\nu_0)\}.$$

Maximum likelihood estimation is possible only of ν_0 and $\kappa\sin\mu_0$. The Dimroth-Watson distribution gives

$$g(\phi|\theta) = \frac{1}{2\pi} C(\kappa,\mu,\theta) \exp\left[-\frac{\kappa}{2}\{\sin^2\mu_0\sin^2\theta\cos^2(\theta-\nu_0) + \sin 2\mu_0\sin 2\theta\cos(\phi-\nu_0)\}\right],$$

where $\{C(\kappa,\mu,\theta)\}^{-1} = I_0(-\frac{\kappa}{2}\sin^2\mu_0\sin^2\theta)I_0(-\frac{\kappa}{2}\sin 2\mu_0\sin 2\theta)$

$$+ 2\sum_{r=1}^{\infty} I_r(-\frac{\kappa}{2}\sin^2\mu_0\sin^2\theta)I_{2r}(-\frac{\kappa}{2}\sin 2\mu_0\sin 2\theta).$$

Unlike the Fisher distribution, provided $\mu_0 \neq 0$, maximum likelihood estimators of all three parameters can be found (Edwards and Mardia, 1980).

4.4 Distribution on a Truncated Sphere from Uniformly Rotating Caps with Uniform Arrivals. Suppose that (μ_0, ν_0) are the coordinates of a recording station at a particular time on the earth, i.e., μ_0 is latitude and fixed whereas ν_0 is an instant of time, i.e., we can take $0<\nu_0\leq 2\pi$ for a day. In various phenomena such as in high energy particles, we can record observations only C_0 degrees from μ_0. Let

$$\alpha_1 = \min(C_0-\mu_0, \mu_0+C_0), \quad \alpha_2 = \max(C_0-\mu_0, \mu_0+C_0).$$

We have assumed $0<C_0<\pi/2$, $0<\mu_0<\pi/2$ without any loss of generality.

The equation of the small circle of the boundary of the cap can be seen as

$$\cos(\phi-\nu_0) = (\sin\mu_0 \sin\theta)^{-1} \cos C_0 - \cot \mu_0 \cot \theta.$$

Above this, the observations will be restricted to this cap. Hence, if there is a uniform distribution on the cap, we have (Edwards and Mardia, 1980)

$$f(\theta,\phi|\mu_0,\nu_0) = \begin{cases} C \sin \theta, & 0<\theta<\alpha_1, \quad 0<\phi<2\pi, \\ C \sin \theta, & \alpha_1<\theta<\alpha_2, \quad \nu_0-a(\theta)<\phi<\nu_0+a(\theta), \end{cases}$$

where $a(\theta) = \cos^{-1}\{(\sin \mu_0 \sin \theta)^{-1} \cos C_0 - \cot\mu_0 \cot\theta\}$. Note that the zone $0<\theta<\alpha_1$, $0<\phi<2\pi$ is the intersection of the caps with respect to rotation along the north axis.

The distribution of uniformly rotating caps is

$$g_{\mu_0}(\theta,\phi|\mu_0) = \int_0^{2\pi} f(\theta,\phi|\mu_0,\nu_0)\, d\nu_0.$$

It is found that

$$g(\theta,\phi|\mu_0) = \begin{cases} 2\pi C \sin\theta, & 0<\theta<\alpha_1, \\ 2Ca(\theta)\sin\theta, & \alpha_1<\theta<\alpha_2, \end{cases}$$

with $C = 1/\{2\pi(1-\cos C_0)\}$. As expected ϕ is uniformly distributed, and the marginal p.d.f., $h(\theta,\mu_0)$, of θ can easily be written down. These provide u_j's for (14) above. This idea can be extended to other distributions, such as the Fisher distribution.

5. FAMILIES OF DISTRIBUTIONS

5.1 Johnson and Wehrly's Bivariate Models. As we have seen, the maximum entropy densities do not in general lead to marginal distributions of types commonly used in directional data. To derive bivariate distributions with given marginals, Johnson and Wehrly (1978), Wehrly and Johnson (1980) use the following construction. Let $f_1(\theta)$ and $f_2(\phi)$ be specified densities on the circle and $F_1(\theta)$ and $F_2(\phi)$ be their distribution functions defined with respect to fixed, arbitrary origins. Also let $g(\cdot)$ be a density on the circle. Then

$$f(\theta,\phi) = 2\pi g\left[2\pi\{F_1(\theta) \pm F_2(\phi)\}\right] f_1(\theta)f_2(\phi),$$

where $0 \leqslant \theta,\phi < 2\pi$ are densities on the torus having the specified marginal densities $f_1(\theta)$ and $f_2(\phi)$. These distributions also lead to a family of distributions for a Markov process.

5.2 Generalized Exponential Models. Beran (1979) proposes as exponential models for directional data,

$$f_h(\underset{\sim}{x}) = \exp\{h(\underset{\sim}{x}) - d(h)\}, \qquad \underset{\sim}{x} \in S_p, \quad h \in M \tag{15}$$

where M is a subspace of $C(S_p)$, the set of all real-valued continuous functions whose domain is S_p, which is invariant under every rotation g in R^p such that $h(\cdot) \in M$ entails $h(g) \in M$. $d(h)$ is chosen to make f_h integrate to one. Both the von Mises-Fisher and the Bingham distributions are included as special cases.

In canonical exponential form (15) can be written in the form

$$f_{\underset{\sim}{\beta}}(\underset{\sim}{x}) = \exp\{\underset{\sim}{\beta}'\underset{\sim}{v}(\underset{\sim}{x}) - C(\underset{\sim}{\beta})\}, \qquad \underset{\sim}{\beta} \in R^q, \ \underset{\sim}{x} \in \underset{\sim}{S}_p,$$

where the $\{v_i : 1 \leqslant i \leqslant q\}$ are functions in $C(S_p)$ such that $\{1, v_1(\underset{\sim}{x}), \ldots, v_q(\underset{\sim}{x})\}$ are linearly independent and $\underset{\sim}{v}(\underset{\sim}{x})$ is the vecor $[v_1(\underset{\sim}{x}), \ldots, v_q(\underset{\sim}{x})]'$ and $\underset{\sim}{\beta} = [\beta_1, \ldots, \beta_q]'$.

Beran discusses both maximum likelihood estimation and also a regression estimator. Briefly, it represents on the circle a model such as

$$\exp\left\{\sum_{i=1}^{k} [\alpha_i \cos i\theta + \beta_i \sin i\theta] - C(\underset{\sim}{\alpha},\underset{\sim}{\beta})\right\}.$$

However, the problem is complicated on the sphere.

5.3 Additive Family. Suppose that $\theta_1, \theta_2, \theta_3$ are independently distributed. Let $\theta = (\theta_1+\theta_2) \bmod 2\pi$, $\phi = (\theta_1+\theta_3) \bmod 2\pi$. An important member is the bivariate von Mises with $\theta_i \sim M(\mu_i,\kappa_i)$, $i=1,2,3$. Using the bivariate normal case, we can define the correlation between θ and ϕ as

$$\rho = \sigma_1^2/\{(\sigma_1^2+\sigma_2^2)(\sigma_1^2+\sigma_3^2)\}^{\frac{1}{2}},$$

where $\sigma_i^2 = A(\kappa_i)$. Another member is the wrapped normal where the marginals are again wrapped normal. For large κ_1, κ_2 and κ_3, we have the bivariate normal situation. These have been used by Holmes and Mardia (1980) in various simulation studies.

5.4 Precision Family. A family which works for Euclidean as well as spherical space is given by

$$\exp\{\kappa\, \phi\left(\frac{x-\mu}{\sigma}\right) - u(\kappa)\},$$

where ϕ is a monotone function and κ is a precision parameter. Thus it contains normal, hyperbolic and von Mises distributions among others.

Under certain conditions, it can be shown that $u(\kappa)$ is strictly convex. For $\sigma = 1$, the maximum likelihood estimators of μ and κ are the solutions of

$$\sum_{i=1}^{n} \phi'(x_i-\hat{\mu}) = 0, \qquad u'(\hat{\kappa}) = \frac{1}{n}\sum_{i=1}^{n} \phi(x_i-\hat{\mu}).$$

These are unique under the condition given. Further, for large n, $\hat{\mu}$ and $\hat{\kappa}$ are independently distributed as $N(\mu,a/n)$ and $N(\kappa,b/n)$ where

$$a^{-1} = \kappa^2 \int \{\phi'(x-\mu)\}^2 f(x;\kappa,\mu)dx, \quad b^{-1} = u''(\kappa).$$

For higher dimensions, we have

$$\exp[\kappa\ \phi\{(\underset{\sim}{x}-\underset{\sim}{\mu})'\ \underset{\sim}{\Sigma}^{-1}(\underset{\sim}{x}-\underset{\sim}{\mu})\} - b(\underset{\sim}{\Sigma},\kappa)].$$

The Fisher and Bingham distributions are members of this class with $\underset{\sim}{x}'\underset{\sim}{x} = 1$. One important member is

$$C(\kappa)|\underset{\sim}{\Sigma}|^{-\frac{1}{2}}\exp[-\tfrac{\kappa}{2}\{(\underset{\sim}{x}-\underset{\sim}{\mu})'\underset{\sim}{\Sigma}^{-1}(\underset{\sim}{x}-\underset{\sim}{\mu}) - 1\}^2] \quad \underset{\sim}{x}\in R^2, \quad \kappa>0,$$

where $C(\kappa) = (\kappa/2\pi)^{\frac{1}{2}}/\{\pi\Phi(\kappa^{\frac{1}{2}})\}$, $\Phi(\cdot)$ = d.f. of $N(0,1)$. This has been used in a critical analysis of megalithic data where elliptic pattern or circular pattern is suspected (see Mardia and Holmes, 1980). Note that $-\phi$ can be replaced by a distance function between any two points x and μ in any general space. For further details, see Mardia (1980b).

6.1 Distributions on a Cylinder. Mardia and Sutton (1978) proposed a model for cylindrical variables (x,θ), $-\infty<x<\infty$, $0<\theta\leq 2\pi$, applications of which are found in rhythmometry, medicine, demography, biology and climatology. For example, one measurement related to wind direction and another to ozone concentration in pollution. The model has p.d.f. given by

$$f(x,\theta) = \{2\pi\ I_0(\kappa)\}^{-1}\ \exp\{\kappa\ \cos(\theta-\mu_0)\}$$

$$\times\ (2\pi\sigma_c^{\ 2})^{-\frac{1}{2}}\exp[-\ (x-\mu_c)^2/2\sigma_c^{\ 2}],$$

$\kappa>0$, $0<\mu_0\leq 2\pi$, $I_0(\kappa)$ is the modified Bessel function of the first kind and order zero and

$$\mu_c = \mu + \sigma\kappa^{\frac{1}{2}}\{\rho_1(\cos\theta - \cos\mu_0) + \rho_2(\sin\theta - \sin\mu_o)\},$$

$$\sigma_c^2 = \sigma^2(1-\rho^2) \quad \text{and} \quad \rho = (\rho_1^2+\rho_2^2)^{\frac{1}{2}}, \qquad 0 \leqslant \rho \leqslant 1.$$

The marginal distribution of θ is von Mises with mean direction μ_0 and concentration parameter κ, but the marginal distribution of x is complicated.

For $\rho = 0$, x and θ are independently distributed as $N(\mu, \sigma^2)$ and $M(\mu_0, \kappa)$ respectively. $\rho = 1$ implies perfect correlation.

Johnson and Wehrly (1977) proposed forming a model from considering a bivariate normal random variable (Y_1, Y_2) with means μ_1, μ_2, variances σ_1^2, σ_2^2 and covariance σ_{12}. Defining $\theta = Y_1 \pmod{2\pi}$, $X = Y_2$, (X,θ) has a characteristic function

$$\phi(p,t) = \exp\{-\tfrac{1}{2}(p^2\sigma_1^2+2pt\sigma_{12}+t^2\sigma_2^2) + i(p\mu_1+t\mu_2)\},$$

where p is an integer and t a real number. The marginal distribution of X is normal and that of θ, wrapped normal.

6.2 Shape Distributions (Triangle). Investigation of central place theory, (Mardia, *et al.*, 1977), and Ley lines (Kendall and Kendall, 1980), lead to appropriate distributions on triangles.

Consider three points $\underset{\sim}{x}_1, \underset{\sim}{x}_2, \underset{\sim}{x}_3$ in R^2. Miles (1970) gives the almost sure distribution of a random Delaunay triangle formed from a Poisson process in the plane. Let α_1, α_2, α_3 be the interior angles of the triangle formed from $\underset{\sim}{x}_1, \underset{\sim}{x}_2, \underset{\sim}{x}_3$. Miles finds the asymptotic distribution of a random triangle to be given by

$$f(\alpha_1,\alpha_2) = \frac{8}{3\pi}\sin\alpha_1 \sin\alpha_2 \sin(\alpha_1+\alpha_2), \; \alpha_1 > 0, \; \alpha_2 > 0, \; \alpha_1+\alpha_2<\pi.$$

Mardia, *et al.* (1977) find the joint distribution of the interior angles of a triangle when $\theta_1, \theta_2, \theta_3$ have independent distributions $M(\mu_j,\kappa)$, j=1,2,3, where the means are spread $2\pi/3$ apart. In this case,

$$f(\alpha_1,\alpha_2) = \frac{1}{\pi^2 I_0^3(\kappa)}\left[I_0[\kappa\{3+2\sum_{j=1}^{3}\cos(2\alpha_j-2\tfrac{\pi}{3})\}^{\frac{1}{2}}] + I_0[\kappa\{3+2\sum_{j=1}^{3}\cos(2\alpha_j+2\tfrac{\pi}{3})\}^{\frac{1}{2}}]\right],\quad \alpha_3=\pi-\alpha_1-\alpha_2.$$

If $\underset{\sim}{x}_1$, $\underset{\sim}{x}_2$, $\underset{\sim}{x}_3$ are independently normally distributed with common mean and covariance matrix $\underset{\sim}{\Sigma}=\sigma^2\underset{\sim}{I}$, then the joint p.d.f. of α_1, α_2, α_3, the interior angles of the triangle thus formed, is given by

$$f(\alpha_1,\alpha_2) = 6S(3-C)^{-2}, \tag{16}$$

where $C=\sum_{j=1}^{3}\cos 2\alpha_j$, $S=\sum_{j=1}^{3}\sin 2\alpha_j$.

If we let $\underset{\sim}{x}_1$, $\underset{\sim}{x}_2$, $\underset{\sim}{x}_3$ be independently normally distributed with means at the vertices of an equilateral triangle and common covariance matrix $\underset{\sim}{\Sigma}$, then

$$f(\alpha_1,\alpha_2) = \frac{3S}{4\pi(3-C)^2}\{1+\frac{3}{4\tau^2}(1+\frac{\sqrt{3S}}{3-C})\}\exp\{-\frac{3}{4\tau^2}(1-\frac{\sqrt{3S}}{3-C})\} + \frac{3S}{4\pi(3-C)^2}\{1+\frac{3}{4\tau^2}(1+\frac{\sqrt{3S}}{3-C})\}\exp\{-\frac{3}{4\tau^2}(1+\frac{\sqrt{3S}}{3-C})\}, \tag{17}$$

where $\tau=\sigma/\rho$, ρ being the radius of the circumcircle on which the means lie. As $\tau\to\infty$, so (17) tends to (16). See Mardia (1980a).

ACKNOWLEDGEMENTS. I wish to express my thanks to Rob Edwards for his help in preparing this paper.

REFERENCES

Beran, R. (1979). Exponential models for directional data. *Annals of Statistics*, 7, 1162-1178.

Bingham, C. (1974). An antipodally symmetric distribution on the sphere. *Annals of Statistics*, 2, 1201-1225.

Bingham, C. and Mardia, K. V. (1978). A small circle distribution on a sphere. *Biometrika*, 65, 379-390.

Cairns, M. B. (1975). *A structural model for the analysis of directional data.* Ph.D. thesis. University of Toronto.

Downs, T. D. (1966). Some relationships among the von Mises distributions of different dimensions. *Biometrika*, 53, 269-272.

Downs, T. D. (1974). Rotational angular correlations. In *Biorhythms and Human Reproduction*, M. Ferrin, F. Halberg, R. M. Richart and L. van der Wiele, eds. Wiley, New York. Pages 97-104.

Edwards, R. and Mardia, K. V. (1980). Some contributions to directional statistics. Research Report 7, Department of Statistics, University of Leeds.

Fraser, D. A. S. (1979). *Inference and Linear Models.* McGraw-Hill, London.

Holmes, D. and Mardia, K. V. (1980). Some statistical problems in megalithic data and directional analysis. Research Report 8, Department of Statistics, University of Leeds.

Johnson, R. A. and Wehrly, T. E. (1977). Measures and models for angular correlation and angular-linear correlation. *Journal of the Royal Statistical Society, Series B*, 39, 222-229.

Johnson, R. A. and Wehrly, T. E. (1978). Some angular-linear distributions and related regression models. *Journal of the American Statistical Association*, 73, 602-606.

Jupp, P. E. and Mardia, K. V. (1979). Maximum likelihood estimators for the matrix von Mises-Fisher and Bingham distributions. *Annals of Statistics*, 7, 590-606.

Jupp, P. E. and Mardia, K. V. (1980). A general correlation coefficient for directional data and related regression problems. *Biometrika*, 67, 163-174.

Kendall, D. G. and Kendall, W. S. (1980). Alignments in two-dimensional random sets of points. *Advances in Applied Probability*, 12, 380-424.

Kent, J. T. (1979). Discussion to "Edgeworth and saddle-point approximations with statistical applications" by O. Barndorff-Nielsen and D. R. Cox. *Journal of the Royal Statistical Society, Series B*, 41, 305-306.

Khatri, C. G. and Mardia, K. V. (1977). The von Mises-Fisher matrix distribution in orientation statistics. *Journal of the Royal Statistical Society, Series B*, 39, 95-106.

Mackenzie, J. K. (1957). The estimation of an orientation relationship. *Acta Crystallographica*, 10, 61-62.

Mardia, K. V. (1972). *Statistics of Directional Data*. Academic Press, London and New York.

Mardia, K. V. (1975a). Statistics of directional data (with discussion). *Journal of the Royal Statistical Society, Series B*, 37, 349-393.

Mardia, K. V. (1975b). Characterizations of directional distributions. In *Statistical Distributions in Scientific Work, Vol. 3*, G. P. Patil, S. Kotz, and J. K. Ord, eds. Reidel, Dordrecht-Holland. Pages 365-385.

Mardia, K. V. (1979). Discussion to "Edgeworth and saddle-point approximations with statistical applications" by O. Barndorff-Nielsen and D. R. Cox. *Journal of the Royal Statistical Society, Series B*, 41, 304.

Mardia, K. V. (1980a). Discussion to "Simulating the Ley Hunter" by S. Broadbent. *Journal of the Royal Statistical Society, Series A*, 123-137.

Mardia, K. V. (1980b). Precision-location family. Research Report 9, Department of Statistics, University of Leeds.

Mardia, K. V., Edwards, R., and Puri, M. L. (1977). Analysis of central place theory. *Bulletin of the International Statistical Institute*, XLVII(2), 93-110.

Mardia, K. V. and Gadsden, R. J. (1977). A circle of best-fit for spherical data and areas of vulcanism. *Journal of the Royal Statistical Society, Series C*, 26, 238-245.

Mardia, K. V. and Holmes, D. (1980). A statistical analysis of megalithic data under elliptic pattern. *Journal of the Royal Statistical Society, Series A*, 3, 293-302.

Mardia, K. V. and Khatri, C. G. (1977). Uniform distribution on a Stiefel manifold. *Journal of Multivariate Analysis*. 7, 468-473.

Mardia, K. V. and Puri, M. L. (1978). A robust spherical correlation coefficient against scale. *Biometrika*, 65, 391-396.

Mardia, K. V. and Sutton, T. W. (1978). A model for cylindrical variables with applications. *Journal of the Royal Statistical Society, Series B*, 40, 229-233.

Mardia, K. V. and Zemroch, P. J. (1977). Table of maximum likelihood estimates for the Bingham distributions. *Journal of Statistical Computation and Simulation*, 6, 29-34.

Miles, R. E. (1970). On the homogeneous planar Poisson point process. *Mathematical Biosciences*, 6, 85-127.

Rao, C. R. (1965). *Linear Statistical Inference and its Applications*. Wiley, New York.

Saw, J. G. (1978). A family of distributions on the m-sphere and some hypothesis tests. *Biometrika*, 65, 69-73.

Selby, B. (1964). Girdle distributions on a sphere. *Biometrika*, 51, 381-392.

Stephens, M. A. (1979). Vector correlation. *Biometrika*, 66, 41-48.

Wehrly, T. E. and Johnson, R. A. (1980). Bivariate models for dependence of angular observations and a related Markov process. *Biometrika*, 67, 255-256.

[*Received July* 1980. *Revised October* 1980]

SIZE DISTRIBUTION OF SUSPENDED PARTICLES — UNIMODALITY, SYMMETRY AND LOGNORMALITY

J. K. GHOSH and B. S. MAZUMDER

Indian Statistical Institute
Calcutta 700 035, India

SUMMARY. The shape of the grain size distribution of suspended material is related to the parameters of flow and the grain size distribution in the bed. It is noted that under certain conditions lognormality may be achieved in suspension even with a hyperbolic distribution in the bed.

KEY WORDS. lognormality, hyperbolic distribution, suspension, diffusion.

1. INTRODUCTION

Grain-size frequencies of naturally occurring sediment populations often follow lognormal (Krumbein, 1938) or hyperbolic distributions (Barndorff-Nielsen, 1977; Bagnold and Barndorff-Nielsen, 1979). Controlled flume experiments indicated that under suitable conditions lognormality can be attained through a process of grain sorting during suspension transportation in water flows, even when the source (bed) materials are not lognormal. These experiments also showed that the suspension load's grain-size distribution is related to flow velocity, height of suspension, and nature of the bed material (Sengupta, 1975, 1979). Lognormality was explained as a transitional phenomenon attained through a process of size sorting within a critical range of velocity and height above a sand bed of a given composition.

The purpose of the present paper is to develop a theoretical framework with a view to relating the suspension load's grain-size distribution, particularly the occurrence of unimodality, symmetry

C. Taillie et al. (eds.), Statistical Distributions in Scientific Work, Vol. 6, 21–32.

and lognormality, to flow parameters and grain-size distribution of the bed material. Grain-size data of two bed materials used for the earlier experiments (bed nos. 2 and 3 of Sengupta, 1979) have been utilized for the present discussion.

The theoretical model that we use here is a simpler version of an earlier model developed from the diffusion equation (Ghosh *et al.*, 1979). The model is briefly explained in Section 2 and compared with some observed data. The statistical consequences relating to shape of the suspension distribution are studied in Section 3, both for the observed data of Section 2 and for general lognormal and hyperbolic beds. Among other things it will be clear from the following discussion that if the bed is lognormal or hyperbolic with a mode lying between 1 and 5ϕ, (where $\phi = -\log_2 D$, D is grain diameter in mm) then under certain flow conditions the suspension distribution follows lognormality.

2. THE MATHEMATICAL MODEL

Consider a steady uniform flow of depth d and longitudinal velocity $u(y)$ at height y above the bed surface. The weight frequency $w_b(\phi)$ is the weight of grains in the sand bed in the range $\phi - .5$ to $\phi + .5$. (We shall also regard it as the weight frequency density at ϕ.) Let $w_b'(\phi) = w_b(\phi)/(\sum w_b(\phi))$. The bed roughness k_s is that value of ϕ for which $\sum_{\phi \geq k_s} w_b'(\phi) = .65$. The average concentration $S_y(\phi)$ at height y is the weight frequency in the range $\phi - .5$ to $\phi + .5$ per unit volume. The average concentration $S_y'(\phi) = S_y(\phi)/(\sum_{\phi} S_y(\phi))$. The average concentration will be assumed to depend only on the space coordinate y.

One needs a model for predicting $S_y'(\phi)$ for $w_b'(\phi)$. In an earlier paper (Ghosh *et al.*, 1979), we advocated a diffusion model for achieving this, and compared numerically our results with other existing approaches due to Einstein (1950) and Gessler (1965). We develop below a simplified version of this. A key step in our earlier approach was the observation that if we fit Hunt's (1954) velocity profile to the observed $u(y)$ and extrapolate it up to k_s, then the fitted value is zero at k_s. To retain this feature but simplify the model in Ghosh *et al.* (1979) we tried the following logarithmic profile

$$u(y) = (u_*/\chi)\log(y/k_s) \tag{1}$$

where u_* is the shear velocity, χ is the von-Kármán constant (0.4) and the constants have been adjusted to make $u(k_s) = 0$. It was found that the fit with observed velocity was excellent. Again for simplicity, we use the single diffusion equation for sediment

$$\frac{\partial S_y(\phi,t)}{\partial t} = \frac{\partial}{\partial y}[c(\phi)S_y(\phi,t)] + \frac{\partial}{\partial y}[\varepsilon(y)\frac{\partial S_y(\phi,t)}{\partial y}] \tag{2}$$

where $c(\phi)$ is the settling velocity of the grain (Terminal fall velocity of quartz spheres in water) and

$$\varepsilon(y) = u_*^2(1-y/d)/\frac{du}{dy}\ .$$

Under equilibrium conditions, the equation (2) leads to, *vide* Rouse (1938),

$$\varepsilon(y)\frac{dS_y(\phi,t)}{dy} + c(\phi)S_y(\phi,t) = 0. \tag{3}$$

Integrating the equation (3) from k_s to y, we get

$$S_y = S_{k_s}\left(\frac{d-y}{y}.\frac{k_s}{d-k_s}\right)^{\frac{c(\phi)}{\chi u_*}} \tag{4}$$

and hence

$$S'_y = S'_{k_s}(\phi)\ e^{-\psi c(\phi) + g(\psi)} \tag{5}$$

$$\text{where}\quad \psi = \frac{1}{\chi u_*}\ \log\left(\frac{y}{d-y}.\frac{d-k_s}{k_s}\right) \tag{6}$$

is a parameter summarizing the effect of flow and

$$e^{-g(\psi)} = \sum_\phi S'_{k_s}(\phi)\ e^{-\psi c(\phi)}. \tag{7}$$

Assuming $S'_{k_s} \simeq w'_b$, we can calculate $S'_y(\phi)$ from equation (5). Calculated values of $S'_y(\phi)$ as well as observed values of $S'_y(\phi)$ for beds 2 and 3 are shown in Figures 1 and 2, respectively.

Formula (5) is frequently used for obtaining $S'_y(\phi)$ from $S_{y_1}(\phi)$, where $0 < y_1 < y$, are both within the so-called suspension zone. This formula is not used for obtaining $S'_y(\phi)$ from w'_b for two reasons: (a) expression (5) breaks down if $y_1 = 0$ and (b) $y_1 = 0$ is outside the suspension region. Our main observation in this context is that (i) effective height of the bed (bed roughness, k_s) is not zero and (ii) at least for the grain sizes $0 \leqslant \phi \leqslant 5$, use of a simple diffusion model does not lead to greater error than the more complicated diffusion models discussed earlier (Ghosh *et al.*, 1979).

It is worth pointing out that equation (2) has an elegant probabilistic interpretation. Consider a particle of size ϕ whose displacement is Markovian with drift $b(y) = \varepsilon'(y) - c(\phi)$ and variance per unit time $a(y) = 2\varepsilon(y)$. Then Kolmogorov's forward differential equation becomes

$$\frac{\partial S_y(\phi,t)}{\partial t} = \tfrac{1}{2}\frac{\partial^2}{\partial y^2}[a(y)S_y(\phi,t)] - \frac{\partial}{\partial y}[b(y)S_y(\phi,t)] \qquad (8)$$

where $S_y(\phi,t)$ is the concentration at time t (before the steady state is reached); clearly equation (8) is identical with (2). Following Dynkin and Yushkevich (1969, Chapter 4), the upper boundary $y = d$ can be taken as repelling and the lower boundary reflecting. If we take the lower boundary to be reflecting, then the stationary distribution is given by

$$p(y,\phi) = \frac{K}{\varepsilon(y)}\exp\{\int_{k_s}^{y}\frac{b(y)}{2\varepsilon(y)}dy\} \qquad (9)$$

where K is chosen such that $\int_{k_s}^{y} p(y,\phi)dy = 1$.

If we assume that the supply $S''_y(\phi)$ of material of size ϕ from the bed is such that S_y is continuous at $y = k_s$, then $S''_{k_s} = S_{k_s}\varepsilon(k_s)/K$ and $S_y(\phi) = S''_{k_s}(\phi)p_y(\phi)$ is the solution (4).

Note that more generally as long as $b(y)$ and $a(y)$ are such that $y = d$ is repelling and $y = k_s$ is reflecting, our

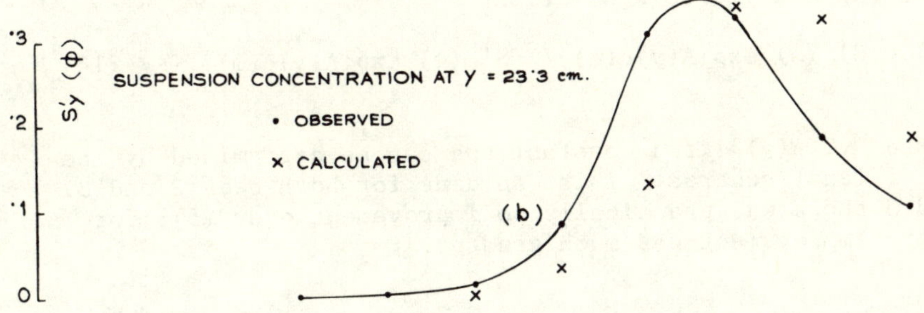

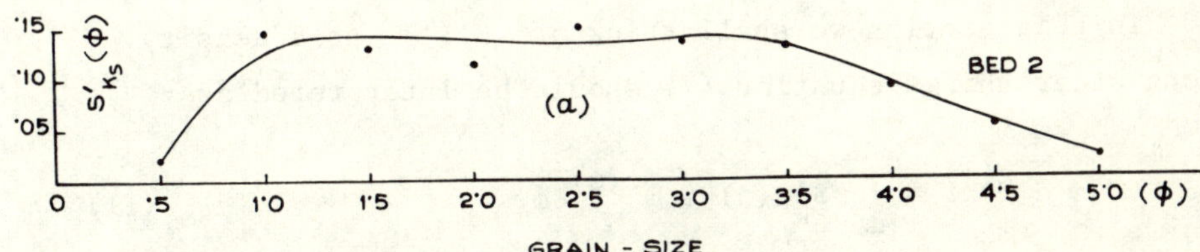

FIG. 1: Bed 2, grain-size distribution (relative concentration) (a) in the bed $S'_{k_s}(\phi)$; *(b) in suspension* $S'_y(\phi)$ *at* y = 23.3 *cm.*

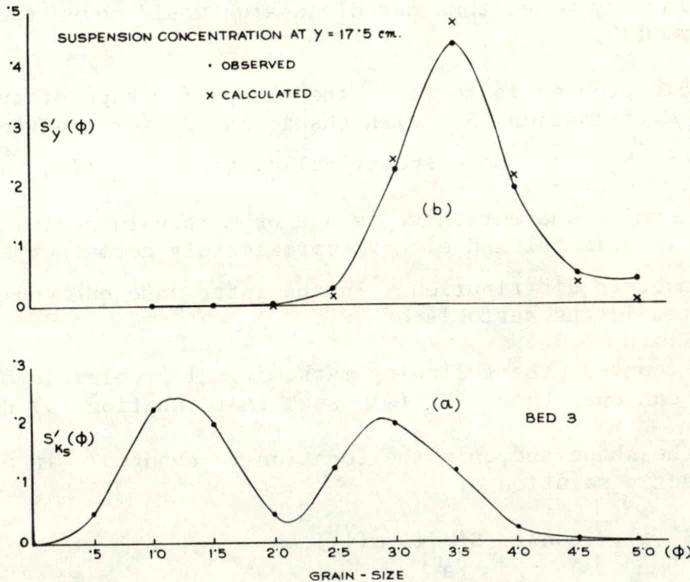

FIG. 2: Bed 3, grain-size distribution (relative concentration) (a) in the bed $S'_{k_s}(\phi)$;*(b) in suspension* $S'_{k_s}(\phi)$ *at* y = 17.5 *cm.*

method works and yields a solution

$$S'_y(\phi) = S'_{k_s}(\phi)\exp\{A(y)c(\phi)\}/\sum_{\phi} S'_{k_s}(\phi)\exp\{A(y)c(\phi)\}. \tag{10}$$

For fixed y, $A(y)$ is a constant and can be determined by the method of least squares. This was done for both beds 2 and 3. For bed 3 there was practically no improvement over (5); for bed 2 the improvement was much greater.

3. SHAPES OF SUSPENSION DISTRIBUTIONS FOR LOGNORMAL AND HYPERBOLIC BEDS

In this section we shall think of $S'_y(\phi)$ as a density. Among other things equation (7) should be interpreted as

$$e^{-g(\psi)} = \int_{-\infty}^{\infty} S'_{k_s}(\phi)\, e^{-\psi c(\phi)} d\phi. \tag{11}$$

In view of equation (5) it is more convenient to write $S'_y(\phi)$ as $S'_\psi(\phi)$. We shall assume throughout that expression (5) is true but in most of the discussion the actual form (6) of ψ will not play any role; thus our discussion would cover the more general form (10).

Our main problem is to study the change in shape of the suspension distribution S'_ψ with change in ψ for a given bed distribution $S'_{k_s}(\phi)$. More specifically, given $S'_{k_s}(\phi)$, we wish to determine whether there is one or more values of ψ for which S'_ψ is unimodal and either approximately normal or at least a symmetric distribution. In the later case one would also be interested in the kurtosis of S'_ψ.

In the context the following mathematical problem is of interest. Can one choose $S'_{k_s}(\phi)$ such that equation (5) does not alter the shape and only the location is changed? In other words we want a solution of

$$S'_\psi(\phi) = \text{const.}\ S'_{k_s}(\phi - \alpha(\psi)). \tag{12}$$

Here $\alpha(\psi)$ would be the difference in the mean or median of $S'_{k_s}(\phi)$ and $S'_\psi(\phi)$. The following proposition contains a complete answer:

Proposition. If $S'_{k_s}(\phi)$ and $S'_{\psi}(\phi)$ satisfy equation (12) and $\alpha(\psi)$ is continuously differentiable for $0 < \psi < \infty$, then $c(\phi)$ is linear in ϕ and $S'_{k_s}(\phi)$ is normal.

This may be proved using Theorems 4 and 4a of Dynkin (1951). We have also a direct proof which is omitted because of lack of space.

Since in our case $c(\phi)$ is not linear this must be regarded as a negative result implying the non-existence of $S'_{k_s}(\phi)$ satisfying equation (12).

As the following discussion shows, a more promising line of enquiry is to relax equation (12) by introducing a scale parameter $B(\psi)$ as well as a location parameter $\alpha(\psi)$, requiring a relation like (12) to hold only approximately and that too only for the range of ϕ of interest (from the point of view of studying suspension), namely $1 \leq \phi \leq 5$, i.e., we want to investigate the possibility of

$$S'_{\psi}(\phi) \simeq \text{const.}\ S'_{k_s}\{[\phi - \alpha(\psi)]/B(\psi)\}, \quad 1 \leq \phi \leq 5. \tag{13}$$

Let us begin by studying the question of unimodality of $S'_{\psi}(\phi)$. It is noted that $\log c(\phi)$ is approximately linear in ϕ (Table 1), i.e., $\log c(\phi) \simeq a + b\phi$, $1 \leq \phi \leq 5$, where $a = 3.5758$, $b = -1.1840$.

TABLE 1

ϕ	1.5	2.0	2.5	3.0	3.5	4.0	4.5	5.0
Observed $\log c(\phi)$	1.617	1.169	.688	.140	-.329	-1.139	-1.796	-2.526
Computed $\log c(\phi)$	1.800	1.208	.616	.024	-.568	-1.160	-1.752	-2.344

Let $C_{\psi}(\phi) = \log S'_{\psi}(\phi)$ and $C_{k_s}(\phi) = \log S'_{k_s}(\phi)$; using (11), equation (5) can be written as

$$C_{\psi}(\phi) = C_{k_s}(\phi) - \psi c(\phi) + g(\psi). \tag{14}$$

Now $C_\psi(\phi)$ may be expected to be unimodal if we can find a unique solution to the equation for the mode ϕ. Then from equation (14)

$$C'_\psi(\hat{\phi}) = C'_{k_s}(\hat{\phi}) - \psi e^{a+b\hat{\phi}} \cdot b = 0, \tag{15}$$

where primes denote derivatives with respect to ϕ. Since $1 \leqslant \phi \leqslant 5$, we are interested in solutions lying in $2 \leqslant \phi \leqslant 4$. Here (I) since $b < 0$, equation (15) can have a solution only when $C'_{k_s}(\hat{\phi}) < 0$, and (II) the solution of (15) is unique if $C''_\psi(\hat{\phi}) < 0$ for $2 \leqslant \hat{\phi} \leqslant 4$. Clearly a sufficient condition for (II) is $C''_{k_s}(\phi) \leqslant 0$ for $2 \leqslant \phi \leqslant 4$.

This condition holds if $S'_{k_s}(\phi)$ is normal or hyperbolic (Barndorff-Nielsen, 1977) with mode to the left of 2. For such $S'_{k_s}(\phi)$ the set of values of ψ for which equation (15) has a solution will determine the region of unimodality. Bed 3 in the range $2 \leqslant \phi \leqslant 5$ is approximately hyperbolic with mode near 3.0, and will be discussed later.

For bed 2, *vide* Figure 3, (I) implies that equation (15) can have a solution only to the right of $\phi = 3.5$; there a straight line describes $C_{k_s}(\phi)$ adequately so that (II) holds, i.e., if (15) has a solution, the solution is unique. Thus the conclusion of above paragraph holds for this bed. Now we will study the normality of suspension distribution above the bed 2 when equation (15) has a solution and (II) holds at $\phi = \hat{\phi}$. It can be checked from Table 1 that

$$c(\phi) \simeq c(\hat{\phi}) + \tfrac{1}{2} c''(\hat{\phi}) \cdot (\phi - \hat{\phi})^2, \qquad |\phi - \hat{\phi}| \leqslant 1 \tag{16}$$

and if we assume

$$C_{k_s}(\phi) \simeq C_{k_s}(\hat{\phi}) + \tfrac{1}{2} C''_{k_s}(\hat{\phi}) \cdot (\phi - \hat{\phi})^2, \qquad |\phi - \hat{\phi}| \leqslant 1, \tag{17}$$

then one gets from equation (14)

$$C_\psi(\phi) \simeq C_\psi(\hat{\phi}) + \tfrac{1}{2} C''_\psi(\hat{\phi}) \cdot (\phi - \hat{\phi})^2, \qquad |\phi - \hat{\phi}| \leqslant 1. \tag{18}$$

The reason for confining attention to $|\phi - \hat{\phi}| \leqslant 1$ in (16) is that most of the mass of S'_y is concentrated here. Clearly equation (17) holds for a lognormal or hyperbolic bed $S'_{k_s}(\phi)$ with mode to the left of $\phi = 2$; it also holds for bed 2. Then $S_\psi(\phi)$ looks approximately like a normal with mean $\hat{\phi}$ and

$$\text{Variance} = -C''_\psi(\hat{\phi}) = -C''_{k_s}(\hat{\phi}) + \psi b^2 e^{a+b\hat{\phi}}. \tag{19}$$

We shall now study in detail the bed 3. In this case we fit a hyperbolic distribution to $S'_{k_s}(\phi)$ for $2 \leqslant \phi \leqslant 5$ as

$$\log S'_{k_s}(\phi) \simeq \nu - \tfrac{1}{2}(\gamma_1+\gamma_2)\{\delta^2+(\phi-\mu)^2\}^{\frac{1}{2}} + \tfrac{1}{2}(\gamma_1-\gamma_2)(\phi-\mu). \tag{20}$$

We determine the parameters γ_1, γ_2 and μ graphically using the geometrical interpretation of these parameters given by Bagnold and Barndorff-Nielsen (1979) and δ from their equation (2.2). The constant ν has been adjusted to agree with the observed frequency in the range $2 \leqslant \phi \leqslant 5$. The estimated values are $\gamma_1 = 1.787$, $\gamma_2 = 4.294$, $\mu = 3.228$, $\delta = .5039$ and $\nu = -.225$.

Here also from Figure 3, it is clear by (I) that equation (15) can have a solution only to the right of $\phi = 3.0$ and (II) also holds for $\phi \geqslant 3.23$. Hence, whenever (15) has a solution, it is unique so that the corresponding $S'_\psi(\phi)$ is unimodal.

We shall now examine when $S'_\psi(\phi)$ can be symmetrical as well as unimodal. Expanding $c(\phi)$ around $\hat{\phi}$ up to the quadratic term, we get

$$\begin{aligned}\log S'_\psi(\phi) \simeq\ & g(\psi) + \nu - \tfrac{1}{2}(\gamma_1+\gamma_2)[\delta^2+(\phi-\mu)^2]^{\frac{1}{2}} \\ & + \tfrac{1}{2}(\gamma_1-\gamma_2)(\phi-\mu) - \psi c(\hat{\phi}) \\ & - \psi b(\phi-\hat{\phi})c(\hat{\phi}) - \tfrac{1}{2}\psi b^2(\phi-\hat{\phi})^2 c(\hat{\phi}).\end{aligned} \tag{21}$$

In order to get symmetry $\hat{\phi}$ must be nearly equal to μ and

$$\tfrac{1}{2}(\gamma_1-\gamma_2) - \psi b c(\hat{\phi}) \simeq 0. \tag{22}$$

(Note that (22) follows from (15) if $\hat{\phi} \simeq \mu$.) Then around $\hat{\phi}$, we have

$$\begin{aligned}\log S'_\psi(\phi) \simeq\ & g(\psi) + \nu - \tfrac{1}{2}(\gamma_1+\gamma_2)[\delta^2+(\phi-\hat{\phi})^2]^{\frac{1}{2}} \\ & - \tfrac{1}{2}\psi b^2(\phi-\hat{\phi})^2 c(\hat{\phi}).\end{aligned} \tag{23}$$

Inspection of (23) suggests that the presence of the term

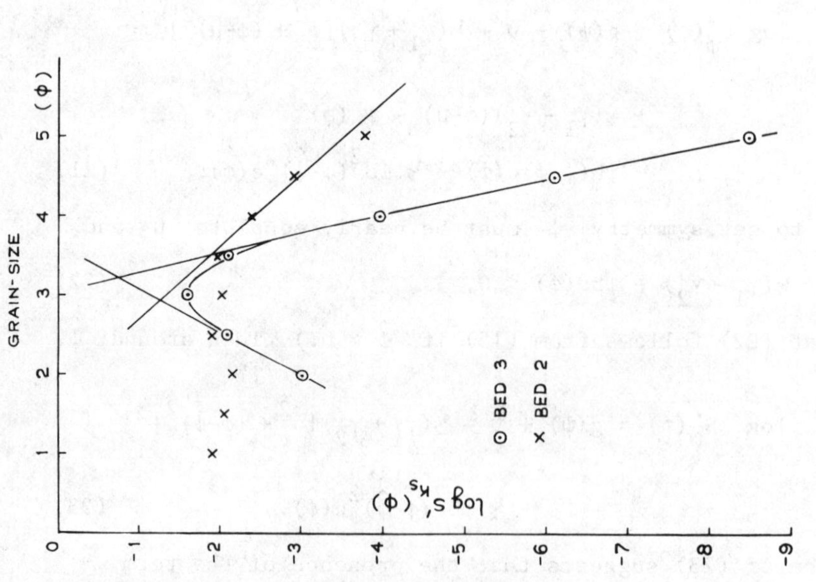

FIG. 3: Grain-size distribution of beds 2 and 3 ($\log S'_{k_s}(\phi)$ against ϕ).

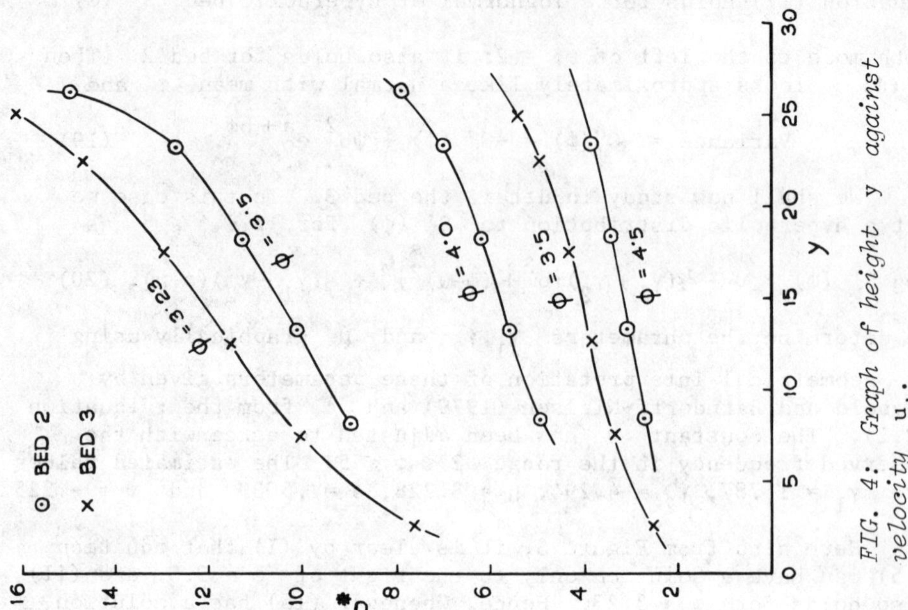

FIG. 4: Graph of height y against velocity u_*.

$[\delta^2 + (\phi-\hat{\phi})^2]^{\frac{1}{2}}$ is likely to lead to less peakedness and hence higher values of the coefficient of kurtosis β_2 than the normal. Confirmation of this expectation is provided in the following numerical calculations.

For the values of $y = 17.5$ cm, $u_* = 6.297$ cm/sec, $d = 27.5$ cm and $k_s = .0451$ cm, we get the value $\psi = 2.7676$. The skewness and kurtosis of the corresponding suspension distribution are, respectively, 0.343 and 4.152 (see Figures 12 and 13 of Sengupta, 1979).

The above analysis illustrates how one can study the unimodality, symmetry and normality of S'_ψ for a given hyperbolic bed distribution $S'_{k_s}(\phi)$.

We now study briefly how one can determine, for a given bed distribution, the flow parameters y and u_* leading to unimodality. Here the relation (6) determining ψ will be used. To fix ideas we work with bed 2. Since by (I) and Figure 3, equation (15) can have a solution only for $\hat{\phi} \geqslant 3.5$ and observed data with $\hat{\phi} > 4.5$ is likely to be scarce, let us work with $3.5 \leqslant \hat{\phi} \leqslant 4.5$. For fixed $\hat{\phi}$ in this range we now solve (15) for ψ; for $\hat{\phi} = 3.5$, $\psi = 1.663$ and for $\hat{\phi} = 4.5$, $\psi = 5.441$. The curves obtained by plotting y against u_* for these two fixed values of ψ in (6) are shown in Figure 4. The zone between these curves gives the values of y and u_* which will give unimodality with peak at some $3.5 \leqslant \hat{\phi} \leqslant 4.5$. As noted before these unimodal distributions will be approximately log-normal.

A similar analysis was made for bed 3. To achieve symmetry we kept $\hat{\phi}$ in the range 3.23 to 3.5. The resulting curves (y against u_*) are also shown in Figure 4. The combinations (y, u_*) obtained this way agree well with our experimental observations for both beds 2 and 3.

ACKNOWLEDGEMENT

We thank Prof. S. Sengupta for suggesting the problem and for providing experimental data.

REFERENCES

Barndorff-Nielsen, O. (1977). Exponentially decreasing distributions for the logarithm of particle size. *Proceedings of the Royal Society, London,* A353, 401-419.

Bagnold, R. A. and Barndorff-Nielsen, O. (1979). The pattern of natural size distributions. Research Report No. 47, Department of Theoretical Statistics, Aarhus University. Denmark.

Dynkin, E. B. (1951). Necessary and sufficient statistics for a family of probability distribution. *Uspehi Mat. Nauk. (N.S.),* 6, 68-90. (Also selected *Translations in Mathematical Statistica and Probability,* 17-40).

Dynkin, E. B. and Yushkevich, A. A. (1969). *Markov Processes.* Plenum Press, New York.

Einstein, H. A. (1950). The bed load function for sediment transportation in open channel flows. Technical Bulletin No. 1026, U. S. Dept. of Agriculture, Soil Conservation Service, Washington, D.C.

Gessler, J. (1965). The beginning of bed load movement of mixtures investigated as natural armoring in channels. Rep. 69, Lab. Hydr. Res. and Soil Mech., ETH, Zurich. (English Trans. by W. M. Keck Lab. Calif. Inst. Tech., California 1967).

Ghosh, J. K., Mazumder, B. S. and Sengupta, S. (1979). Methods of computation of suspended load from bed materials and flow parameters. Flume Project, Technical Report No. #1/79, Indian Statistical Institute, Calcutta.

Hunt, J. N. (1954). The turbulent transport of suspended sediment in open channels. *Proceedings of the Royal Society, London,* A224, 322-335.

Krumbein, W. C. (1938). Size frequency distributions of sediments and the normal Phi curve. *Journal of Sedimentary Petrology,* 8, 84-90.

Rouse, H. (1938). *Fluid Mechanics for Hydraulic Engineers* (Dover edition, 1961). Dover, New York.

Sengupta, S. (1975). Size-sorting during suspension transportation-lognormality and other characteristics. *Sedimentology,* 22, 257-273.

Sengupta, S. (1979). Grain-size distribution of suspended load in relation to bed materials and flow velocity. *Sedimentology,* 26, 63-82.

[*Received September* 1980]

OFFSHORE OIL/GAS LEASE BIDDING AND THE WEIBULL DISTRIBUTION

DANNY DYER

Department of Mathematics
The University of Texas at Arlington
Arlington, Texas 76019 USA

SUMMARY. Competitive sealed bids for offshore oil and gas leases issued by the federal government of the United States are usually taken to be lognormally distributed. However, there are known discrepancies under the lognormal model between the observed and theoretical "money left on the table" -- one of the important indicators of how competitive is the bidding process. By using multi-sample goodness-of-fit tests, the Weibull distribution is shown to be a more viable bid model than the lognormal distribution for tracts leased from 1954 to 1978 which received 11 or more non-low, noise bids. Under the Weibull-bids hypothesis, the concept of "money left on the table" is examined. In addition, tests for homogeneity of variances of log-bids are discussed.

KEY WORDS. Weibull distribution, extreme value distribution, goodness-of-fit tests, Bartlett's test, competitive bidding, oil and gas leases.

1. INTRODUCTION

Periodically since 1954 the federal government of the United States has offered for lease tracts of land off the United States coastline. There is frequent and spirited competition among oil and gas companies for the leases on certain tracts because of the possibility that hydrocarbons might be found therefrom. When an offshore tract is offered for lease, competitive sealed bids are received by the Bureau of Land Management and the submitter of the highest bid is awarded the lease. The leasee then has exclusive rights for a period of time to explore for hydrocarbons

C. Taillie et al. (eds.), Statistical Distributions in Scientific Work, Vol. 6, 33–45.

on that tract. Since the first federal offshore oil and gas lease sale on October 13, 1954 through a sale held on December 19, 1978, over 3300 tracts have been leased by the federal government to oil and gas companies. Each lease may be identified by its outer continental shelf (OCS) number. The amount of money paid to the federal government for these leases has been in excess of $20 billion. Approximately half of the leases received only one or two bids and are usually excluded from statistical analyses. Approximately 75% of the leases received five or fewer bids. The maximum number of bids received on a lease was eighteen.

The competitive sealed bids for an individual offshore oil and gas lease are often assumed to follow a lognormal distribution (Brown, 1969; Crawford, 1970; Pelto, 1971). Under the lognormality assumption, however, there are known discrepancies between the observed and theoretical "money left on the table" -- one of the more important indicators of how competitive is the bidding process (Dougherty and Lohrenz, 1976, 1977). In addition, it is sometimes believed that for all tracts within a sale, the standard deviations of the logarithms of bids (log-bids) are not significantly different -- thus implying that large bids do not tend to be proportionately more or less precise than small ones (Brown, 1969, p. 37). Based on results from Bartlett's test for homogeneity of variances (Dyer and Keating, 1980), Dougherty and Lohrenz (1977) have argued against such a belief. However, Bartlett's test is highly sensitive to the lognormality assumption in the sense that if the bids are not lognormally distributed, the hypothesis of equal variances may well be rejected far more often, on the average, than would otherwise occur under the nominal significance level (Box, 1953). It has therefore become rather judicious to test the hypothesis that the bids on various groups of leases follow lognormal distributions. To carry out such a test, a multi-sample Shapiro-Wilk W-test may be implemented (Wilk and Shapiro, 1968). Based on this test procedure, it has been found that the lognormal-bids hypothesis is rejected on certain groups of leases (for example, the 15-bid leases) but not rejected on other groups of leases (for example, the 16-bid leases). These results along with a study by Genter and Bruckner (1978) would indicate, at least on a large scale basis, the lognormal distribution should *not* be assumed as the underlying distribution of the bids.

As an alternative to the lognormal distribution, the Pareto distribution has been recently considered due to its frequent use as a statistical model in certain economic situations. And although the Pareto-bids hypothesis may be tenable on certain small groups of leases (Dyer, 1979), it has been shown (Keating and Bruckner, 1979) that the lognormal distribution is a much more suitable model on a large scale basis. Thus the problem remains as to an acceptable model for those groups of leases for which lognormality has been rejected. Of course, it may also be the case that even for those groups of leases for which lognormality was not rejected, yet some other distribution might also be a plausible model (Bruckner and Johnson, 1978).

It is the purpose of this paper to study the use of the Weibull distribution as an acceptable model for the distribution of bids on a lease. A similar study has been made by Smiley (1979) who advocates the adoption of the Weibull distribution in lieu of the lognormal distribution as a model for the lease bids. Unfortunately, the test procedures used are somewhat questionable. In addition to Smiley's work, we are motivated by (a) a discussion by Rothkopf (1969) on the use of the Weibull distribution in modeling a competitive bidding process and (b) the fact that for many of the leases, the histograms for the log-bids have a negatively skewed appearance -- the relevancy of which will become clear in the next section. We discuss multi-sample test procedures for discriminating between a lognormal distribution and a Weibull distribution and for testing the Weibull-bids hypothesis. Under the Weibull-bids hypothesis, the concept of "money left on the table" is examined. In addition, tests for homogeneity of variances of log-bids are discussed.

The data base used throughout this study consists of all federal offshore oil and gas leases which received 11 or more sealed bids (low, noise bids are excluded) -- a total of 120 leases. The data are compiled by the General Services Administration Data Services Division, Fort Worth, Texas. Low, noise bids (bids that do not appear to have been made in the same competitive spirit as the other bids) have been identified by the Dougherty and Lohrenz (1976) "30-30 algorithm". When low, noise bids are excluded, the maximum number of bids received on a lease becomes sixteen.

2. DISCRIMINATION BETWEEN THE LOGNORMAL AND WEIBULL DISTRIBUTIONS

On the basis of n independent observations $x_1, x_2, \cdots, x_n$, we wish to choose either the lognormal distribution with density function

$$f_L(x;\mu,\sigma) = (\sqrt{2\pi}\,\sigma x)^{-1} \exp[-(\ln x - \mu)^2/2\sigma^2], \quad -\infty < x < \infty \qquad (1)$$

or the Weibull distribution with density function

$$f_W(x;b,c) = (c/b)(x/b)^{c-1} \exp[-(x/b)^c], \quad x > 0 \qquad (2)$$

as a model. By transforming the data by taking logarithms, say $z_i = \ln x_i$, we equivalently would be choosing between the normal distribution with location and scale parameters μ and σ or the Type I extreme value distribution (for minimum values) with density function

$$f_{EV}(z;\xi,\delta) = (1/\delta)e^{(z-\xi)/\delta}\exp[-e^{(z-\xi)/\delta}], \quad -\infty < z < \infty \qquad (3)$$

where $\xi = \ln b$ and $\delta = 1/c$ are location and scale parameters. The density function given by (3) is *negatively skewed*.

The ratio of maximized likelihoods (RML) test of H_0: $X \sim f_L(x;\mu,\sigma)$ against $H_1: X \sim f_W(x;b,c)$ rejects H_0 (see Dumonceaux and Antle, 1973) whenever

$$\mathrm{RML} \geqslant \mathrm{RML}_\alpha, \qquad (4)$$

where the test statistic

$$\mathrm{RML} = (2\pi e\hat{\sigma}^2)^{\frac{1}{2}}\left[\prod_{i=1}^{n} X_i f_W(X_i;\hat{b},\hat{c})\right]^{1/n} \qquad (5)$$

$$\text{with} \quad \hat{\sigma}^2 = \sum_{i=1}^{n} (\ln X_i - \hat{\mu})^2/n, \quad \hat{\mu} = \sum_{i=1}^{n} (\ln X_i)/n, \qquad (6)$$

and the maximum likelihood estimators of b and c, $\hat{b}$ and $\hat{c}$, are determined by solving the equations

$$n/\hat{c} - n\sum_{i=1}^{n} X_i^{\hat{c}} \ln X_i \Big/ \sum_{i=1}^{n} X_i^{\hat{c}} + \sum_{i=1}^{n} \ln X_i = 0 \qquad (7)$$

$$\hat{b} = \left(\sum_{i=1}^{n} X_i^{\hat{c}}/n\right)^{1/\hat{c}}. \qquad (8)$$

Since the distribution of RML is independent of parameters (Antle and Bain, 1969), the size α critical values RML_α may be obtained by Monte Carlo techniques.

In multi-sample situations we do not require the critical value but instead determine the attained significance level (ASL). For the j*th* sample,

$$\mathrm{ASL}(j) = \mathrm{Prob}\{\mathrm{RML} \geqslant \mathrm{RML}_0 \text{ when lognormality is present}\},$$

where RML_0 is the calculated value of RML. Littell and Folks (1973) have shown that Fisher's method (1950, pp. 99-101) is optimal (in the sense of Bahadur relative efficiency) among essentially all methods of combining independent tests. Specifically, we reject the hypothesis that k samples follow lognormal distributions as opposed to Weibull distributions at the level of significance α if

$$- 2 \ln\left(\prod_{j=1}^{k} ASL(j)\right) > \chi^2_{2k}(1-\alpha), \tag{9}$$

where $\chi^2_{2k}(1-\alpha)$ is the $100(1-\alpha)th$ percentile of the χ^2-distribution with 2k degrees of freedom. The percentiles of the distribution of RML have been determined empirically for sample sizes of $n = 11(1)16$ and are available from the author. For a given value of n, a random sample of size n was generated from a lognormal distribution with $\mu = 0$ and $\sigma = 1$ using the IMSL random lognormal deviate generator subroutine. Based on the sample data, the value of RML was then calculated. This procedure was followed 5000 times. The 5000 values of RML were tabulated in increasing order. The jth percentile of the distribution of RML was taken to be the $50jth$ ordered value.

For each lease in the data base, the value of the test statistic was calculated by using equations (5)-(8) and the bids on the lease. The value RML_0 was then approximately located in the tabulated percentiles and the corresponding ASL was determined. Based on these determinations, a summary of the test results is given in Table 1. The lognormal distribution as opposed to the Weibull distribution as a model for the bids in each group of leases as well as in the group of all leases considered is overwhelmingly rejected. For example, for the group of 12-bid leases, the value of the test statistic to test H_0: lognormal *vs.* H_1: Weibull is 115.895. A value at least as large as this would be obtained less than once in 10,000 times, on the average, if lognormal distributions as opposed to Weibull distributions were the true underlying bid models.

3. TESTING THE WEIBULL-BIDS HYPOTHESIS

The test results of Section 2 indicate that Weibull distributions provide better fits, collectively speaking, for the lease bids than do lognormal distributions. This does not infer, however, the acceptability of the Weibull model. We now examine an omnibus goodness-of-fit test for the Weibull-bids hypothesis.

TABLE 1: Test results for groups of leases.

Group of Leases	lognormal *vs.* Weibull		Weibull *vs.* arbitrary	
	$-2\ \ln(\Pi\ \text{ASL}(j))$	ASL	$-2\ \ln(\Pi\ P_j)$	ASL
11-bid	124.331	$< 10^{-4}$	84.693 68.386[a]	.110 .335[a]
12-bid	115.895	$< 10^{-4}$	65.074 54.254[b]	.445 .688[b]
13-bid	101.624	$< 10^{-4}$	68.551 59.865[c]	.120 .275[c]
14-bid	41.160	.014	28.468	.245
15-bid	49.555	$< 10^{-4}$	16.292	.581
16-bid	17.184	.024	12.234	.141
all	449.749	$< 10^{-10}$	275.312 239.499[abc]	.057 .289[abc]

Note: (a) OCS Nos. 981, 1743, and 2020 excluded.
(b) OCS Nos. 2018 and 3481 excluded.
(c) OCS No. 2238 excluded.

Mann, Scheuer, and Fertig (1973) have developed a test procedure for the Weibull or Type I extreme value distribution which exploits the fact that the right tail of the Type I extreme value distribution is "shorter" than that of usual appropriate alternative distributions, while the left tail is "longer." Let $Z_{(i)} = \ln X_{(i)}$ denote the logarithm of the *ith* ordered observation. The test statistic is

$$S = \frac{\sum_{i=[n/2]+1}^{n-1} (Z_{(i+1)} - Z_{(i)})/[E(V_{(i+1)}) - E(V_{(i)})]}{\sum_{i=1}^{n-1} (Z_{(i+1)} - Z_{(i)})/[E(V_{(i+1)}) - E(V_{(i)})]} , \qquad (10)$$

where $E(V_{(i)}) = E[(Z_{(i)} - \xi)/\delta]$ is the expected value of the *ith* reduced extreme value order statistic. The hypothesis that the sample, i.e., X_i's, was drawn from a Weibull distribution is rejected if the calculated value of S, say S_0, exceeds the tabulated percentile (see Mann, Scheuer, and Fertig, 1973). As a multi-sample test procedure, we require the attained significance level (ASL) for the *jth* sample as given by

$$P_j = \text{Prob}\{S > S_0 \text{ when the Weibull distribution is present}\}$$

$$= \sum_{i=0}^{[\![(n-3)/2]\!]} \binom{n-2}{i} S_0^i (1-S_0)^{n-2-i} . \qquad (11)$$

Using Fisher's method, we reject the hypothesis that Weibull distributions model the collection of k independent samples at level of significance α if

$$- 2 \ln\left(\prod_{j=1}^{k} P_j\right) > \chi^2_{2k}(1-\alpha) . \qquad (12)$$

A power study by Mann, Scheuer, and Fertig (1973) found the S-test to be the most powerful (among those tests considered) for discriminating between a Weibull distribution and a lognormal distribution. Littell, McClave, and Offen (1979) have shown the S-test also performed rather well for various other alternative distributions.

Testing that the bids for a particular offshore oil and gas lease follow a Weibull distribution is equivalent to testing that the log-bids follow a Type I extreme value distribution.

For each lease, we calculated the value of the test statistic, S_0, and the corresponding ASL. A summary of the test results is given in Table 1. The hypothesis that the bids follow Weibull distributions is not rejected at the 10% level of significance for *each* group of 11-, 12-, 13-, 14-, 15-, and 16-bid leases. The hypothesis that the bids on all leases considered follow Weibull distributions is not rejected at the 5% level of significance. The somewhat low overall group ASL of .057 is, however, due primarily to very low ASL's on 6 leases (OCS Nos. 981, 1743, 2018, 2020, 2238, and 3481). Based on the ASL for the RML-test for these leases, it would appear that the bids on these leases are substantially more likely to follow lognormal distributions than Weibull distributions. If these leases are removed from consideration, then based on the remaining 114 leases (95% of 120 total leases), $-2 \ln(\prod_j P_j) = 239.499$ and the corresponding ASL is .289. In this case, the Weibull-bids hypothesis is not rejected at even the 25% level of significance.

4. MONEY LEFT ON THE TABLE

The phrase "money left on the table" refers to the amount by which the winning highest bid exceeds the second highest bid. Since each bid is sealed and therefore unknown (supposedly) to all other bidders at the time the bids are made, money left on the table is the amount of money in excess of what would have been necessary to win had the bidding process been open. Such an amount is often used to indicate how competitive is the bidding process.

For a particular lease for which n bids were received, let $X_{n(j)}$ be the j*th* largest bid. The fraction of money left on the table, denoted by LOT_n, is the amount of money left on the table normalized by the winning bid, that is

$$LOT_n \equiv (X_{n(n)} - X_{n(n-1)})/X_{n(n)}. \tag{13}$$

By taking expectations under the Weibull-bids hypothesis, it follows that

$$E[\ln(1 - LOT_n)] = -\delta[E(V_{n(n)}) - E(V_{n(n-1)})]$$

or, equivalently, $- c\ E[\ln(1-LOT_n)] = E(V_{n(n)}) - E(V_{n(n-1)})$, (14)

where $V_{n(j)} = (\ln X_{n(j)} - \xi)/\delta$ is the j*th* reduced extreme value order statistic. The left hand side of (14) is termed the LOT-transform and is a function of n alone. For given n, its value may be determined from tables given by Mann, Scheuer, and Fertig (1973). The LOT-transform is shown as a solid curve in Figure 1 and can be used to compare with actual bid data. To make such a comparison, the observed value of the LOT-transform for a particular lease for which n bids were received is calculated by

$$\text{LOT-transform (observed)} = -\hat{c}\ln(1-\text{LOT}_n) = \hat{c}(\ln x_{n(n)} - \ln x_{n(n-1)}), \quad (15)$$

where $\hat{c}$ is the maximum likelihood estimate of the Weibull parameter c based on the bids for that lease. For each lease, the observed LOT-transform was calculated. For all leases for which n bids were received, the average observed value of the LOT-transform is plotted in Figure 1 (except the case where $n = 16$ for which there are only 4 leases). There is good agreement with the theoretical curve (average absolute error is about .09) which is consistent with the conclusion of the Dougherty and Lohrenz (1977) study as to possible reasons for the disparity between the observed and theoretical values of the LOT-transform (average absolute error is about .25) assuming lognormally distributed bids on a lease. Under the lognormal-bids assumption, less money is left on the table than predicted. The conclusion was that due to "bidders' tendencies to restraint when the stakes are extraordinarily high (as they tend to be when the number of bids on a lease is high)," the bids apparently depart from lognormality.

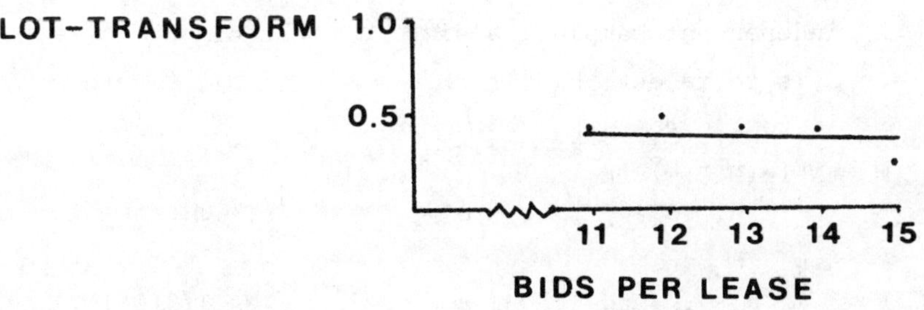

FIG. 1: A comparison of the observed and theoretical LOT-*transform.*

5. STANDARD DEVIATIONS OF LOG-BIDS

Variability of the logarithms of bids has often been examined (Brown, 1969; Pelto, 1971; Dougherty and Lohrenz, 1976, 1977) -- the importance being that the standard deviations of log-bids may be used to measure the precision or imprecision of the bidding process. For example, if the standard deviation of a 5-bid lease is not significantly different from that of a 15-bid lease, one implication is that large bids do not tend to be proportionately more or less precise than small ones. There have been arguments both for as well as against the hypotheses that the standard deviations of log-bids on individual leases are constant (a) for all leases and (b) for all leases within a particular sale. We examine these hypotheses under the assumption that the log-bids follow Type I extreme value distributions.

The test procedure we use is given in Bain (1978, p. 288) and is designed to test equality of shape parameters among several supposedly Weibull populations. It is a variation of Bartlett's test for homogeneity of variances. For a given lease for which n bids were received, we calculated an estimate of δ given by

$$\tilde{\delta} = \frac{-\sum_{i=1}^{s} \ln x_{n(i)} + [s/(n-s)] \sum_{i=s+1}^{n} \ln x_{n(i)}}{nk_n} ,$$

where k_n is an unbiasing constant tabled in Bain (1978, p. 268) and $s = [\![.84n]\!]$. This estimate of δ is linear unbiased and has high efficiency relative to the maximum likelihood estimate of δ. An estimate of the lease standard deviation is $1.283\,\tilde{\delta}$ which, for each lease, is available from the author. Approximately $h\tilde{\delta}/\delta \sim \chi^2(h)$, where the values of h are tabled in Bain (1978, p. 268).

For k independent samples, a size α test of $H_0: \delta_1 = \delta_2 = \cdots = \delta_k$ is to reject H_0 if $M/c > \chi^2_{k-1}(1-\alpha)$, where

$$M = N \ln(N^{-1} \sum_{j=1}^{k} h_j \tilde{\delta}_j) - \sum_{j=1}^{k} h_j \ln \tilde{\delta}_j ,$$

$$N = \sum_{j=1}^{k} h_j , \quad \text{and} \quad c = 1 + (\sum_{j=1}^{k} h_j^{-1} - N^{-1})/3(k-1).$$

If H_0 is not rejected, a combined samples estimate of δ which has minimum variance among the class of unbiased estimators of δ which are linear functions of the $\tilde{\delta}_i$'s is given by

$$\tilde{\delta}_c = \sum_{j=1}^{k} h_j \tilde{\delta}_j / N .$$

The corresponding combined samples estimate of the common standard deviation is $1.283\ \tilde{\delta}_c$.

The results of the implementation of the above test to the data base are given in Table 2. The hypothesis of homogeneity of standard deviations of log-bids on all leases is rejected. All leases were then grouped according to sale dates; however, only sale dates with three or more leases were tested. In all cases, the hypothesis of homogeneity of standard deviations of log-bids on leases within a sale date is not rejected. Moreover, there is some indication of a possible time trend -- sale dates in the 1960's and early 1970's, generally speaking, appear to have somewhat larger standard deviations than the more recent sale dates. Many more leases would, however, be needed to examine this possibility.

TABLE 2: Test results for H_0: *homogeneity of variances of log-bids (* indicates* H_0 *is rejected).*

Sale Date	No. of Leases	M/c	ASL	Combined Samples Estimate of Standard Deviation
3/16/62	3	.851	.665	1.357
6/13/67	16	4.865	.988	1.133
5/21/68	9	5.630	.697	1.592
12/15/70	34	29.717	.590	1.180
9/12/72	4	4.748	.199	1.107
12/19/72	18	21.653	.201	.889
6/19/73	18	19.053	.334	.902
5/29/74	3	.936	.640	.755
6/23/77	3	.521	.761	.884
all leases	120	167.284	.002*	---

ACKNOWLEDGEMENTS

This research was supported in part by the Conservation Division, U. S. Geological Survey, Denver, Colorado under the auspices of John Lohrenz. The author gratefully acknowledges Frank W. Conner, General Services Administration Data Services Division, Fort Worth, Texas and Krishna Mudunuri, Kentron International, Hurst, Texas for many helpful discussions in addition to providing the data base and other pertinent information. Paul C. Chiou, Mathematics Department, University of Texas at Arlington performed all programming.

REFERENCES

Antle, C. E. and Bain, L. J. (1969). A property of maximum likelihood estimators of location and scale parameters. *SIAM Review*, 11, 251-253.

Bain, L. J. (1978). *Statistical Analysis of Reliability and Life-Testing Models*. Marcel Dekker, New York.

Box, G. E. P. (1953). Non-normality and tests on variances. *Biometrika*, 40, 318-335.

Brown, K. C. (1969). Bidding for offshore oil: toward an optimal strategy. *Journal of the Graduate Research Center*, 38, Nos. 1 and 2, Southern Methodist University Press.

Bruckner, L. A. and Johnson, M. M. (1978). On the probability distribution of bids on outer continental shelf oil and gas leases. LA-7190-MS, Los Alamos Scientific Laboratory.

Crawford, P. B. (1970). Texas offshore bidding patterns. *Journal of Petroleum Technology*, 22, 283-289.

Dougherty, E. L. and Lohrenz, J. (1976). Statistical analyses of bids for federal offshore leases. *Journal of Petroleum Technology*, 28, 1377-1390.

Dougherty, E. L. and Lohrenz, J. (1977). Money left on the table in sealed, competitive bidding: federal offshore oil and gas lease bids. Paper SPE 6501, *Proceedings Economics and Evaluation Symposium*, Dallas, Texas, February 21-22, 1977.

Dumonceaux, R. and Antle, C. E. (1973). Discrimination between the log-normal and the Weibull distributions. *Technometrics*, 15, 923-926.

Dyer, D. (1979). A survey of certain k-sample test procedures with applications to LPR-5 data. Technical Report No. 109, University of Texas at Arlington.

Dyer, D. and Keating, J. P. (1980). On the determination of critical values for Bartlett's test. *Journal of the American Statistical Association*, 75, 313-319.

Fisher, R. A. (1950). *Statistical Methods for Research Workers* (11*th* ed.). Oliver and Boyd, London.

Genter, F. C. and Bruckner, L. A. (1978). Investigation of the lognormality of offshore oil and gas lease bidding data. LA-7339-MS, Los Alamos Scientific Laboratory.

Keating, J. P. and Bruckner, L. A. (1979). The Pareto distribution as an economic model. LA-8128-MS, Los Alamos Scientific Laboratory.

Littell, R. C. and Folks, J. L. (1973). Asymptotic optimality of Fisher's method of combining independent tests II. *Journal of the American Statistical Association,* 68, 193-194.

Littell, R. C., McClave, J. T. and Offen, W. W. (1979). Goodness-of-fit tests for the two parameter Weibull distribution. *Communications in Statistics,* B8(3), 257-269.

Mann, N. R., Scheuer, E. M. and Fertig, K. W. (1973). A new goodness-of-fit test for the two-parameter Weibull or extreme-value distribution with unknown parameters. *Communications in Statistics,* 2(5), 383-400.

Pelto, C. R. (1971). The statistical structure of bidding for oil and mineral rights. *Journal of the American Statistical Association,* 66, 456-460.

Rothkopf, M. H. (1969). A model of rational competitive bidding. *Management Science,* 15, 362-373.

Smiley, A. K. (1979). *Competitive Bidding Under Uncertainty.* Ballinger, Cambridge, Massachusetts.

Wilk, M. B. and Shapiro, S. S. (1968). The joint assessment of normality of several independent samples. *Technometrics,* 10, 825-839.

[*Received June* 1980. *Revised October* 1980]

STATISTICAL DISTRIBUTIONS OCCURRING IN PHOTOELECTRON PHENOMENA, RADAR AND INFRARED APPLICATIONS

FRANK McNOLTY

Lockheed Palo Alto Research Laboratory
Palo Alto, California 94304 USA

ELDON HANSEN

Lockheed Missles and Space Company
Sunnyvale, California 94086 USA

SUMMARY. Photoelectron counting distributions are derived for the detection of stochastic electromagnetic fields, at optical frequencies, which are characterized by either unknown or fluctuating intensity. In the general case the analysis applies to Helstrom modal decomposition of aperture fields in terms of the eigenfunctions of an integral equation whose kernel is the mutual coherence function of the incident radiation (a Karhunen-Loeve expansion). Special cases assume a short term counting condition and a point-detector photoresponsive surface. Statistical distributions are also derived for fluctuating radar cross section when N pulses are noncoherently integrated during each antenna scan and the target amplitude is independent from scan to scan. Distributions for the random output voltage of an infrared scanner are provided in the general case of a fluctuating target of unknown size.

KEY WORDS. photoelectron counting distributions, modal decomposition, fluctuating radar cross section, short-term counting, special functions, coherent radiation, coherence separable, doubly stochastic.

1. PHOTOELECTRON STATISTICS

In the analysis which follows the emphasis is placed upon statistical problems related to stochastic electromagnetic

C. Taillie et al. (eds.), Statistical Distributions in Scientific Work, Vol. 6, 47–77.

fields whose intensity is either fluctuating or constant and unknown. Mixed Laguerre distributions are derived in which the electron counting probability is often expressible in terms of Jacobi polynomials when the underlying assumptions imposed upon the incident radiation vary from a point detector with short term counting to the multi-mode case for coherence separable fields. In each case the attendant generalized probability generating functions are also provided.

We define the instantaneous intensity $I(t,\underset{\sim}{r})$ of the incident polarized wave as the square of the modulus of the analytic signal representation $g(t,\underset{\sim}{r})$ of the field $f(t,\underset{\sim}{r})$. For a monochromatic field $f(t,\underset{\sim}{r})$ the analytic signal is Goodman (1981)

$$g(t,\underset{\sim}{r}) = a(\underset{\sim}{r}) \exp(i\omega t) \tag{1}$$

$$f(t,\underset{\sim}{r}) = \mathrm{Re}\{g(t,\underset{\sim}{r})\} \tag{2}$$

and $$I(t,\underset{\sim}{r}) = |g(t,\underset{\sim}{r})|^2 = |a(\underset{\sim}{r})|^2 \tag{3}$$

where ω is the frequency of the received radiation, t is time in seconds, $\underset{\sim}{r}$ is a two-vector on the photodetector surface, $a(\underset{\sim}{r})$ is the complex envelope and $I(t,\underset{\sim}{r})$ has the physical units watts/cm^2. In the multi-mode case we will not always carefully identify the domain of $\underset{\sim}{r}$ as being either the receiver lens (aperture plane) or the detector (focal plane), since it is known that the Karhunen-Loeve expansion in either plane has the same eigenvalues. In the case of a nonmonochromatic signal $f(t)$ which has Fourier transform $\mathscr{F}(\nu)$ the analytic signal representation $g(t)$ is given by Goodman (1981).

$$g(t) = 2 \int_{-\infty}^{\infty} \mathscr{F}(\nu) \exp(-j2\pi\nu t)d\nu$$

$$= f(t) + \frac{j}{\pi} \int_{-\infty}^{\infty} \frac{f(s)ds}{(s-t)} + f(t) + j\mathscr{H}(t) \tag{4}$$

where $\mathscr{H}(t)$ is the Hilbert transform of $f(t)$.

1.1 Short-term Counting Intervals and Coherent Electromagnetic Fields. We are interested in the electron counting statistics for photodetection phenomena involving doubly stochastic Poisson processes, a field $f(t,\underset{\sim}{r})$ which is coherent over the detector surface $\mathscr{A}$ and a short-term counting interval T. The number of electrons emitted from the surface $\mathscr{A}$ of the photodetector during a time interval T as a result of the incident radiation intensity is a discrete random variable K defined over the nonnegative integers and described by a Poisson distribution whose level m is itself a realization of a stochastic process, (Cox and Lewis, 1966). Following Gagliardi *et al.* (1970) the level is given by

$$m = \alpha \cdot \int_{\mathcal{A}} \int_{t}^{t+T} I(\rho,\underset{\sim}{r})\, d\rho d\underset{\sim}{r} = \alpha \cdot \int_{\mathcal{A}} \int_{t}^{t+T} |a(\rho,\underset{\sim}{r})|^2 d\rho d\underset{\sim}{r} \quad (5)$$

where $I(t,\underset{\sim}{r})$ is the electromagnetic field intensity defined in (3) and T is the counting interval. In order to define the level as a dimensionless quantity, the factor α is assigned the physical units $(\text{watt-secs})^{-1}$.

From a statistical point of view, expression (5) doesn't provide an expedient relationship between the received field intensity and the level m. This dilemma can be mitigated by imposing two constraints upon the incident field $f(t,\underset{\sim}{r})$ -- one spatial and the other temporal. The first constraint requires that $f(t,\underset{\sim}{r})$ be coherent over $\mathcal{A}$ for any time t whence $I(t,\underset{\sim}{r}) = I(t)$, while the second constraint restricts the length of the counting interval T so that

$$m = \alpha A \cdot \int_{t}^{t+T} I(\rho) d\rho \simeq \alpha A I(t) T \quad (6)$$

where the quantity A denotes area and has the physical units cm^2. Expression (6) provides a simple algebraic relationship, between m and the square $|a(t,\underset{\sim}{r})|^2$ of the complex envelope of the field, which has been extensively exploited in the literature.

It is known (Gagliardi *et al.*, 1970; Fowles, 1967) that the detector must absorb a quantity of energy from the field equal to hf (the energy of a photon with frequency f) in order to free an electron where f is the frequency of the field and h is Planck's constant. By introducing the detector quantum efficiency η (ratio of absorbed energy to incident energy) one can identify the constant α appearing in (5) and (6) as $\alpha = \eta/hf$.

The point of departure for the discussion in this section is the known (Gagliardi *et al.*, 1970) short-term counting, point-detector probability P(K) for an incident electromagnetic field which is the sum of a randomly (uniform over $0,2\pi$) phased, coherent monochromatic field with constant intensity I and a stationary, zero mean, narrow-band Gaussian noise field with variance σ^2. The envelope of this stochastic field has the Rice distribution and the resulting (unconditional) counting probability is given by Gagliardi *et al.* (1970).

$$P(K) = \left[\frac{1}{1 + 2\alpha\sigma^2 TA}\right] \cdot \left[\frac{2\alpha\sigma^2 TA}{1 + 2\alpha\sigma^2 TA}\right]^K \cdot \exp\left[-\frac{\alpha VAT}{1 + 2\alpha\sigma^2 TA}\right] \cdot$$

$$\cdot L_K\left[\frac{-V/2\sigma^2}{1 + 2\alpha\sigma^2 TA}\right] \qquad (7)$$

where $K = 0,1,2,3,\cdots$; $L_K(x)$ is the Laguerre polynomial; T is the counting interval in secs; $\alpha = \eta/hf$ and A is the integrated detector area. In (7) we are now using the symbol V to denote intensity. Literally, $P(K)$ = probability that exactly K electrons will be released from the photoresponsive surface of integrated area A by the impinging electromagnetic field during the time interval T.

In the vernacular of the detection theory a photodetector exposed to a completely coherent field is called a point detector, since the electron emission rate is independent of any particular region within A. That is, the resulting electron flow may be said to effectively emanate from a single point on the detector surface. In terms of the Karhunen-Loeve modal expansion, to be discussed below, expression (7) applies to the case of a single spatial mode.

The probability generating function corresponding to (7) is given by

$$G(S|V) = [1 + 2\alpha\sigma^2 TA(1 - S)]^{-1} \cdot \exp\left[\frac{\alpha VAT(S - 1)}{1 + 2\alpha\sigma^2 TA(1 - S)}\right] \qquad (8)$$

where again $\alpha, \sigma, T, A > 0$ and $0 \leqslant S \leqslant 1$.

From McNolty and Hansen (1979) and McNolty *et al.* (1975) we now define four distributions which are suitable descriptors of randomly fluctuating field intensity. These distributions variously characterize target objects which are either tumbling, precessing or which simply have an unknown orientation with respect to the sensor line of sight.

$$g_1(V)dV = (2/\pi)(\pi^2 r^4 d^2 - V^2)^{-1/2} \cdot dV \qquad (9)$$

where $0 \leqslant V \leqslant \pi r^2 d$; d has physical units watts/cm^4; r is in cm; V is in watts/cm^2; $r,d > 0$ and in McNolty and Hansen (1979) r is defined as the radius of a thin circular flat plate.

$$g_2(V)dV = (4/\pi^3 r^2 d) \cdot F\left[\pi/2,\ (1-V^2/\pi^2 r^4 d^2)^{1/2}\right] \qquad (10)$$

where $0 \leqslant V < \pi r^2 d$; $F(x,k)$ is the elliptic integral of the first kind and $r,d > 0$.

$$g_3(V)dV = b^{\lambda}[\Gamma(\lambda)]^{-1}\, V^{\lambda-1} \cdot \exp(-bV)dV \qquad (11)$$

where $0 \leqslant V < \infty$; λ is a dimensionless number; b is in cm^2/watts and $b,\lambda > 0$.

$$g_4(V)dV = (2/\beta)^{P-1} \cdot V^{(P-1)/2} \cdot q^P \exp(-\beta^2/4q)\exp(-qV) \cdot$$

$$\cdot\ I_{P-1}(\beta V^{1/2})dV \qquad (12)$$

where $0 \leqslant V < \infty$; P is a pure number; q is in cm^2/watts; β^2 is cm^2/watts; $q,P > 0$; $\beta \geqslant 0$ and $I_{P-1}(x)$ is the modified Bessel function of the first kind, order $P - 1$.

It is now appropriate to recall the language of Patil and Boswell (1970) and henceforth refer to expressions (9), (10), (11), and (12) as mixing distributions or, better, intensity mixing distributions. Since these expressions are to be applied in consort with (7) and (8), we will again borrow from Patil's terminology in order to coin the term "mixed Laguerre distribution" for the resulting (unconditional) counting probability P(K).

Expression (9) applies to a situation in which an optical sensor views a tumbling, thin flat circular plate of radius r whose tumbling axis is always perpendicular to the line of sight from the sensor. More generally, expression (10) describes a case where the tumbling axis may assume an arbitrary (random) angle with the sensor line of sight. Similarly (11) and (12) are discussed in McNolty and Hansen (1979), McNolty *et al.* (1975a,b).

1.2 Mixed Laguerre and Mixed Generalized Laguerre Distributions. Expression (7) will now be taken as a conditional probability $P(K|V)$ and by employing the distribution $g_1(V)$ for fluctuating intensity the unconditional counting probability becomes

$$P_1(K) = \int_0^{\pi r^2 d} g_1(V)P(K|V)dV = \frac{(2\alpha\sigma^2 TA)^K K!}{\sqrt{\pi}(1 + 2\alpha\sigma^2 TA)^{K+1}} \cdot \sum_{j=0}^{\infty} \frac{\left(\frac{j-1}{2}\right)!}{(j/2)!} \cdot$$

$$\cdot \left[\frac{\pi r^2 d}{2\sigma^2(1 + 2\alpha\sigma^2 TA)}\right]^j \cdot \sum_{n=0}^{j} \frac{(-2\sigma^2\alpha AT)^n}{n!\,[(j-n)!]^2 (K - j + n)!}$$

$$= \frac{(2\alpha\sigma^2 TA)^K}{(1 + 2\alpha\sigma^2 TA)^{K+1}} \sum_{j=0}^{\infty} \frac{(-1)^j}{[(j/2)!]^2} \cdot \left(\frac{\pi r^2 d}{4\sigma^2}\right)^j \cdot$$

$$\cdot P_j^{(0,K-j)}\left[\frac{-(1 - 2\alpha\sigma^2 AT)}{(1 + 2\alpha\sigma^2 AT)}\right] \tag{13}$$

where $K = 0,1,2,\cdots$ and $P_n^{(a,b)}(x)$ is the Jacobi polynomial.

A cursory inspection of (13) indicates that the factor preceding the summation sign is the Bosé-Einstein count probability corresponding to a zero mean, narrow band stationary Gaussian field and no signal.

The unconditional probability generating function corresponding to (13) is

$$G_1(S) = \int_0^{\pi r^2 d} g_1(V)G(S|V)dV = [1 + 2\alpha\sigma^2 TA(1 - S)]^{-1} \cdot$$

$$\cdot \sum_{n=0}^{\infty} 2^{-n} [(n/2)!]^{-2} \cdot \left[\frac{\alpha AT\pi r^2 d(S - 1)}{1 + 2\alpha\sigma^2 TA(1 - S)}\right]^n$$

$$= [1 + 2\alpha\sigma^2 TA(1 - S)]^{-1} \cdot [I_0(x) + \mathscr{L}_0(x)], \quad 0 \leqslant S \leqslant 1 \tag{14}$$

where $x = \alpha AT\pi r^2 d(S - 1)\,[1 + 2\alpha\sigma^2 TA(1 - S)]^{-1}$; $I_0(x)$ is the modified Bessel function of the first kind of order zero, $\mathscr{L}_0(x)$ is the modified Struve function and $G(S|V)$ is given in expression (8).

Similarly from expressions (7) and (10) one can write

$$P_2(K) = \int_0^{\pi r^2 d} g_2(V)P(K|V)dV = (1/\sqrt{\pi})(2\alpha\sigma^2 TA)^K (1 + 2\alpha\sigma^2 TA)^{-K-1} \cdot$$

$$\cdot \sum_{i=0}^{\infty} \frac{(i/2 - 1/2)!}{[(i/2)!]^3} \cdot \left(-\frac{\pi r^2 d}{4\sigma^2}\right)^i \cdot P_i^{(0,K-i)}\left[\frac{-(1 - 2\alpha\sigma^2 AT)}{1 + 2\alpha\sigma^2 AT}\right] \quad (15)$$

where $K = 0,1,2,\cdots$ and again $P_n^{(a,b)}(x)$ is the Jacobi polynomial.

From expressions (8) and (10) the probability generating function corresponding to (15) is given by

$$G_2(S) = \int_0^{\pi r^2 d} g_2(V)\, G(S|V)dV$$

$$= (1/\pi)[1 + 2\alpha\sigma^2 TA(1 - S)]^{-1} \sum_{K=0}^{\infty}\sum_{n=0}^{\infty} \binom{-1/2}{K} \cdot$$

$$\cdot B\left(\frac{n+1}{2}, K+1\right)(-1)^K (2K)! \cdot [2^{2K}(K!)^2 n!]^{-1}$$

$$\cdot \left[\frac{\alpha AT\pi r^2 d(S - 1)}{1 + 2\alpha\sigma^2 TA(1 - S)}\right]^n$$

$$= (1/\sqrt{\pi})\ [1 + 2\alpha\sigma^2 TA(1 - S)]^{-1} \cdot \sum_{n=0}^{\infty} \frac{(n/2 - 1/2)!}{2^n[(n/2)!]^3} \cdot$$

$$\cdot \left[\frac{\alpha AT\pi r^2 d(S - 1)}{1 + 2\alpha\sigma^2 TA(1 - S)}\right]^n \quad (16)$$

where $0 \leqslant S \leqslant 1$ and $B(x,y)$ is the beta function.

The unconditional counting probability corresponding to expressions (7) and (11) becomes

$$P_3(K) = \int_0^{\infty} g_3(V)P(K|V)dV = b^{\lambda}(2\alpha\sigma^2 TA)^K \frac{(1 + 2\alpha\sigma^2 TA)^{\lambda-K-1}}{[\alpha AT + b(1 + 2\alpha\sigma^2 TA)]^{\lambda}} \cdot$$

$$\cdot \sum_{j=0}^{K} (-K)_j(\lambda)_j(j!)^{-2}(-1/2\sigma^2)^j\ [\alpha AT + b(1 + 2\alpha\sigma^2 TA)]^{-j}$$

$$= \frac{b^{\lambda}(\alpha TA)^K(1 + 2b\sigma^2)^K}{(1 + 2\alpha\sigma^2 TA)^{1-\lambda}} [b(1 + 2\alpha\sigma^2 TA) + \alpha AT]^{-\lambda-K} \cdot$$

$$\cdot P_K^{(0,-\lambda-K)}\left[\frac{2b\sigma^2(1 + 2\alpha\sigma^2 TA) + 2\alpha\sigma^2 TA - 1}{2b\sigma^2(1 + 2\alpha\sigma^2 TA) + 2\alpha\sigma^2 TA + 1}\right] \quad (17)$$

where again $P_n^{(a,b)}(x)$ is the Jacobi polynomial.

The probability generating function corresponding to (17) is

$$G_3(S) = \int_0^\infty g_3(V)G(S|V)dV = b^{\lambda}[1 + 2\alpha\sigma^2 TA(1 - S)]^{-1} \cdot$$

$$\cdot [1 + 2\alpha\sigma^2 TA(1 - S)]^{\lambda} \cdot [b - \alpha TA(S - 1)(2b\sigma^2 + 1)]^{-\lambda} \quad (18)$$

where $0 \leqslant S \leqslant 1$ and $G(S|V)$ is given by (8).

From expressions (7) and (12) the unconditional counting probability may be written as

$$P_4(K) = \int_0^\infty g_4(V)P(K|V)dV = \frac{(2\alpha\sigma^2 TA)^K K! q^P}{(1 + 2\alpha\sigma^2 TA)^{K-P+1}\Gamma(P)} \cdot \exp(-\beta^2/4q) \cdot$$

$$\cdot \sum_{j=0}^{K} \frac{\Gamma(P + j)}{(j!)^2(K - j)!} \cdot (1/2\sigma^2)^j [q(1 + 2\alpha\sigma^2 TA) + \alpha AT]^{-P-j} \cdot$$

$$\cdot {}_1F_1\left\{P + j; P; \frac{\beta^2(1 + 2\alpha\sigma^2 TA)}{4[q(1 + 2\alpha\sigma^2 TA) + \alpha AT]}\right\}$$

$$= \frac{(2\alpha\sigma^2 TA)^K q^P \exp(-\beta^2/4q)}{(1 + 2\alpha\sigma^2 TA)^{K+1-P}[q(1 + 2\alpha\sigma^2 TA) + \alpha AT]^P} \cdot \sum_{n=0}^{\infty} (1/n!)(\beta^2/4)^n \cdot$$

$$\cdot \left[\frac{1 + 2\alpha\sigma^2 TA}{q(1 + 2\alpha\sigma^2 TA) + \alpha AT}\right]^n \cdot P_K^{(0,P+n-K-1)}\left(1 + \frac{2C_1}{q + C_2}\right) \quad (19)$$

by using Hansen (1975, 10.9.6). Here C_1 and C_2 are used to denote $C_1 = (1/2\sigma^2)(1 + 2\alpha\sigma^2 TA)^{-1}$ and $C_2 = \alpha AT(1 + 2\alpha\sigma^2 TA)^{-1}$. In (19), ${}_1F_1(a;b;x)$ is the confluent hypergeometric function.

Throughout this paper the authors will derive, wherever possible, the probability generating function (or characteristic function) corresponding to each distribution. These pairs of functions should provide a convenient mathematical description of the underlying stochastic process.

The generating function $G_4(S)$ corresponding to (19) becomes

$$G_4(S) = \int_0^\infty g_4(V)G(S|V)dV = [1 + 2\alpha\sigma^2 TA(1 - S)]^{-1} \cdot q^P \exp(-\beta^2/4q) \cdot$$

$$\cdot \exp\left\{\frac{\beta^2 [1 + 2\alpha\sigma^2 TA(1 - S)]}{4 [q + 2q\alpha\sigma^2 TA(1 - S) - \alpha AT(S - 1)]}\right\} \cdot$$

$$\cdot \left[\frac{1 + 2\alpha\sigma^2 TA(1 - S)}{q + 2q\alpha\sigma^2 TA(1 - S) - \alpha AT(S - 1)}\right]^P, \quad 0 \leqslant S \leqslant 1 \quad (20)$$

From expressions (11) and (12) it is seen that (17) and (18) are special cases of (19) and (20), respectively.

We now consider a more general case of short-term counting and point-detector photoelectron emission when the received radiation is the sum of M independent randomly phased (uniform over $0, 2\pi$), coherent (over the detector surface $\mathscr{A}$) monochromatic fields each with constant intensity V_i and each accompanied (additively) by a stationary zero mean, narrow-band Gaussian noise field with variance σ^2. The envelope of the combined field has a distribution of the same form as (12) with $P = M$, $q = (2\alpha\sigma^2 TA)^{-1}$ and $\beta = \sqrt{I}(\sigma^2\sqrt{\alpha TA})^{-1}$. In this case the probability is given by Gagliardi, *et al.* (1970).

$$P^M(K) = \frac{(2\alpha\sigma^2 TA)^K}{(1 + 2\alpha\sigma^2 TA)^{K+M}} \exp\left[\frac{-\alpha VAT}{1 + 2\alpha\sigma^2 TA}\right] \cdot L_K^{(M-1)}\left[\frac{-V/2\sigma^2}{1 + 2\alpha\sigma^2 TA}\right] \quad (21)$$

where $K = 0,1,2,\cdots$; $V = \sum_{i=1}^{M} V_i$; $L_n^\alpha(x)$ is the generalized Laguerre polynomial and the physical dimensions of all parameters are the same as in expression (7) which is a special case of (21) for $M = 1$.

The probability generating function corresponding to (21) is

$$G^{M}(S) = [1 + 2\alpha\sigma^{2}TA(1 - S)]^{-M} \cdot \exp\left[\frac{\alpha VAT(S - 1)}{1 + 2\alpha\sigma^{2}TA(1 - S)}\right] \tag{22}$$

where $\alpha,\sigma,T,A > 0$; $0 \leqslant S \leqslant 1$; $M = 1,2,3,\cdots$ and $V > 0$.

The mixed generalized Laguerre distributions corresponding to (13) - (20) are given below. For expedience $g_3(V)$ will now be omitted, since it is a special case of $g_4(V)$. Again in Karhunen-Loeve terms each case will correspond to a single spatial mode and a single temporal mode.

The unconditional counting probability corresponding to (9) and (21) becomes,

$$P_1^{M}(K) = \int_0^{\pi r^2 d} g_1(V)P^{M}(K|V)dV = \frac{(2\alpha\sigma^{2}TA)^{K}}{\sqrt{\pi}K!\,(1 + 2\alpha\sigma^{2}TA)^{K+M}} \cdot$$

$$\cdot \sum_{n=0}^{\infty} \frac{(\frac{n-1}{2})!\,(K + M - 1)!\,(-1)^{n}}{(n/2)!(n + M - 1)!} \cdot$$

$$\cdot \left(\frac{\pi r^{2}d}{2\sigma^{2}}\right)^{n} P_n^{(M-1,K-n)}\left[\frac{-(1 - 2\alpha\sigma^{2}AT)}{(1 + 2\alpha\sigma^{2}AT)}\right] \tag{23}$$

and from expression (14) we immediately obtain the corresponding generating function

$$G_1^{M}(S) = [1 + 2\alpha\sigma^{2}TA(1 - S)]^{-M} \cdot [I_0(x) + \mathscr{L}_0(x)] \tag{24}$$

where $0 \leqslant S \leqslant 1$, $x = \alpha AT\pi r^{2}D(S - 1)\,[1 + 2\alpha\sigma^{2}TA(1 - S)]^{-1}$ and again $I_0(x)$ and $\mathscr{L}_0(x)$ are the modified Bessel and Struve functions respectively.

From expressions (10) and (21) we have

$$P_2^{M}(K) = \int_0^{\pi r^2 d} g_2(V)P^{M}(K|V)dV$$

$$= \frac{(2\alpha\sigma^{2}TA)^{K}}{\pi(1 + 2\alpha\sigma^{2}TA)^{K+M}} \sum_{m=0}^{\infty} \sum_{i=0}^{K} \sum_{n=0}^{\infty} \frac{(-1)^{m}}{m!} \cdot$$

$$\cdot \frac{(\alpha AT)^{m}(\pi r^{2}d)^{m+i}}{(\sigma^{2})^{i}(1 + 2\alpha\sigma^{2}AT)^{m+i}} \cdot \frac{(K + m - 1)!\left(\frac{m + i - i}{2}\right)!}{(i + M - 1)!(K - i)!i!} \cdot$$

$$\cdot \frac{(1/2)_n (2n)!}{\left(\frac{m+i+1}{2}\right)!(n!)^2 \left(\frac{m+i+3}{2}\right)_n 2^{2n}}$$

$$= \frac{(2\alpha\sigma^2 TA)^K (K+M-1)!}{\pi(1+2\alpha\sigma^2 TA)^{K+M} K!} \sum_{i=0}^{\infty} \left[\frac{(i/2-1/2)!}{(i/2)!}\right]^2 \cdot$$

$$\cdot \frac{(-1)^i}{(i+M-1)!} \left(\frac{\pi r^2 d}{2\sigma^2}\right)^i P_i^{(M-1,K-i)} \left[\frac{-(1-2\alpha\sigma^2 AT)}{(1+2\alpha\sigma^2 AT)}\right] \tag{25}$$

The generating function $G_2^M(S)$ corresponding to (25) is easily obtained from (16) by changing the factor $[1 + 2\alpha\sigma^2 TA(1 - S)]^{-1}$ in that expression to $[1 + 2\alpha\sigma^2 TA(1 - S)]^{-M}$.

The unconditional photoelectron count probability corresponding to (21) and (12) is

$$P_4^M(K) = \int_0^\infty g_4(V) P^M(K|V)\,dV = \frac{(2\alpha\sigma^2 TA)^K q^P}{(1+2\alpha\sigma^2 TA)^{K+M-P}\,\Gamma(P)} \cdot \exp(-\beta^2/4q) \cdot$$

$$\cdot (K+M-1)! \sum_{i=0}^{K} \frac{(P+i-1)!(1/2\sigma^2)^i}{i!(K-i)!(i+M-1)!} \cdot$$

$$\cdot [q(1+2\alpha\sigma^2 TA) + \alpha AT]^{-i-P} \cdot$$

$$\cdot {}_1F_1 \left\{P+i; P; \frac{\beta^2(1+2\alpha\sigma^2 TA)}{4[q(1+2\alpha\sigma^2 TA)+\alpha AT]}\right\} \tag{26}$$

where $K = 0,1,2,\cdots$ and ${}_1F_1(a;b;x)$ is the confluent hypergeometric function.

Again $G_4^M(S)$ may be obtained by simply multiplying expression (20) by $[1 + 2\alpha\sigma^2 TA(1 - S)]^{-M+1}$.

1.3 Karhunen-Loeve Expansion of the Stochastic Electromagnetic Field into Modal Space. In Sections 1.1 and 1.2 the short-term, point-detector condition was imposed in order to extract a statistically convenient relationship between the Poisson level m and the squared modulus $|a(t,r)|^2$ of the complex envelope. In this section that constraint is relaxed by employing the

Karhunen-Loeve (K-L) expansion. While this expansion permits a greater mathematical generality, it often obscures the physical simplicity of our earlier results. The situation is almost exactly analogous to a principal components analysis of classical statistics in which the new coordinate system can eliminate undesirable multi-collinearity and yet yield coordinates which are not easily interpreted in a physical sense.

When the received stochastic field is coherence separable Helstrom (1970), Papoulis (1968), Gagliardi *et al.* (1970) the mutual coherence function factors into a spatial coherence function and a temporal correlation function, i.e.,

$$R(t_1,\underset{\sim}{r}_1;t_2,\underset{\sim}{r}_2) = R_s(\underset{\sim}{r}_1,\underset{\sim}{r}_2)R_t(t_1,t_2) \quad . \tag{27}$$

The attendant Karhunen-Loeve integral equation is similarly factorable, providing eigenfunctions of the form

$$\phi_j(t,\underset{\sim}{r}) = g_j(t)\cdot w_j(\underset{\sim}{r}) \tag{28}$$

and eigenvalues γ_j in the form

$$\gamma_j = \gamma_{jt} \cdot \gamma_{js} \quad . \tag{29}$$

That is,

$$\int_t^{t+T} R_t(t_1,t_2)g_j(t_2)dt_2 = \gamma_{jt}g_j(t_1) \tag{30}$$

$$\int_{\mathscr{A}} R_s(\underset{\sim}{r}_1,\underset{\sim}{r}_2)w_j(\underset{\sim}{r}_2)d\underset{\sim}{r}_2 = \gamma_{js}w_j(\underset{\sim}{r}_1) \tag{31}$$

so that the K-L expansion becomes

$$f(t,\underset{\sim}{r}) = \sum_{i=0}^{\infty}\sum_{j=0}^{\infty} b_{ij}g_i(t)\cdot w_j(\underset{\sim}{r}) \tag{32}$$

where each term in (32) represents a separate mode (coordinate) of the stochastic electromagnetic field over time and space.

By means of the Fresnel-Kirchoff approximation relating the source irradiance and the spatial coherence over the receiver one can show Helstrom (1970), Gagliardi (1970) that for circular or rectangular apertures the number of spatial modes is equal to

$$D_s = (A_0/L^2) \cdot (\lambda^2/A)^{-1} + 1 \tag{33}$$

where A_0 is the source irradiance area, L is the distance from source to receiver, λ is the wavelength of the incident field, A is the receiver area and (33) holds true when the source solid angle lies within the receiver field of view. The term λ^2/A of (33) is the diffraction limited field of view of the sensor. In this case we can say that the number of significant spatial modes is equal to the number of solid angles Ω_{dL} needed to cover the source solid angle Ω_s.

In the case of a Gaussian white noise process (having a flat spectrum N_{0b}) and band limited to $\pm B$ Hz, the number of significant temporal modes is approximately

$$D_t = 2BT + 1 \tag{34}$$

where T is again the counting interval.

Now if in expression (21) we let $2\alpha\sigma^2 TA = \mu_{b0} = \alpha N_{0b} = \alpha \cdot$ (average noise energy per mode), $\alpha IAT = \mu_s$ = total signal energy collected over all modes and $M = D_t D_s$ then our results (23) - (26) apply to a situation involving the sum of a monochromatic signal field and a coherence-separable stationary Gaussian white noise field with a flat intensity spectrum where both signal and and noise are bandlimited to B Hz. In applying (23) - (26) to this case, we implicitly set $\alpha AT = 1$ and let $V = \mu_s$ in (9), (10), (11), (12), (21) and (22). That is the six parameters in (21) are replaced by the three parameters μ_s, μ_{b0}, D where $P^M(K|V)$ is transformed to $P^M(K|\mu_s)$. While the algebra proceeds in an orderly manner, it is seen that the physically interpretable fluctuating intensity V has been replaced by a less intuitive quantity μ_s. The latter is the sum of D elemental energies μ_{si} where μ_s has actually been rendered dimensionless by the factor α. In the K-L modal space it is now more natural to think of our distributions (9) - (12) as providing a first order description of a randomly varying scene, rather than relating them to a particular fluctuating or fixed (with unknown orientation) target source as in the single mode case.

2. FLUCTUATING RADAR CROSS SECTION

The objective of this section is to provide further contributions to that branch of radar cross section theory which applies statistical distributions. The earliest systematic treatment of this general subject appears to be the work of Minoru Nakagami beginning in the early 1940's. Nakagami was primarily concerned with the statistical fluctuations in the intensity of long distance, high frequency propagation. Much of his work is summarized in Nakagami (1960). Studies relating directly to fluctuating radar cross section began with Swerling (1954, 1957, 1960) who exploited his experience as a radar consultant at the Rand Corporation in formulating a meaningful approach to the problem. Later studies by McNolty and Hansen (1974) and McNolty *et al.* (1975a,b) extended the earlier statistical structure of random radar cross section. The work by McNolty began in the early 1960's with extensive classified investigations related to actual VHF and UHF radar time history (pulse returns) data. These data included returns from random scatterers (chaff), hard bodies alone and chaff plus hard body. The time histories were studied from the viewpoint of their first order statistical distributions and also in terms of their time series structure. In some of these cases remarkable text-book style fits to large sample, real world, radar cross section histograms were obtained by McNolty using gamma, Swerling, inverse Gaussian, log normal, Maxwell-Boltzmann and other distributions. Second order Markov processes also proved useful in fitting and simulating radar cross section time histories. It is evident that the statistical distribution has provided a means for gaining much insight into this problem area.

Radar cross section σ can be defined in a simplistic manner simply by solving for σ (an unfortunate choice of symbols from a statistician's perspective) in the radar-range equation

$$P_r = P_t G_t \sigma G_r \lambda^2 [(4\pi)^3 r_t^2 r^2 L_t L_{mt} L_{mr} L_r L_p]^{-1} \qquad (35)$$

where P_t is the transmitter power in watts, G_t is the gain of the transmitting antenna in the direction of the target, L_t is a numerical factor to account for losses in the transmitting system, L_r is a similar factor for the receiving system, r_t is the range between the transmitting antenna and the target, σ is the radar cross section, L_{mt} and L_{mr} are numerical factors which allow the propagating medium to have loss, r is the range between the target and the receiving antenna, G_r is the gain of the receiving antenna in the direction of the target, λ is the radar wavelength and L_p is a numerical factor to account for polarization losses. It is assumed that the target is far enough from the antenna so that the far-field condition prevails and the incident electromagnetic field takes the form of a plane wave. It is seen from (35) that σ has physical units of square meters.

Intrinsically, σ depends upon the target shape and material, the aspect (view) angle, the radar frequency and the polarization of the radar transmitting and receiving antennas. In the case of a linearly polarized wave there is general (in terms of trends) agreement between horizontal and vertical polarization, but not necessarily at particular aspect angles. From (35) it is seen that $\sigma = k_1 P_r$ where k_1 is an appropriate constant representing the remaining four parameters.

When the target is a perfectly conducting sphere of radius 'a' then three distinct regions of radar backscattering can be discerned: (1) the Rayleigh region in which the greatest dimension of a target object is much smaller than a wavelength, (2) the Mie region in which $1 \leqslant 2\pi a/\lambda < 10$, and (3) the high frequency region in which $2\pi a/\lambda > 10$. In the latter region σ simply approaches the projected area πa^2 of the sphere; thus, it can be called the optical region. For instance, rough calculations tell us that for a VHF radar (150 MHz) with $a = 2/\pi$, $\sigma \simeq 1.0 m^2$ while $\pi a^2 \simeq 1.25$ m^2 and for an S-band radar (3000 MHz) with $a = 1/10\pi$, $\sigma \simeq .0025$ m^2 while $\pi a^2 \simeq .003$ m^2 where these values hold true for a certain point in the Mie or resonance region. The sphere is obviously independent of the viewing angle, but our point here is that it is not unreasonable to relate σ directly (via an appropriate scale factor) to the projected area of the perfectly conducting sphere in the high frequency region. Now recalling our comments at the end of Section 1.1, let us make it clear that (9) and (10) were transformed from projected area distributions, McNolty and Hansen (1979) to distributions for fluctuating intensity. In this section it suits our purposes to relate (9) and (10) to the random variable A which now denotes signal amplitude and not projected area. Emphasizing the high frequencies near the optical region we write

$$\sigma = k_1 P_r = k_2 A_m^2 = k_3 A_r \tag{36}$$

temporarily using A_m and A_r to explicitly distinguish between signal amplitude and projected area, respectively. Then

$$A_r = (k_2/k_3) A_m^2 = K A_m^2 \tag{37}$$

and expression (37) may be used to define the signal-amplitude analog to the probability densities (9) and (10). That is, writing A for A_m

$$g_5(A)dA = (4A/\pi)[(\pi^2 r^4/K^2) - A^4]^{-1/2} \cdot dA, \qquad 0 \leqslant A \leqslant r\sqrt{\pi/K} \tag{38}$$

and

$$g_6(A)\,dA = (8KA/\pi^3 r^2)F\left\{\pi/2,\ (K/\pi r^2)\,[(\pi^2 r^4/K^2) - A^4]^{-1/2}\right\}dA,$$
$$0 \leqslant A \leqslant r\sqrt{\pi/K} \quad (39)$$

where in (38) and (39) the variable A^2 is in watts, K is in cm^2/watts and $F(x,k)$ is the elliptic integral of the first kind.

2.1 Pulse Integration. Scan-to-Scan Amplitude Independence. Marcum (1960) derived the pdf for the sum y of N noncoherently integrated, randomly (uniformly distributed over $0,2\pi$) phased signal-plus-noise pulses each of constant signal amplitude out of a square-law detector in the form

$$f(y)\,dy = (2y/NR_p)^{(N-1)/2}\cdot \exp(-y-NR_p/2)\cdot I_{N-1}(\sqrt{2NR_p y})\,dy, \quad y \geqslant 0 \quad (40)$$

where R_p is the ratio of the single-pulse maximum instantaneous signal power to the average noise power out of the matched filter preceding the square-law device. Expression (40) is the pdf of the quantity $y = \sum_{i=1}^{N} r_i^2/2$ where r_i is the envelope of the *ith* signal-plus-noise (zero mean, narrowband Gaussian) pulse. Di Franco and Rubin (1968) use the relationship

$$R_p = 2E_p/N_0 = 2A^2\cdot\mathscr{E}_p/N_0 = A^2 \quad (41)$$

where in (41) the energy of the rf quadrature components has been normalized via $2\mathscr{E}_p/N_0 = 1$; E_p is the energy of a single pulse and equals $A^2\mathscr{E}_p$; $N_0/2$ is the two-sided noise power spectrum and A is the signal amplitude. The amplitude of all N signal pulses is assumed to remain constant in (40) and the corresponding characteristic function is given by

$$\phi_N(t) = (1 - it)^{-N}\cdot \exp\left[\frac{itNA^2}{2(1 - it)}\right]. \quad (42)$$

Expressions (40) and (42) are completely dimensionless and accordingly we effectively normalize (38) and (39) by transforming K to a new constant having units cm^2.

Now writing the characteristic function (42) in the form $\phi_N(t|A)$ and using the prior pdf (38), the unconditional characteristic function becomes

$$\Theta_5(t) = \int_0^{r\sqrt{\pi/K}} \phi_N(t|A) g_5(A)\,dA$$

$$= (4/\pi)(1 - it)^{-N} \cdot \int_0^{r\sqrt{\pi/K}} \exp\left[\frac{itNA^2}{2(1 - it)}\right] \cdot A[(\pi^2 r^4/K^2) - A^4]^{-1/2} \cdot dA$$

$$= (a\pi/4) \cdot [I_0(b) + \mathscr{L}_0(b)] \tag{43}$$

where $I_0(b)$ and $\mathscr{L}_0(b)$ are the modified Bessel and Struve functions respectively. In (43) the quantities a and b are given by

$$a = (4/\pi)(1 - it)^{-N} \quad \text{and} \quad b = itN\pi r^2[2K(1 - it)]^{-1}. \tag{44}$$

It is easily seen that $\Theta_5(0) = 1$ as should be the case.

The unconditional pdf corresponding to (43) is obtained from (40) and (38)

$$f_5(y) = (4/\pi)(2y/N)^{(N-1)/2} \exp(-y) \cdot \int_0^{r\sqrt{\pi/K}} A^{2-N} \exp(-NA^2/2) \cdot$$

$$\cdot [(\pi^2 r^4/K^2) - A^4]^{-1/2} \cdot I_{N-1}(\sqrt{2NY}\,A)\,dA$$

$$= (a/2cd)(d/2)^N \sqrt{\pi} \cdot \sum_{n=0}^{\infty} (-1)^n b^n (1/n!) \sum_{j=0}^{\infty} \left(\frac{d}{2}\right)^{2j} \cdot \frac{\Gamma\left(\frac{n+j+1}{2}\right)}{j!\,\Gamma(N+j)\,\Gamma\left(\frac{n+j}{2}+1\right)}$$

$$= (a\sqrt{\pi}/4c)(d/2)^{N-1} \sum_{n=0}^{\infty} \frac{(-b)^n \Gamma\left(\frac{n+1}{2}\right)}{(N+n-1)!\,\Gamma\left(\frac{n}{2}+1\right)} \cdot L_n^{(N-1)}(d^2/4b) \quad , \tag{45}$$

where $y \geqslant 0$,

$$a = (4/\pi)(2y/N)^{\frac{N-1}{2}} \cdot \exp(-y), \quad b = N\pi r^2/2K \tag{46}$$

$$c = r^{N-1}(\pi/K)^{\frac{N-1}{2}}, \quad d = r\sqrt{2Ny\pi/K} \tag{47}$$

and $L_n^{\alpha}(x)$ is the generalized Laguerre polynomial.

From (39) and (42) the N-pulse unconditional characteristic function is given by

$$\Theta_6(t) = \int_0^{r\sqrt{\pi/K}} \phi_N(t|A)g_6(A)dA = 8K[\pi^3r^2(1 - it)^N]^{-1} \cdot$$

$$\cdot \int_0^{r\sqrt{\pi/K}} A \exp\left[\frac{itNA^2}{2(1 - it)}\right] \cdot F\left\{\pi/2, (K/\pi r^2)\cdot[(\pi^2r^4/K^2) - A^4]^{1/2}\right\}dA$$

$$= (a\pi/8) \sum_{n=0}^{\infty} \sum_{m=0}^{\infty} \frac{(-1)^n\Gamma(n + 1/2)b^m\Gamma\left(\frac{m + 1}{2}\right)}{n!(-1/2 - n)!m!\Gamma(n + m/2 + 3/2)}$$

$$= (a/8)\pi^{3/2} \sum_{m=0}^{\infty} \frac{\left(\frac{m - 1}{2}\right)!}{\left[\left(\frac{m}{2}\right)!\right]^3} \cdot \left(\frac{b}{2}\right)^m \quad (48)$$

where $F(x,k)$ is again the elliptic integral of the first kind and Hansen (1975, 7.4.14) has been used to obtain the last equality. In (48) the quantities a and b are identified as follows

$$a = 8[\pi^2(1 - it)^N]^{-1} \quad \text{and} \quad b = itN\pi r^2[2K(1 - it)]^{-1}. \quad (49)$$

The unconditional pdf corresponding to (48) becomes

$$f_6(y) = (2y/N)^{\frac{N-1}{2}} \cdot (8K/\pi^3r^2) \exp(-y) \cdot \int_0^{r\sqrt{\pi/K}} A^{2-N} \exp(-NA^2/2)$$

$$\cdot I_{N-1}(\sqrt{2NY} \cdot A) \cdot F\left\{\pi/2, (K/\pi r^2)\cdot[(\pi^2r^4/K^2) - A^4]^{1/2}\right\}dA$$

$$= (a\pi/8)(c/2)^{N-1} \sum_{j=0}^{\infty} \sum_{n=0}^{\infty} \sum_{m=0}^{\infty} \frac{(c/2)^{2j}(-1)^n\Gamma(n+1/2)(-1)^m b^m\Gamma\left(\frac{m+j+1}{2}\right)}{j!\Gamma(N+j)n!(-1/2-n)!m!\Gamma\left(\frac{m+j}{2}+n+\frac{3}{2}\right)}$$

$$= (a\pi/8)(c/2)^{N-1} \cdot \sum_{m=0}^{\infty} \left[\frac{\left(\frac{m - 1}{2}\right)!}{\left(\frac{m}{2}\right)!}\right]^2 \cdot \frac{(-b)^m}{(N + m - 1)!} L_m^{(N-1)}(c^2/4b) \quad (50)$$

where $y \geqslant 0$ and $L_n^{\alpha}(x)$ is the Laguerre polynomial.

$$a = 8(2yK/N)^{\frac{N-1}{2}} \cdot \left[r^{N-1}\pi^{\frac{N+3}{2}}\right]^{-1} \cdot \exp(-y) \quad , \quad (51)$$

$$b = N\pi r^2/2K \quad \text{and} \quad c = \sqrt{2Ny\pi/K} \cdot r \quad . \quad (52)$$

The interpretation of the density functions (45) and (50) is that they each correspond to the sum of N noncoherently integrated pulses out of a square-law detector, i.e., $Y = \sum_{i=1}^{N} r_i^2/2$ where r_i is the envelope of the *ith* signal-plus-noise pulse and the noise is stationary, zero mean, narrowband and Gaussian. The sequence of N pulses is called a pulse train and it is assumed that the orientation of the target changes slowly in comparison with the duration of the pulse train. The radar makes a complete search of the surveillance volume in one scan and on, say, the *ith* scan it receives an N-pulse echo from a given target which might be tumbling, precessing, wobbling or simply translating with respect to the radar. The target signal amplitude on each pulse of the *ith* scan is a random variable $\mathscr{A}$ which might fall in the interval $(A_i, A_i + dA)$ and is constant for all pulses in the train, while on the *jth* scan $\mathscr{A}$ might assume a value in $(A_i, A_i + dA)$, and so on. In expression (45) the first order stochastic behavior of $\mathscr{A}$ is characterized by the pdf (38), while expression (50) was derived by employing the distribution (39).

3. FLUCTUATING INFRARED TARGETS

3.1 Introduction. The analysis in this section focuses on the self-emissive radiation component for radiators (targets) having a Lambertian surface including cases in which the earthshine and solar reflectivity contributions to the irradiance received at the sensor are comparatively weak. We must request the reader's indulgence while a modest amount of introductory material is presented in this section. Hopefully this discussion will help motivate and clarify the statistical results which are provided in Section 3.2.

In the case of exoatmospheric propagation, the self-emission component of the spectral irradiance incident at the collector of the sensor from a perfectly diffuse radiator (lambertian surface) may be approximated by McNolty and Clow (1980a),

$$H_\lambda = \text{watts}/\mu\cdot\text{cm}^2 = \varepsilon(\lambda)AW_\lambda(T)/\pi R^2 = AM_\lambda/\pi R^2 \qquad (53)$$

where

R = slant range from object to sensor,

ε = radiant emissivity of object.

$W_\lambda(T)$ = spectral radiant emittance of an object's surface

$= c_1\{\lambda^5[\exp(c_2/\lambda T) - 1]\}^{-1}$ (Planck's law),

T = temperature of object's surface in degrees Kelvin,

λ = wavelength of radiation in microns ($\mu = 10^{-4}$ cm)

$c_1 = 3.7413 \times 10^4\ W\mu^4/cm^2$, $c_2 = 1.4388 \times 10^4\ \mu^oK$.

If we now denote the power per unit λ delivered to the collector by $P_{col,\lambda}$ then

$$P_{in} = \int P_{col,\lambda} d\lambda = A_{col} \cdot \overline{L(\lambda)} \int H_\lambda d\lambda = H \cdot A_{col} \cdot L \quad (54)$$

where A_{col} is the collector area in cm^2, $\overline{L(\lambda)}$ is the average value of $L(\lambda)$ over the optical passband and $L(\lambda)$ denotes the losses due to: absorption in the transmissive optical components, incomplete reflection by any mirror, spurious reflections (Fresnel reflections) at the surfaces of transmissive optical components and scattering due to imperfections in optical components.

From (53) and (54)

$$P_{in} = A_{col} \cdot L \cdot (A/\pi R^2) \cdot \int M_\lambda d\lambda = LA_{col} \cdot (A/\pi R^2) \cdot M \quad (55)$$

Using the matched filter as a convenient example, the normalized (i.e. for a 1 - W optical power input) noiseless signal voltage out of the filter is given by McNolty *et al.* (1972a,b)

$$s(t) = \int_{-\infty}^{\infty} S(f) \exp(j2\pi ft)df = \int_{-\infty}^{\infty} C(f)H(f) \exp(j2\pi ft)df$$

$$= K \int_{-\infty}^{\infty} |C(f)|^2 \cdot [\beta(f)]^{-1} \cdot \exp[-j2\pi f(\Delta - t)]df$$

$$= (1/K) \int_{-\infty}^{\infty} N_m(\omega) \exp[-j\omega(\Delta - T)]d\omega/2\pi \quad (56)$$

where

S(f) = signal voltage density spectrum at the filter output,

C(f) = signal voltage density spectrum into the matched filter,

H(f) = transfer function for matched filter,

$\beta(f)$ = colored noise power spectrum into filter,

$N_m(\omega)$ = colored noise power spectrum out of filter,

K = filter gain factor, and

Δ = realizability delay.

Thus, s( ) = normalized noiseless peak signal voltage out of filter and

S_p = actual noiseless peak signal voltage at the filter output due to the radiating object where

$$S_p = P_{in} s(\Delta) = L \cdot A_{col} \cdot M \cdot As(\Delta)/\pi R^2 \tag{57}$$

so that

$$S_p = A\left\{L \cdot A_{col} \cdot Ms(\Delta)/\pi R^2\right\} = AD \tag{58}$$

the quantity D denoting the expression in braces.

The integrands in expression (56) include such sensor parameters as the detector responsivity, various time constants, and noise parameters corresponding to photon noise, thermal noise, generation-recombination noise and modulation $(1/f^\alpha)$ noise.

In McNolty and Hansen (1979) the pdf for the projected area A of a tumbling, thin, circular plate of radius r is derived under several conditions. The tumbling axis is assumed to pass through the center of the plate and without loss of generality the reader may assume that the axis coincides with a horizontal line on this page. The circular plate rotates about this axis through an angle ϕ so that the viewer perceives a projected area $A = \pi r^2 \cos\phi$. That is, the area varies from zero $(\phi = 90^\circ)$ to a maximum of πr^2 when $\phi = 0^\circ$. There is also a second independent orientation angle θ which refers to the position of the tumbling axis. It can either lie in the plane of the page (as indicated above) or it can be normal to the page pointing toward the reader or take any angle in between -- always passing through the center of the plate. We do not permit the plate to have a wobbling motion. In McNolty and Hansen (1979) the distribution of A was derived under the following general conditions:

(a) ϕ and θ are each uniformly distributed over $(0,\pi/2)$,
(b) ϕ is uniform over $(0,\pi/2)$ and θ is nonuniform (or vice versa),

(c) ϕ and θ are each nonuniformly distributed over $(0,\pi/2)$ and
(d) a special simplified case in which the tumbling axis is always perpendicular to the viewer's (sensor's) line of sight.

The pdf corresponding to condition (a) above is obtained from expression (10) by letting $V = Ad$; the pdf for special case (d) is obtained from expression (9) by again letting $V = Ad$.

In case (b) above, in which ϕ and θ are distributed according to

$$p_1(\phi)d\phi = (2/\pi)d\phi, \quad 0 \leqslant \phi \leqslant \pi/2 \tag{59}$$

$$p_2(\theta)d\theta = 4\Gamma(\lambda)[\Gamma^2(\lambda/2)2^{\lambda}]^{-1} \cdot (\sin 2\theta)^{\lambda-1}d\theta, \quad 0 \leqslant \theta \leqslant \pi/2 \tag{60}$$

the pdf for the projected area becomes

$$\begin{aligned} g(a)da = {} & 2\Gamma(\lambda)a^{\lambda/2-1/2} \cdot [\Gamma(\lambda/2)\Gamma(\lambda - 1/2)]^{-1} \cdot (1 - a^2)^{\lambda/4-1/4} \cdot \\ & \cdot [(2/\pi)\sin(\pi\lambda/2) \cdot Q_{\lambda/2-1}^{\lambda/2-1/2}(1 - 2a^2) - \cos(\pi\lambda/2) \cdot \\ & \cdot P_{\lambda/2-1}^{\lambda/2-1/2}(1 - 2a^2)]da \end{aligned} \tag{61}$$

where $a = A/\pi r^2$, $0 \leqslant a \leqslant 1$ and $P_n^m(x)$, $Q_n^m(x)$ are the associated Legendre functions of degree n and order m of the first and second kinds, respectively.

For the case (c) in which ϕ and θ are each distributed according to (60) the pdf for projected area becomes $(a = A/\pi r^2)$

$$g(a)da = 4a^{\lambda-1}\Gamma^2(\lambda)[\Gamma^4(\lambda/2)]^{-1} \cdot (1 - a^2)^{\lambda/2-1} \cdot \cdot Q_{\lambda/2-1}\left(\frac{1 + a^2}{1 - a^2}\right)da \tag{62}$$

where $0 \leqslant a \leqslant 1$ and $Q_n(x)$ is the Legendre function of the second kind. As in the case of expression (61) the derivation of (62) is quite lengthy and involves an analytic continuation of the parameter λ.

From McNolty and Hansen (1979, 1974) we will present and describe the following four density functions which are to be used in Section 3.2 below.

$$\tau(x) = (\pi\sigma_n\sqrt{2})^{-1} \cdot \exp(-x^2/2\sigma_n^2) \cdot \sum_{K=0}^{\infty} \alpha^K (2\sigma_n\sqrt{2})^{-K} \cdot H_K(x/\sigma_n\sqrt{2}) \cdot$$

$$\cdot\ \Gamma(K/2 + 1/2) \cdot [\Gamma(K/2 + 1)]^{-3} \tag{63}$$

where $-\infty < x < \infty$, $\alpha = \pi D r^2$ and $H_n(x)$ is a Hermite polynomial.

Expression (63) is the pdf for the output voltage x of an unspecified filter in the scanner receiver when the output consists of the peak S_p of the signal pulse plus mean zero, stationary Gaussian noise with variance σ_n^2 and when the pulse arrival time is known. In this case it is assumed that we were able to sample the output x at the exact time of the pulse peak arrival. Further, S_p (in volts) is distributed according to (10) with $V = (d/D)S_p$ where D has physical units volts/cm^2. Then

$$\gamma(x) = \ell(\pi\sqrt{2\pi}\ \sigma_n)^{-1} \exp(-x^2/2\sigma_n^2) \cdot \sum_{m=0}^{\infty} \alpha^m H_m(x/\sqrt{2}\ \sigma_n)$$

$$\cdot\ [m!(m + \ell^2)^{1/2} \cdot (\sqrt{2}\ \sigma_n)^m]^{-1} \cdot \Gamma^2(m/2 + 1/2) \cdot$$

$$\cdot\ [\Gamma^2(m/2 + 1)]^{-1} \tag{64}$$

where again $\alpha = \pi D r^2$, $-\infty < x < \infty$ and $H_m(u)$ is a Hermite polynomial.

Expression (64) is the pdf for the output voltage x of an unspecified filter in the scanner when the output consists of the signal plus mean zero, stationary Gaussian noise and now when the pulse peak arrival time is unknown. Again as in (63) the pulse peak S_p (in volts) is fluctuating and distributed according to expression (10) with $V = (d/D)S_p$, D in volts/cm^2. Unlike (63) expression (64) presumes that the pulse peak location time has been estimated and that the resulting sampled amplitude s of the pulse has a pdf McNolty *et al.* (1974a) given by

$$h(s) = (\ell/\sqrt{\pi})S_p^{-\ell^2} \cdot [\ell n(S_p/s)]^{-1/2} \cdot s^{\ell^2-1}, \quad 0 \leqslant s \leqslant S_p \tag{65}$$

where σ_1 is the dispersion parameter corresponding to the deterministic Gaussian shaped signal pulse, σ_2 is the error sigma associated with the unbiased pulse peak estimation (location) procedure and the parameter $\ell = \sigma_1/\sigma_2$ is a measure of the estimation effectiveness.

The mean and variance of s are given by

$$E(s) = \ell S_p/(\ell^2 + 1)^{1/2} \tag{66}$$

$$\sigma_s^2 = \ell S_p^2[(\ell^2 + 2)^{-1/2} - \ell/(\ell^2 + 1)] \tag{67}$$

where $E(s) \to S_p$ and $\sigma_s^2 \to 0$ as $\ell \to \infty$, i.e. as the pulse-peak-location estimation error approaches zero.

The last distribution in this section is

$$\beta(x) = \ell(\sqrt{2\pi}\,\sigma_n)^{-1} \cdot \exp(-x^2/2\sigma_n^2) \cdot \sum_{m=0}^{\infty} H_m(x\sqrt{c}) \cdot (\alpha\sqrt{c}/2)^m \cdot$$

$$\cdot\, [(m + \ell^2)^{1/2} \cdot \Gamma^2(m/2 + 1)]^{-1} \tag{68}$$

where $-\infty < x < \infty$, $\alpha = \pi D r^2$, $c = 1/2\sigma_n^2$ and $H_m(u)$ is again a Hermite polynomial. Expression (68) has the same physical rationale as (64), except that now the pulse peak S_p is distributed according to expression (9) with $V = (d/D)S_p$.

3.2 Fluctuating Infrared Targets of Unknown Size. The objective in this section is to provide distributions which apply to infrared targets which are both fluctuating (tumbling) and of unknown size. The results apply exactly to thin, flat circular target objects and in an approximate way to radiators having similar shapes.

The following two prior distributions for the radius r of the flat plate will be applied to expressions (63), (64) and (68) above. This problem was motivated by a study in which one of the authors tediously contended with a large ensemble of tumbling target objects having different sizes, but similar shapes. The first prior pdf $g_7(r)$ has the beta form, while the second $g_8(r)$ is a Bessel distribution (McNolty, 1973),

$$g_7(r)dr = 2\Gamma(p + q)[\Gamma(p)\Gamma(q)a^{2(p+q-1)}]^{-1}\, r^{2p-1}(a^2 - r^2)^{q-1}dr \tag{69}$$

where $0 < r < a$; $p,q < 0$.

$$g_8(r)dr = \beta(2/\beta)^P r^P q^P \exp(-\beta^2/4q) \exp(-qr^2) I_{P-1}(\beta r)dr \quad (70)$$

where $0 \leqslant r \leqslant \infty$; $\beta \geqslant 0$; $q, P > 0$ and $I_a(x)$ is the modified Bessel function of the first kind of order a.

Now regarding (63), (64) and (68) as conditional distributions we write them as $\tau(x|r)$, $\gamma(x|r)$ and $\beta(x|r)$, respectively. As in Section 2 the resulting unconditional distributions will be denoted by a subscript to indicate which prior was used.

From (63) and (69),

$$\tau_7(x) = \frac{\Gamma(p+q)}{\pi\sigma_n\sqrt{2}\,\Gamma(p)} \cdot \exp(-x^2/2\sigma_n^2) \sum_{K=0}^{\infty} \left(\frac{\pi Da^2}{2\sigma_n\sqrt{2}}\right)^K \cdot$$

$$\cdot \frac{\Gamma(p+K)\Gamma(K/2+1/2)}{\Gamma^3(K/2+1)\Gamma(p+q+K)} \cdot H_K(x/\sigma_n\sqrt{2}), \quad (71)$$

where $-\infty < x < \infty$ and from (64) and (69)

$$\gamma_7(x) = \frac{\ell\Gamma(p+q)}{\pi\sqrt{2\pi}\sigma_n\Gamma(p)} \cdot \exp(-x^2/2\sigma_n^2) \cdot$$

$$\cdot \sum_{m=0}^{\infty} \left(\frac{\pi Da^2}{\sqrt{2}\sigma_n}\right)^m \cdot H_m(x/\sigma_n\sqrt{2}) \cdot \frac{\Gamma^2(m/2+1/2)\Gamma(m+p)}{m!(m+\ell^2)^{1/2}\Gamma^2(m/2+1)\Gamma(m+p+q)},$$

$$-\infty < x < \infty \quad (72)$$

The characteristic function corresponding to (72) is given by

$$\theta(t) = \ell\Gamma(p+q)[\pi\Gamma(p)]^{-1} \cdot \exp(-\sigma_n^2 t^2/2) \cdot \sum_{m=0}^{\infty} \frac{(it\pi Da^2)^m}{m!(m+\ell^2)^{1/2}} \cdot$$

$$\cdot \frac{\Gamma^2(m/2+1/2)\Gamma(p+m)}{\Gamma^2(m/2+1)\Gamma(p+q+m)}, \quad -\infty < t < \infty \quad (73)$$

where $\theta(0) = 1$ as should be the case.

From (68) and (69) we have

$$\beta_7(x) = \ell\Gamma(p+q)[\sqrt{2\pi}\,\sigma_n\Gamma(p)]^{-1} \cdot \exp(-x^2/2\sigma_n^2) \sum_{K=0}^{\infty} H_K(x\sqrt{c}) \cdot$$

$$\cdot (\pi Da^2\sqrt{c}/2)^K\Gamma(p+K)[(K+\ell^2)^{1/2} \cdot \Gamma^2(K/2+1)\Gamma(p+q+K)]^{-1}$$

$$-\infty < x < \infty \qquad (74)$$

and the characteristic function corresponding to (74) is

$$\theta(t) = \ell\Gamma(p+q)[\Gamma(p)]^{-1} \cdot \exp(-t^2/4c) \sum_{K=0}^{\infty} (i\pi tDa^2/2)^K \cdot$$

$$\cdot \frac{\Gamma(p+K)}{(K+\ell^2)^{1/2}\Gamma^2(K/2+1)\Gamma(p+q+K)}, \qquad (75)$$

where $-\infty < t < \infty$ and $\theta(0) = 1$.

4. RADIAL, COMPONENT AND NONUNIFORM PHASE DISTRIBUTIONS

4.1 Trivariate Case. We begin by defining a two-sided Bessel distribution for the x-component of the form

$$f(x)dx = 2^{(Q_1/4 - 3/2)} \cdot \gamma_1^{-(Q_1/2)+1} |x|^{(Q_1/2)} \lambda^{(Q_1/2)} \cdot$$

$$\cdot \exp -[\tfrac{1}{2}(\gamma_1^2/2\lambda + \lambda x^2)] \cdot I_{(Q_1/2-1)}\left(\gamma_1 \frac{|x|}{\sqrt{2}}\right) dx \qquad (76)$$

where $-\infty < x < \infty$ and the distributions of y and z have the same form as (76), but do not necessarily have the same values assigned to the parameters γ and Q. These distributions are

$$f(y)dy = f(y;\lambda,\gamma_2,Q_2)dy, \quad -\infty < y < \infty \qquad (77)$$

$$f(z)dz = f(z;\lambda,\gamma_3,Q_3)dz, \quad -\infty < z < \infty \qquad (78)$$

where the random variables x,y,z are independent and the parameters are constrained as follows: $\lambda > 0; \gamma_1,\gamma_2,\gamma_3 \geq 0; \quad Q_1,Q_2,Q_3 > 0$. Here $I_\nu(x)$ is the modified Bessel function of the first kind of order ν.

In a spherical coordinate system the variables x,y,z are related to θ,ϕ,r by

$$x = r \sin\theta \cos\phi, \quad y = r \sin\theta \sin\phi, \quad z = r\cos\theta \tag{79}$$

where $0 \leqslant r < \infty$, $0 \leqslant \phi \leqslant 2\pi$, $0 \leqslant \theta \leqslant \pi$. Also we have $x = \rho\cos\phi$ and $y = \rho\sin\phi$ as is well known.

One can show that the pdf for ϕ is given by

$$f(\phi)d\phi = (1/2)\sum_{i=0}^{\infty}\sum_{j=0}^{\infty} H(i,j;\lambda;\gamma_1,\gamma_2;Q_1,Q_2)|\sin\phi|^{Q_2+2j-1} \cdot \cdot |\cos\phi|^{Q_1+2i-1} \cdot d\phi, \tag{80}$$

where $0 \leqslant \phi \leqslant 2\pi$.

In (80) the $H(\ )$ notation is used to denote

$$H(i,j;\lambda;\gamma_1,\gamma_2;Q_1,Q_2) = \exp[-(\gamma_1^2 + \gamma_2^2)/4\lambda] \cdot (\gamma_1/2)^{2i} \cdot \cdot (\gamma_2/2)^{2j} \cdot [\lambda^{i+j} i!j!B(Q_1/2 + i, Q_2/2 + j)]^{-1} \tag{81}$$

where $B(x,y)$ is the beta function.

Without derivation the distribution of θ is given by

$$f(\theta)d\theta = \sum_{i=0}^{\infty}\sum_{j=0}^{\infty} H\left(i,j;\gamma_3,\sqrt{\gamma_1^2 + \gamma_2^2};Q_3,Q_1 + Q_2\right) \cdot \cdot (\sin\theta)^{Q_1+Q_2+2j-1} \cdot |\cos\theta|^{Q_3+2i-1}\, d\theta, \quad 0 \leqslant \theta \leqslant \pi \tag{82}$$

where $H(\ )$ has the same form as in (81).

The distribution of $\rho = \sqrt{x^2 + y^2}$ is given by expression (70) with $\beta = \gamma/\sqrt{2}$, $q = \lambda/2$, $P = Q/2$; $Q = Q_1 + Q_2$ and $\gamma^2 = \gamma_1^2 + \gamma_2^2$.

Similarly the pdf for $r = \sqrt{x^2 + y^2 + z^2}$ has the same form as (70) with $\beta = \gamma/\sqrt{2}$, $q = \lambda/2$, $P = Q/2$ and now $Q = Q_1 + Q_2 + Q_3$, $\gamma^2 = \gamma_1^2 + \gamma_2^2 + \gamma_3^2$.

4.2 A Generalized Quadratic Form Distribution. We define a K dimensional quadratic form by

$$s = \sum_{i=1}^{K} (x_{0i} + x_{1i})^2 \tag{83}$$

where the distribution of each x_{0i} has the same form as (76), but with parameters λ, Q_i and γ_i; while each x_{1i} is normally distributed with mean zero and variance σ^2. All of the random variables x_{0i} and x_{1i} are statistically independent and it can be shown by means of characteristic functions that the distribution of s is given by

$$f(s)ds = \sigma^{q-K}\lambda^{q/2}s^{K/2-1}[2^{K/2}\Gamma(K/2)(1+\lambda\sigma^2)^{q/2}]^{-1} \cdot$$

$$\cdot \exp(-s/2\sigma^2 - \Gamma^2/4\lambda) \cdot \sum_{m=0}^{\infty} \sigma^{2m}\Gamma^{2m}[4^m m!(1+\lambda\sigma^2)^m]^{-1} \cdot$$

$$\cdot {}_1F_1[q/2 + m; K/2; s/2\sigma^2(1+\lambda\sigma^2)]ds, \quad s \geqslant 0, \tag{84}$$

where

$$q = Q_1 + Q_2 + \cdots + Q_K \quad \text{and} \quad \Gamma^2 = \gamma_1^2 + \gamma_2^2 + \cdots + \gamma_K^2 .$$

4.3 Photoelectron Statistics. Associated Legendre Mixing Distributions. In this section expression (61) will be transformed to a pdf for fluctuating intensity V by noting that expression (53) for the incident spectral irradiance may be integrated over λ to yield

$$V = (A/\pi R^2) \int M_\lambda d\lambda = (M/\pi R^2)A = Ad . \tag{85}$$

Thus, making the transformation $a = V/d\pi r^2$ yields the desired pdf for V from (61) when ϕ is uniformly distributed as in (59) and θ is nonuniformly distributed according to (60). It is easily seen that (80) yields (59) via $\gamma_1 = \gamma_2 = 0$ and $Q_1 = Q_2 = 1$; while (60) is obtained from (80) by letting $\gamma_1 = \gamma_2 = 0$ and $Q_1 = Q_2 = \lambda$. Now denoting (61) by $g(V)$ and writing expression (7) as $P(K|V)$ we will obtain

$$P(K) = \int_0^{\pi d r^2} g(V)P(K|V)dV \tag{86}$$

where (86) is the short-term, point detector (single spatial mode, single temporal mode) unconditional counting probability. Like many of the other integrations performed in this paper, (86) is definitely a non-trivial calculation. Eventually one can show that

$$P(K) = \frac{\Gamma(\lambda)(2\alpha\sigma^2 TA)^K}{\sqrt{\pi}\,\Gamma(\lambda/2)(1+2\alpha\sigma^2 TA)^{K+1}} \cdot \sum_{n=0}^{\infty} (-1)^n \left(\frac{\alpha AT\pi r^2 d}{1+2\alpha\sigma^2 TA}\right)^n \cdot$$

$$\cdot \frac{\Gamma^2(n/2+1/2)}{n!\,\Gamma(\lambda+n/2)\Gamma(n/2+1)} \cdot {}_2F_1\,[-n,-K;1;(2\alpha\sigma^2 AT)^{-1}] \qquad (86)$$

where $K = 0,1,2,\cdots$.

The probability generating function corresponding to (86) is given by

$$G(S) = \frac{\Gamma(\lambda)}{\sqrt{\pi}\,\Gamma(\lambda/2)[1+2\alpha\sigma^2 TA(1-S)]} \cdot \sum_{n=0}^{\infty} \frac{\Gamma(\lambda/2+n/2)\Gamma(n/2+1/2)}{n!\,\Gamma(\lambda+n/2)\Gamma(1+n/2)} \cdot$$

$$\cdot \left[\frac{\alpha AT\pi r^2 d(S-1)}{1+2\alpha\sigma^2 TA(1-S)}\right]^n, \quad 0 \leqslant S \leqslant 1 \qquad (87)$$

and it is gratifying to observe that $G(1)$ does equal one.

REFERENCES

Cox, D. R. and Lewis, P. A. W. (1966). *The Statistical Analysis of Series of Events.* Methuen Monograph, Wiley, New York.

DiFranco, J. V. and Rubin, W. L. (1968). *Radar Detection.* Prentice-Hall, Englewood Cliffs, N. J.

Fowles, G. R. (1968). *Introduction to Modern Optics.* Holt, Rinehart and Winston, New York.

Gagliardi, R. M., Karp, S. and O'Neill, E. L. (1970). Communication theory for the free space optical channel. *Proceedings of the IEEE,* 58, 1611-1625

Goodman, J. W. (1981). *Statistical Optics.* Wiley, New York.

Gradshteyn, I. S. and Ryzhik, I. M. (1965). *Tables of Integrals, Series and Products.* Academic Press, New York.

Hansen, E. R. (1975). *A Table of Series and Products.* Prentice-Hall, Englewood Cliffs, N. J.

Helstrom, C. W. (1970). Modal decomposition of aperture fields in detection and estimation of incoherent objects. *Journal of the Optical Society of America,* 60, 521-530.

McNolty, F. W., Clow, R. and Hansen, E. R. (1972a). Some matched filter configurations for infrared systems. *IEEE Transactions on Aerospace and Electronic Systems,* AES-8, 428-438.

McNolty, F. W., Clow, R. and Hansen, E. R. (1972b). Some properties of the output of an integrator in an infrared system. *IEEE Transactions on Aerospace and Electronic Systems*, AES-8, 552-558.

McNolty, F. W. (1973). Some probability density functions and their characteristic functions. *Mathematics of Computation*, 27, 495-504.

McNolty, F. W., Clow, R. and Hansen, E. R. (1974a). Bayesian density functions for Gaussian pulse shapes in Gaussian noise. *Proceedings of the IEEE*, 62, 134-136.

McNolty, F. W. and Hansen, E. R. (1974b). Some aspects of Swerling models for fluctuating radar cross section. *IEEE Transactions on Aerospace and Electronic Systems*, AES-10, 218-285.

McNolty, F. W., Huynen, R. and Hansen, E. R. (1975a). Certain statistical distributions involving special functions and their applications. In *Statistical Distributions in Scientific Work, Vol. I,* G. P. Patil, S. Kotz and J. K. Ord, eds. Reidel, Dordrecht-Holland. Pages 131-160.

McNolty, F. W., Huynen, R. and Hansen, E. R. (1975b). Component distributions for fluctuating radar targets. *IEEE Transactions on Aerospace and electronic systems*, AES-11, 1316-1332.

McNolty, F. W. and Hansen, E. R. (1979). Probability densities and characteristic functions for fluctuating targets. *IEEE Transactions on Aerospace and Electronic Systems*, AES-15, 474-480.

McNolty, F. W. and Clow, R. (1980a). Methodology for target discrimination. *Applied Optics*, 19, 984-999.

McNolty, F. W., Doyle, J. and Hansen, E. R. (1980b). Properties of the mixed exponential failure process. *Technometrics*, to appear in November issue.

Nakagami, M. (1960). The m-distribution--a general formula of intensity distribution of rapid fading. Section I of *Statistical Methods in Radio Wave Propagation*, 3-36, Pergamon Press, New York.

Papoulis, A. (1968). *Systems and Transforms with Applications in Optics*. McGraw-Hill, New York.

Patil, G. P. and Boswell, M. T. (1970). Chance mechanisms generating the negative binomial distribution. In *Random Counts in Models and Structures*, G. P. Patil, ed. The Pennsylvania State University Press, University Park and London.

Rihaczek, A. W. (1969). *Principles of High Resolution Radar*. McGraw-Hill, New York.

Swerling, P. (1954). Probability of detection for fluctuating targets. Rand Corporation Research Memo, RM-1217, Santa Monica, Calif.

Swerling, P. (1957). Detection of pulsed signals in the presence of noise. *IEEE Transactions on Information Theory*, IT-3, 175-178.

Swerling, P. (1960). Probability of detection for fluctuating targets. *IEEE Transactions on Information Theory*, IT-6, 269-308.

[*Received June* 1980. *Revised October* 1980]

APPLICATION OF DISCRETE DISTRIBUTIONS FOR ESTIMATING THE NUMBER OF ORGANIC COMPOUNDS IN WATER

K. G. JANARDAN*

Sangamon State University
Springfield, Illinois 62708 USA

D. J. SCHAEFFER
Illinois Environmental Protection Agency
Springfield, Illinois 62708 USA

SUMMARY. Shackelford and Keith (1976) have compiled data on the organic compounds which have been identified in water. A complex sample consists of a large number of distinct compounds. Of the compounds potentially identifiable by a given method, some are actually identified. Methods for estimating the number of additional compounds which could have been identified are presented. These data are used to provide estimates of the number of observed compounds (350) which are expected to appear in a similar list developed from nonoverlapping sources. The expected number of unobserved (but observable) compounds (about 1200) which could be identified using techniques comparable to those giving rise to the data in Shackelford and Keith's list is obtained. These methods are then used to determine if the rate at which we are identifying new compounds (in a deliberate search for new compounds) is greater than the rate determined from Shackelford and Keith's data base.

KEY WORDS. Unseen species, Poisson, negative binomial, generalized geometric distributions, aquatic environment, estimation of organic compounds.

---

*Presently on sabbatical leave at the University of Pittsburgh.

C. Taillie et al. (eds.), Statistical Distributions in Scientific Work, Vol. 6, 79–94.

1. INTRODUCTION

Almost every aspect of environmental control depends on identifying and quantifying chemical pollutants in the environment. Although it is known that an estimated 60,000 organic compounds are produced and used by various industries in the USA, accurate estimates of the number of organic compounds that are being discharged are not available. Present techniques enable analysts to capture and identify 5-20 percent of the isolable compounds in the sample (Janardan *et al.*, 1980).

In the past, environmental efforts in the United States have been concerned with well known, readily characterized, stable organic pollutants or broad groups of materials such as phenols, biochemically oxidizable substances, etc. Recent U.S. Federal legislation, such as the Public Drinking Water Act of 1976, and the Toxic Substances Control Act, and the successful law suit by the Environmental Defense Fund (EDF) are forcing environmental efforts in new directions. As a result of the EDF suit, 129 organic and inorganic substances have been classified as priority pollutants. These are subject to new regulations for controlling industries. These regulations are also supplemented to pretreatment requirements limiting the introduction of toxic wastes and the treatment residues such as sludge, into sewers. The wastes which are generated at the source either by removal or by accumulation during treatment, are regulated by the Resource Conservation and Recovery Act. This plethora of U.S. Federal regulations is not supported by the essential scientific information which is needed to define the nature of these substances; the quantities in and being discharged into the environment, the method of analysis, and human health consideration, the control strategies and so on. Further, no good estimate of the number of compounds which might ultimately have to be regulated exists.

Shackelford and Keith (1976) have compiled data from government reports, unpublished studies, and published works on the organic compounds found and identified in 33 different water types. One thousand two hundred fifty-eight distinct compounds for which a definite chemical structure was assigned by the original authors were found in these reports. Five hundred and three compounds were observed once, 238 twice, 133 three times, etc. The total 5,720 separate identifications of the 1,258 distinct compounds is given in Table 1. In this table the left and top margins give the frequency of occurrence (k = 1,2,•••,110), while the body of the table give the number of distinct compounds appearing with that frequency.

The illinois EPA has initiated a study (Somani *et al.*, 1980) of trace organics contained in discharges to the State's waters. This study has identified many organic compounds that are not referenced in Shackelford and Keith's report. Thus, one would like to know how many more new organic compounds there are in the (aquatic environment which have not yet been discovered. In

TABLE 1: Number of Organic Compounds Listed in Shackelford and Keith (1976). *Entry* k *is* n_k, *the number of organic compounds found exactly* k *times such that* $\sum n_k = 1258$ *and* $\sum kn_k = 5720$.

k	1	2	3	4	5	6	7	8	9	10
0+	503	238	133	80	56	46	20	14	15	18
10+	15	16	10	10	9	4	12	6	7	4
20+	4	1	4	0	2	3	1	5	4	3
30+	0	1	1	1	1	0	0	0	0	0
40+	0	0	1	0	0	1	1	0	0	1
50+	0	0	0	0	0	0	0	0	1	0
60+	1	0	0	0	0	2	0	0	0	0
70+	0	0	0	0	0	1	0	0	0	0
80+	0	0	0	0	0	0	0	0	0	0
90+	0	0	0	0	0	0	0	1	0	0
100+	0	0	1	0	0	0	0	0	0	0

particular, if additional samples are collected so that t additional new data bases each containing 5,720 entries are produced, how many more new compounds, in addition to the 1,258 distinct compounds already tabulated, would we expect to find? The present study provides an answer to this question.

Estimating the number of unseen species, in this case the number of unidentified organic compounds, is a familiar problem in ecological studies. R. A. Fisher (1943) developed a parametric model, and Good and Toulmin (1956) a nonparametric model, for estimating the number of unseen species. Efron and Thisted (1976) extended these models from the theory of empirical Bayes estimation. In the present study we have examined the relevance of these and other approaches to the problem of estimating the number of 'unidentified' organic compounds in water. The results provide an answer to the question posed earlier.

2. FISHER'S BASIC POISSON MODEL

Fisher (1943) developed his model with the basic assumption that members of each species in the environment enter the trap according to a Poisson law. We make a similar assumption that organic compounds in the water are captured and identified according to a Poisson law.

Suppose there exist S organic compounds (such as trihalomethanes) continuously present in a source, such as drinking water (Smith *et al.*, 1980). Our ability to detect a compound which is actually present depends on a number of factors such as the actual concentration of the material in the source, the reproducibility and efficiency of the collection and concentration steps, analytical methodology, etc. (Janardan and Schaeffer, 1979). Then, if replicate samples are collected from a given source, these factors will act randomly and we will either be able or unable to identify a given compound in a given replicate. Let X_j be the number of times the j*th* compound will be found, then X_j follows Poisson distribution:

$$P(X_j) = e^{-\lambda_j} \lambda_j^{X_j} / X_j! \quad j = 1,2,\cdots,S, \quad X_j = 0,1,2,\cdots \qquad (1)$$

For example, Janardan *et al.*, (1980) have recently reported data on the compounds identified in simultaneously collected replicate samples for each of two distinct sources. For source 1, for example, 24 compounds were identified once, 18 twice, zero thrice, and 1 four times. Twenty-seven compounds observed in source 1 were not identified in source 2 (Table 2).

The results in Table 2 show that the assumption of the Poisson distribution for the random varaible X_j is not unrealistic.

Having demonstrated that the Poisson assumption is justified for the type of data being used here, we closely follow Efron and Thisted's (1976) approach (which the reader should consult for a

TABLE 2: Distribution of Number of Compounds Observed in Two Distinct Sources: (1) Illinois River, Peoria; (2) Petroleum refinery, Lockport.

	Source 1		Source 2	
Occurrence	Obs. No. of Compounds	Expected from Poisson	Obs. No. of Compounds	Expected from Poisson
0	20	22.8	27	26.6
1	24	23.2	22	19.3
2	18	11.8	3	7.0
3	0	4.0	1	1.7
4	1	1.0	1	0.3
5	0	0	1	0
Chi-square (2 D.F.)		0.603		1.38

fuller discussion of the mathematics). It is convenient mathematically to assume that the observed data base of 5720 entries corresponds to the interval [-1,0] and that the future t sets (each of size 5720) correspond to the interval [0,t]. In order to estimate the number of organic compounds which are likely to be present in the water we extrapolate the results of the interval [-1,0] to the interval [0,t].

Let η_k be the expected number of organics which appear exactly k times in [-1,0] so that

$$\eta_k = S \int_0^\infty e^{-\lambda} \lambda^k / k! \; dG(\lambda). \tag{2}$$

Let $\Delta(t)$ be the expected number of new organic compounds to be found in [0,t]. Since the average rates of occurrence, λ_j in (1), vary from compound to compound, the λ's will have their own distribution. We assume that the distribution of λ's is given by $G(\lambda)$. Thus $\Delta(t) \equiv S$ times (probability an organic compound is not discovered in [-1,0] but that it is discovered at least once in [0,t]). Then

$$\Delta(t) \equiv S \int_0^\infty e^{-\lambda}(1-e^{-\lambda t}) \; dG(\lambda). \tag{3}$$

Our purpose is to estimate $\Delta(t)$. Expanding the expression $(1-e^{-\lambda t})$ on the right hand side of (3) in powers of t and using (2) we obtain

$$\Delta(t) = \eta_1 t - \eta_2 t^2 + \eta_3 t^3 - \cdots. \tag{4}$$

The right hand side of (4) need not converge. As Efron and Thisted (1976) showed, by assuming it does, an unbiased estimate for $\Delta(t)$ is given as:

$$\hat{\Delta}(t) = n_1 t - n_2 t^2 + n_3 t^3 - \cdots \tag{5}$$

For Shackelford and Keith's data in Table 1 the number of previously unobserved compounds appearing in a new data base (t = 1) is:

$$\hat{\Delta}(1) = 503 - 238 + 122 - 80 + 56 - \cdots = 347. \tag{6}$$

Calculations show that for values of t larger than one however, meaningless estimates ($\hat{\Delta}(2) = 9.825E30$, $\hat{\Delta}(3) = 1.386E49$) are obtained because the higher powers of t produce large oscillations in the series.

Since Σn_k is Poisson distributed with the parameter $\Sigma\eta_k$ the variance of $\hat{\Delta}(1)$ is given by $Var(\hat{\Delta}(1)) = \Sigma\eta_k = 1258$; standard deviation is 35.

3. GOOD AND TOULMIN'S MODEL

Fisher's estimator diverges for large values of t. Good and Toulmin (1956) use Euler's transformation to force convergence in the oscillating series (5) by making the transformation $t = u/(2-u)$, $0 \leqslant u \leqslant 2$. Then $\Delta(t)$ in (5) becomes

$$\Delta(t) = \sum_{k=1}^{\infty} (-1)^{k+1}\eta_k t^k = \sum_{k=1}^{\infty} (-1)^{k+1}\eta_k u^k[(1+u/2+u^2/4+\cdots)/2]^k$$

$$= \sum_{y=1}^{\infty} \xi_y u^y \tag{7}$$

where $$\xi_y = \sum_{k=1}^{y} \binom{y-1}{k-1} (-1)^{k+1} \eta_k/2^y. \tag{8}$$

Let $\Delta^p(t) \equiv \sum_{k=1}^{p} (-1)^{k+1} t^k\eta_k$, and $\Delta^p(u) \equiv \sum_{y=1}^{p} \xi_y u^y$, then

$$\Delta(t) = \lim_{p\to\infty} \Delta^p(t) \quad \text{and} \quad \Delta(u) = \lim_{p\to\infty} \Delta^p(u).$$

For positive η_k, the partial sums $\Delta^p(u)$ will converge more rapidly to the common limit than the sums $\Delta^p(t)$. Thus, Good and Toulmin (1956) suggest estimating $\Delta(t)$ by

$$\hat{\Delta}^p(u) = \sum_{y=1}^{p} \hat{\xi}_y u^y, \tag{9}$$

where $u = 2t/(1+t)$ and $\hat{\xi}_y$ is obtained by replacing η_k in (8) by the estimators n_k. Some values of $\hat{\xi}_y$ are given in Table 3.

Efron and Thisted (1976) stated that there is no good theoretical criterion for deciding the proper choice of the number of Euler coefficients. These authors choose $p = 9$ for their data, since higher values of the coefficients were negative. They noted that this choice was based on ". . . (admittedly weak) theoretical reasons . . ." which were not given.

TABLE 3: Estimated Euler's Coefficients for the Organic Data.

y	ξ_y	y	ξ_y
1	251.50	6	1.20
2	66.26	7	0.42
3	20.00	8	-0.20
4	6.75	9	-0.68
5	2.66	10	-0.99

Good (1953) approached the problem of selecting p by using graphical smoothing of $\sqrt{n_k}$ and by the fitting of ". . . one or another of nine special hypotheses . . ." which he presents. We approached the problem by using Monte Carlo simulation.

Our "population" for the purposes of simulation consisted of the 5,720 observations and frequencies given in Table 1. A sample of size 1,000 (or 2,000) was drawn from this population using uniform random numbers and the η_k's for these samples were obtained. This constituted the "working sample" (analogous to the sample given in Table 1). Then, t additional samples of size 1,000 (or 2,000) were generated from the population of 5,720 observations in the same fashion. The number of new compounds present in the subsequent samples were obtained by comparing the "working sample" with the *tth* set. The cumulative number of new compounds was calculated for $t = 1,2,\cdots$. Euler coefficients were computed for each value of t, and the estimated number of new compounds were obtained from (9) for $p = 7$ to 10. As negative values of ξ_y $(p > 7)$ were included in the sum, the estimated number of compounds, varied erratically as a function of t, and eventually became negative. Thus, we rejected the inclusion of negative Euler coefficients since their inclusion decreased the estimates, which eventually became negative. An approach based on computing η_k values for the frequencies obtained by fitting the data in Table 1 to various discrete distributions will be presented subsequently.

Computation of $\hat{\Delta}^7$ (1) from Table gives a value of 349. Table 4 gives the estimates of the number of new compounds which are expected to appear in t $(= 1,2,\cdots,24)$ additional sets using $p_0 = 7$.

TABLE 4: Estimates of Unseen Compounds Using Non-parametric Model (9).

t	$\Delta^7(u)$	t	$\Delta^7(u)$
1	349	13	1044
2	543	14	1057
3	669	15	1068
4	757	16	1078
5	822	17	1087
6	873	18	1095
7	913	19	1102
8	944	20	1109
9	972	21	1115
10	994	22	1120
11	1013	23	1125
12	1030	24	1130

Initially, estimates were computed for values of t up to 100. Applying graphical differentiation techniques, it was found that the differences in the estimates for $t > 25$ were not significant.

Good and Toulmin's model does not enable us to obtain a standard error on the estimate. Standard errors can be computed from the model (Efron and Thisted, 1976) given next.

4. EFRON AND THISTED'S GENERAL LINEAR MODEL

Efron and Thisted show that the Euler's transformation $\Delta^p(u)$ is just the average of the oscillating series $\Delta^x(t)$ over values of x binomially distributed with mean $p/(1+t)$ and variance $pt/(1+t)^2$. The averaging process is what smooths out the oscillations. The estimator $\hat{\Delta}^p(u)$, with $\eta_k = n_k$ is equivalent to the form:

$$\hat{\Delta}(t) = \sum_{x=1}^{\infty} h_x n_x \tag{11}$$

where

$$h_x = (-1)^{x+1} t^x \sum_{k=x}^{p} \binom{p}{k} \left(\frac{1}{1+t}\right)^p \left(\frac{t}{1+t}\right)^{p-k} \tag{12}$$

for $x = 1,2,\cdots,p$ and $h_x = 0$ for $x > p$.

The estimator (11) is called a general linear estimator whose variance is given by:

$$\mathrm{Var}(\hat{\Delta}(t)) = \mathrm{Var}\left(\sum_{x=1}^{\infty} h_x n_x\right) = \sum_{x=1}^{\infty} h_x^2 \eta_x. \tag{13}$$

The estimates of $\Delta(t)$ obtained from (11) are biased. Since the series (11) is an infinite series, some arbitrary truncation point, p, must be selected for computational purposes. "The choice of p must take into account both bias and variance" (Efron and Thisted, 1976). Following the computational procedure given by these authors (which we will not describe here), the bias was calculated for various values of λ and t. With $p = 7$ and $t = 1$ the bias ranges from -8.6 to $+1.18$ and the variance is 806.92. For $p = 12$ and $t = 1$ the bias goes from -0.14 to 0.92, with a variance of 993.19. Taking $p = 7$, $t = 24$, the bias varies from -22675.79 to 760.75, while the variance is 259,903.17; for $p = 12$, and $t = 24$ the bias goes from -18485.10 to 931.46 with a variance of 48,916,036. Thus, as p increases, the range of bias decreases, while the variance increases substantially. The choice of $p = 7$ is therefore reasonable. The estimated variance is obtained by replacing η_x's in (13) by their estimates n_x:

$$\mathrm{Var}(\hat{\Delta}(t)) = \sum_{x=1}^{\infty} h_x^2 n_x.$$

So, $\mathrm{Var}(\hat{\Delta}(10)) = 139864.78$; standard deviation = 374. Table 5 gives the general linear estimates of the expected number of previously unseen organic compounds which will appear in t future lists of identified compounds for $t = 1,2,\cdots,24$ and $p = 7$, along with their standard deviations.

5. MODELS BASED ON OTHER DISTRIBUTIONS

The previous sections used the Poisson assumption to develop estimates of the number of new compounds which will be observed from t additional data bases. The models give similar estimates of the number of compounds which are expected to be observed. Here, we consider models based on other discrete distributions namely, Fisher's Negative Binomial Distribution (FNBD), the Logarithmic Series Distribution (LSD) (Pielou, 1977), and a Generalized Geometric Distribution (GGD). Although these distributions provide direct estimates of the total number of compounds, the standard errors of these estimates cannot be obtained. Thus, Efron and Thisted's (1976) approach may be preferable.

TABLE 5: General Linear Estimates and Standard Deviations.

t*	$\hat{\Delta}$	S.D.($\hat{\Delta}$(t))	Lower Bound**	Upper Bound
1	349	28	321	377
2	543	78	465	621
3	669	135	534	804
4	758	187	571	945
5	823	232	591	1055
6	873	270	603	1143
7	913	302	611	1215
8	945	330	615	1275
9	972	354	618	1326
10	994	374	620	1368
11	1013	392	621	1405
12	1030	408	622	1464
13	1044	422	622	1466
14	1056	434	622	1490
15	1068	445	623	1513
16	1078	455	623	1533
17	1087	464	623	1551
18	1095	472	623	1567
19	1102	480	622	1582
20	1109	487	622	1596
21	1115	493	622	1608
22	1121	499	622	1620
23	1126	505	621	1631
24	1131	510	621	1641

*t = number of data bases   ** 1 standard deviation limits

5.2 Fisher's Negative Binomial Model. Let the distribution function $G(\lambda)$ in Section 2 be approximated by a gamma distribution with density function

$$g(\lambda;\ \alpha,\beta) = C\ \lambda^{\alpha-1} e^{-\lambda/\beta} \qquad \alpha > 0,\ \beta > 0. \tag{14}$$

where $C = 1/\beta^{\alpha}\Gamma(\alpha)$. From (2) we have

$$\eta_k = S\ C \int_0^{\infty} [\lambda^{k+\alpha-1} e^{-\lambda(1+1/\beta)}/k!]d\lambda$$

$$= S\ C\ \Gamma(k+\alpha)\ \gamma^{k+\alpha}/k! \tag{15}$$

$$\eta_1 = S\ C\ \Gamma(1+\alpha)\ \gamma^{1+\alpha}. \tag{16}$$

Substituting (8) in (7) we get

$$\eta_k = \eta_1 \, \Gamma(k+\alpha) \, \gamma^{k-1}/k! \, \Gamma(1+\alpha) \tag{17}$$

where $\gamma = \beta/(1+\beta)$. Equation (17) is proportional to the negative binomial distribution with parameters α and γ. It may be noted that the expression (17) allows the parameter α to assume values also in the range -1 to 0 giving finite values of η_k $(k = 1,2,3,\cdots)$. The density (14) becomes improper at the origin for values of α less than 1, and the constant C becomes meaningless. Fisher liked $\alpha = 0$, which makes the expression (17) a density function of the logarithmic distribution discussed in Section 5.2.

We can write the expression (3) in the form

$$\Delta(t) = \eta_1 \int_0^\infty e^{-\lambda}(1-e^{-\lambda t})dG(\lambda) / \int_0^\infty \lambda \, e^{-\lambda} \, dG(\lambda) \tag{18}$$

in order to avoid the ambiquities in the case where G is improper. Substituting (14) in (18) gives

$$\Delta(t) = \begin{cases} \eta_1[1-(1+\gamma t)^{-\alpha}]/\gamma\alpha & \text{if } \alpha > 0 \\ \eta_1 \log(1+\gamma t)/\gamma & \text{if } \alpha = 0 \\ \eta_1[(1+\gamma t)^{-\alpha}-1]/(-\gamma\alpha) & \text{if } \alpha < 0 \end{cases} \tag{19}$$

When $\alpha > 0$, $\Delta(t)$ in (19) approaches the limit η_1/α as t goes to infinity. Using Method B of Brass (1958), the estimates of the parameters α and γ were obtained as 0.2076 and 0.4891. Substituting the values $\hat{\eta}_1 = 503$, $\hat{\alpha} = 0.2076$ and $\hat{\gamma} = 0.4891$ into the first line of (19) gives the estimates $\hat{\eta}_k$ remarkably close to the observed n_k (see Table 6, column 3). The total number of unobserved organic compounds in the aquatic environment is

$$\text{limit } \Delta(t) = (\eta_1/\alpha) = 503/0.2028 = 2480.$$

Hence, the total number of distinct compounds is 2490 + 1258 = 3738, in excellent agreement with the estimates obtained in the earlier sections.

TABLE 6: Frequency of Occurrence of 1,258 *Organic Compounds*

Frequency of occurrence	Observed No. of compounds	Expected no. of Compounds FNBD	LSD	GGD
1	503	503	473.2	544.0
2	238	209	215.6	212.3
3	133	123	131.0	117.1
4	80	83	89.5	75.3
5	56	60	65.3	52.9
6	46	46	49.6	39.3
7	20	36	38.7	30.4
8	14	29	30.9	24.2
9	15	24	25.0	19.7
10	18	20	20.5	16.3
11	15	17	17.0	13.7
12	16	14	14.2	11.6
13	10	12	11.9	10.0
14	10	10	10.1	8.7
15	9	9	8.6	7.6
16	4	8	7.3	6.7
17	12	7	6.3	5.9
18	6	6	5.4	5.2
19	7	5	4.7	4.7
20 or more	46	37	33.2	52.4

5.2 Logarithmic Series Distribution (LSD) Model. Pielou (1977) gives the LSD as $P(x) = \alpha\,\theta^x/x$, where $\alpha = -[(\ln(1-\theta)]^{-1}$, θ was estimated using a maximum likelihood iterative procedure (Pielou 1975) which gave $\theta = 0.911212$. Treating each new data base $(t = 1,2,\cdots)$ as a "quadrat" we can apply the procedure described by Pielou (1977) to estimate the total number of compounds. Thus, let S_t be the total number of distinct compounds expected to be identified in a large data base of t times that of Shackelford and Keith. When t is large

$$S_t = S\alpha[\ln(N/S\alpha)] + S\alpha[\ln(t)]$$

where $N = S\theta\alpha/(1-\theta)$. Using $S = 1258$ and $\theta = 0.911212$ gives $S_t = 2860$, and the number of new, distinct compounds is $2860 - 1258 = 1602$.

5.3 Generalized Geometric Distribution (GGD) Model. The GGD was developed by Plunkett and Jain (1975). The probability function is given by

$$P(x) - \frac{\Gamma(1 + \beta x)}{x!\Gamma(2 + \beta x - x)} \alpha^{x-1} (1-\alpha)^{1+(\beta-1)x}$$

where α and β are estimated by the method of moments. Here

$$\hat{\alpha} = 1 - S^2/(m^2(m-1)) = 0.135846$$

$$\hat{\beta} = m(m-1)^2/(m^2(m-1)-S^2) = 5.74231.$$

The resulting fit of the data using these values is given in Table 6. Then, the total number of compounds is given as $S^* = S/\alpha = 1258/0.135846 = 9260$ compounds. The number of unseen compounds is 8002.

6. RESULTS AND DISCUSSION

Recently, several authors have suggested statistical approaches for estimating the quantities of unseen objects in a population. These methods have now been applied to the problems of estimating the number of organic pollutants likely to be in the aquatic environment.

In developing these estimates it is assumed that:

1. The data base (Shackelford and Keith) which is used is accurate and unbiased; that is, it correctly reports what has been found, and the findings reflect the whole population (of sample types).

2. The data used were all obtained in essentially the same manner, and future data will be similarly obtained. Thus, the development of new, more sensitive procedures are not directly anticipated by these estimates.

3. The estimates refer only to the characterizable compounds: that is, compounds for which a definite structure and name can be assigned. Separable compounds, (those which produce a unique peak on chromatography or molecular weight by mass spectrometry, etc.) are not included.

Under these constraints it is estimated that if a large number $(t = 24)$ of data bases equivalent in size and character to Shackelford and Keith's are collected, an additional 1131 (621-1641) new compounds will be identified. Since their list includes 1258 separate compounds, the 95% confidence bounds on the total number of identifiable compounds is 2389 ± 1020.

The development of these estimates was motivated by the need to identify and quantitate the organic compounds being discharged to, and present in, Illinois' waters.

A specific objective of this effort was the identification of as many components in the sample as possible. Evaluation of the data from samples simultaneously collected by different methods showed that no single method or combination of methods effectively captured/recovered the same set of compounds from a source (Janardan *et al.*, 1980). This finding provided experimental evidence that if we were going to successfully identify as many compounds as possible in each sample, an estimate of the number of compounds present in each sample was required. If these estimates were then totaled for all possible samples then an estimate of the magnitude of the effort could be made.

At least three quality control problems related to knowing the total number of compounds were identified. First, realistic estimates of the magnitude of the proposed effort were required. Second, as new compounds relative to Shackelford and Keith's base were identified, a method for determining if the rate of these identifications was lower than, higher than, or at the expected rate was needed. This required an estimate of the expected rate. Third, would the expected rate of identifications plateau, since if this occurred management goals could be established.

At the end of the IEPA program's first year, 208 identifications of 160 distinct compounds, of which 124 were "new", had been made. From the simulations described earlier, the expected number of new compounds in a base of 208 identifications is found to be ~106 ± 24 (2 SD) = (82, 130). Thus, in spite of IEPA's efforts to maximize the identification of new compounds, there is no evidence that their rate of identification of "new" compounds exceeds the expected rate.

This finding provides a basis for speculating on the real severity of some of the initial assumptions. Briefly, the IEPA tried to maximize the rate of discovery of "new" compounds by: selecting industrial sources which engineering review had identified as discharging complex organic wastes such as specialty chemical and petroleum refinery wastes; using, and subsequently modifying, the best general extraction and analysis techniques currently available; matching the mass spectral fragmentation patterns of the unknown against computer based libraries of compounds ; examining even minor constituents in the sample. In spite of the fact that the IEPA tried to optimize their procedures, and that these procedures (and equipment) were "modern" in comparison to most of the studies in Shackelford and Keith, the rate of identification agress with the predicted rate. Unless there is a significant flaw in our experimental work we infer that rate at which compounds will be identified is in agreement with the estimates reported in the present paper.

7. CONCLUSIONS

Several different methods have been examined for obtaining estimates of the number of compounds which might be identified in the environment using currently available analytical methods. The total number of identifiable distinct compounds estimated using the various statistical procedures described herein is between about 2500 to about 9300.

The procedures described here have been used by IEPA for quality control purposes in at least two ways. First, as new data bases are formed, the actual rates of discovery of new compounds are compared with the theoretical rates. If compounds accumulate at a higher than expected rate, this may signal: a change to more sensitive or precise methods, changes in sample characteristics, bias (error) in identification, etc. If the rate of accumulating new compounds is significantly below expectation, lack of analytical sensitivity, mis-identification, changes in waste characteristics, etc. may be indicated.

Second, new data can be pooled with existing files, and revised estimates of the number of expected compounds obtained. For regulatory agencies concerned with the potential impact of non-specific regulations, these estimates provide a realistic picture of the magnitudes of the control problems.

ACKNOWLEDGEMENT

The authors thank Marla Gregory and Judy Hornyak for typing the manuscript, and M. V. Srinivas for initial programming work.

REFERENCES

Brass, W. (1958). Simplified methods of fitting the truncated negative binomial distribution. *Biometrika*, 45, 59-68.

Efron, B. and Thisted, R. (1976). Estimating the number of new species: How many words did Shakespeare know? *Biometrika*, 63, 435-447.

Fisher, R. A., Corbet, A. S. and Williams, C. B. (1943). The relation between the number of species and the number of individuals in a random sample of an animal population. *Journal of Animal Ecology*, 12, 42-58.

Good, I. J. (1953). The population frequencies of species and the estimation of population parameters. *Biometrika*, 40, 23-264.

Good, I. J. and Toulmin, G. H. (1956). The number of new species, and the increase in population coverage, when a sample is increased. *Biometrika*, 43, 45-63.

Janardan, K. G. and Schaeffer, D. J. (1979). Propagation of errors in estimating the levels of trace organics in environmental sources. *Analytical Chemistry*, 51, 1024-1026.

Janardan, K. G., Schaeffer, D. J., and Somani, S. M. (1980). Efficiencies of liquid-liquid extraction carbon, and XAD-2 absoprtion in isolating organic compounds from environmental sources. *Bulletin Environmental Contamination and Toxicology*, 24, 145-151.

McLafferty, F. W. (1980). Tandem mass spectromerty (MS/MS): A promising new analytical technique for specific component determination in complete mixtures. *Accounts of Chemical Research*, 13, 33-39.

Pielou, E. C. (1975). *Ecological Diversity*. Wiley-Interscience, New York.

Peilou, E. C. (1977). *Mathematical Ecology*. Wiley-Interscience, New York.

Plunkett, I. G. and Jain, G. C. (1975). Three generalized negative binomial distributions. *Biometrika*, 17, 286-302.

Shackelford, W. M. and Keith, L. H. (1976). Frequency of Organic Compounds Identified in Water. Athens, GA: US Environmental Protection Agency, EPA 600/4-76-062.

Smith, V. L., Cech, I., Brown, J. H. and Bogdan, G. F. (1980). Temporal variations in trihalomethane content of drinking water. *Environmental Science and Technology*, 14, 190-196.

Somani, S. M., Teece, R. G. and Schaeffer, D. J. (1980). Identification of co-carcinogens and promoters in industrial discharges into, and in the Illinois River. *Journal of Toxicology and Environmental Health*, 6, 317-333.

[*Received June* 1980. *Revised October* 1980]

SOME BIVARIATE PROBABILITY MODELS APPLICABLE TO TRAFFIC ACCIDENTS AND FATALITIES

RAMALINGAM SHANMUGAM

University of South Alabama
Mobile, Alabama 36688 USA

JAGBIR SINGH

Temple University
Philadelphia, Pennsylvania 19122 USA

SUMMARY. Following Leiter and Hamdan (1973) two bivariate probability models are generalized to better explain the phenomenon of highway traffic accidents and fatalities. Both the models are then studied for fitting to a data set given in Leiter and Hamdan. It is noticed that the generalized models provide a significantly improved fit when compared with the Leiter and Hamdan's model.

KEY WORDS. Bivariate probability models, generalized Poisson-quasi binomial model, estimation, goodness-of-fit, traffic accidents, fatalities data.

1. INTRODUCTION

Leiter and Hamdan (1973) considered two bivariate probability models to study the pheonmenon of highway traffic accidents and fatalities. A basic assumption underlying their models was the constancy of the accident rate in a given interval of time. In reality, however, the accident rate varies depending upon many hazards such as the traffic intensity, the road conditions, and so on. Therefore, their models proved inadequate because they did not account for the variability as a consequence of highway hazards.

C. Taillie et al. (eds.), Statistical Distributions in Scientific Work, Vol. 6, 95–103.

In this paper we study two generalized bivariate probability models as alternatives to those of Leiter and Hamdan. Our models take into consideration the presence of accident hazard factors which may cause accidents. Naturally we consider estimation and show that Leiter and Hamdan's results are special cases of ours. The fit of the generalized models is illustrated using the data given in Leiter and Hamdan.

2. GENERALIZED POISSON-QUASI BINOMIAL MODEL

Consider the number of accidents X recorded at a specific location in a given time interval follows a generalized Poisson probability law,

$$P(X=x) = \lambda_1(\lambda_1+x\lambda_2)^{x-1}\exp[-(\lambda_1+x\lambda_2)]/x!, \tag{1}$$

where the parameters $\lambda_1 > 0$ and $-1 < \lambda_2 < 1$ represent the accident rate and accident-hazard level, respectively, such that $(\lambda_1+x\lambda_2) > 0$ for $x = 0,1,2,\cdots$. The probability distribution in (1), also called Lagrangian distribution in the literature (Consul, 1975), is known to be a member of the modified power series distributions(Gupta, 1974). Some statistical properties of this distribution have been investigated by Consul and Jain (1973). Notice that when there is no accident hazard, that is $\lambda_2 = 0$, the model in (1) reduces to the Poisson probability law.

Define the indicator random variable (r.v.) Y_i associated with the *ith* accident as follows: $Y_i = 1$ with probability $0 < p < 1$ if the *ith* accident is fatal, and $Y_i = 0$ otherwise with probability $q = 1-p$. Let Y denote the total number of fatal accidents out of X accidents which follow the distribution in (1). So $Y = Y_1 + Y_2 +\cdots+ Y_X$. If an accident-hazard exists in a given interval of time, that is $\lambda_2 \neq 0$, then r.v.'s $Y_1, Y_2,\cdots,Y_X$ cannot be assumed mutually independent. We assume, however, that the conditional distribution of the total number of fatal accidents Y, given that there were $X = x$ number of accidents, would be a quasi-binomial probability law with parameters x and p. That is,

$$p[Y=y|X=x] = \binom{x}{y}\left(\frac{\lambda_1 pq}{\lambda_1+x\lambda_2}\right)\left(\frac{\lambda_1 p+y\lambda_2}{\lambda_1+x\lambda_2}\right)^{y-1}\left(\frac{\lambda_1 q+(x-y)\lambda_2}{\lambda_1+x\lambda_2}\right)^{x-y-1} \tag{2}$$

where $y = 0,1,2,\cdots,x$. If $\lambda_2 = 0$, then the quasi=binomial model reduces to the binomial law.

We may point out that Consul (1975) uses the quasi-binomial as a damage process to characterize the generalized Poisson in (1).

Now the joint probability distribution $p(x,y)$ of the total number of accidents X and the total number of fatal accidents Y is

$$\lambda_1^2 pq(\lambda_1 p+y\lambda_2)^{y-1}[\lambda_1 q+(x-y)\lambda_2]^{x-y-1}\exp[-(\lambda_1+x\lambda_2)]/y!(x-y)! \quad (3)$$

where $x = 0,1,2,\cdots$; $y = 0,1,2,3,\cdots,x$; $\lambda_1 > 0$; $-1 < \lambda_2 < 1$; $0 < p = 1 - q < 1$ and $(\lambda_1+x\lambda_2) > 0$. It can be shown that the marginal distribution of Y is a generalized Poisson probability law with parameters $\lambda_1 p$ and λ_2. That is:

$$p(Y=y) = \lambda_1 p(\lambda_1 p+y\lambda_2)^{y-1} \exp[-(\lambda_1 p+y\lambda_2)]/y!$$

$y = 0,1,2,\cdots$; $\lambda_1 p > 0$; $-1 < \lambda_2 < 1$ such that $(\lambda_1 p+y\lambda_2) > 0$. Further, the conditional distribution of X given $Y = y$ can be easily shown to be

$$p[X=x|Y=y] = \lambda_1 q[\lambda_1 q+(x-y)\lambda_2]^{x-y-1} \exp[-(\lambda_1 q+(x-y)\lambda_2)]/(x-y)!$$

for $x \geqslant y$, and zero otherwise. This is also a generalized Poisson distribution with parameters $\lambda_1 q$ and λ_2 but shifted y units towards the right. Using the Abel identities (Riordan, 1968) and some properties of the generalized Poisson distribution we obtain, after algebraic simplifications, the following:

$$E[Y|X=x] = xp \ , \ E[X|Y=y] = y + \lambda_1 q/(1-\lambda_2)$$

$$\mathrm{Var}[Y|X=x] = x^2 pq - \frac{x(x-1)\lambda_1 pq}{(\lambda_1+x\lambda_2)} \sum_{s=0}^{x-2} \frac{(x-2)(s)\lambda_2^s}{(\lambda_1+x\lambda_2)^s}$$

$$\mathrm{Var}[X|Y=y] = \lambda_1 q/(1-\lambda_2)^3 .$$

From the above results, it is easy to see that

$$\mathrm{Var}(X) = \lambda_1/(1-\lambda_2)^3$$

$$\mathrm{Cov}[X,Y] = \lambda_1 p/(1-\lambda_2)^3 \quad \text{and} \quad \mathrm{Corr}[X,Y] = +\sqrt{p} .$$

3. ESTIMATION OF THE PARAMETERS λ_1, λ_2 AND p

Suppose that (X_i, Y_i), $i = 1,2,\cdots,n$, are independent and identically distributed following the joint probability law $p(x,y)$ given in (3). The maximum likelihood estimators are difficult to obtain since the maximum likelihood equations are tedious indeed. So, we choose the method of moments to estimate the parameters. The moment estimates are:

$$\hat{\lambda}_2 = 1 - \sqrt{\bar{x}}/S_x, \ \hat{\lambda}_1 = \bar{x}(1-\hat{\lambda}_2), \ \hat{p} = \bar{y}/\bar{x}$$

where S_x denotes the standard deviation of X values.

4. GENERALIZED POISSON - GENERALIZED POISSON MODEL

Consider again the number of accidents X in a given time interval to be governed by the generalized Poisson probabilistic law given in (1). Let Z_i, $i = 1,2,\cdots,n$ represent the number of fatalities in the *ith* accident and that it obeys a generalized Poisson probabilistic law.

$$p(Z_i=z) = \lambda_3(\lambda_3+a\lambda_2)^{a-1}\exp[-(\lambda_3+a\lambda_2)]/a!$$

$$a = 0,1,2,\cdots; \ -1 < \lambda_2 < 1$$

such that $(\lambda_3+a\lambda_2) > 0$ where $\lambda_3 > 0$ represents the 'fatal-rate'. The total number of fatalities in X accidents are $Z = Z_1 + Z_2 + \cdots + Z_x$. The conditional distribution of Z is also a generalized Poisson with parameters $x\lambda_3$ and λ_2. That is,

$$p[Z=z|X=x] = x\lambda_3(x\lambda_3+z\lambda_2)^{z-1}\exp[-(x\lambda_3+z\lambda_2)]/z!$$

where $(x\lambda_3 + z\lambda_2) > 0$. The joint distribution $g(x,z)$ of X and Z is

$$g(x,z) = \lambda_1\lambda_3(\lambda_1+x\lambda_2)^{x-1}(x\lambda_3+z\lambda_2)^{z-1}\exp[-(\lambda_1+x(\lambda_2+\lambda_3)+z\lambda_2)]/(x-1)!z! \quad (4)$$

where $x = 1,2,\cdots,$ and $x = 0,1,2,\cdots.$

Closed expressions for the marginal probability distribution of the total number of fatalities Z and the conditional distribution of X given $Z = z$ are difficult to obtain. However, one can show that

$$E(Z) = \lambda_1\lambda_3/(1-\lambda_2)^2,\ Var(Z) = \lambda_1\lambda_3(1-\lambda_2+\lambda_3)/(1-\lambda_2)^5$$

$$E[Z|X=x] = x\lambda_3/(1-\lambda_2),\ Var[Z|X=x] = x\lambda_3/(1-\lambda_2)^3$$

$$Var(X) = \lambda_1/(1-\lambda_2)^3,\ Cov[X,Z] = \lambda_1\lambda_3/(1-\lambda_2)^4$$

$$Corr[X,Z] = \sqrt{\lambda_3/(1-\lambda_2+\lambda_3)}.$$

5. ESTIMATION OF THE PARAMETERS λ_1, λ_2 AND λ_3

Suppose that (X_i, Z_i), $i = 1,2,\cdots,n$, are n independent pairs of r.v.'s from the generalized Poisson-generalized Poisson bivariate probability model in (4) where Z_i represents the number of fatalities in X_i number of accidents which happened in the *ith* time interval. The maximum likelihood estimators are difficult to compute. Hence, we choose to estimate the parameters by the method of moments. The estimates are:

$$\hat{\lambda}_2 = 1 - \sqrt{\bar{x}}/S_x,\ \hat{\lambda}_1 = \bar{x}(1-\hat{\lambda}_2),\text{ and } \hat{\lambda}_3 = \bar{z}(1-\hat{\lambda}_2)^2/\hat{\lambda}_1.$$

6. GOODNESS -OF-FIT

We shall fit the generalized models to an accidents-fatalities data (Leiter and Hamdan, 1973) which were collected by the Virginia State Police from January 1, 1969 to October 31, 1970. The fit of each model is judged by the chi-square goodness-of-fit criterion. We have fitted these models to each of the six data sets. There are six tables; one for each data set. The first entry in each cell of the table is the actual observation; the second entry is the fitted value using the generalized models while the third is the fitted value from Leiter and Hamdan's models. Notice that the generalized models consistently fit quite well compared to the Leiter-Hamdan models.

ACKNOWLEDGMENTS

This research was partially supported by the Data Analysis Laboratory of Temple University.

TABLE 1: Observed and fitted distributions for the number of injury accidents (X) and the number of fatal accidents (Y) for the entire study.

X	Y = 0	Y = 1+	TOTAL
	286		286
0	284.89		284.89
	269.78		269.78
	198	18	216
1	202.31	13.72	216.03
	217.85	14.78	232.63
	82	10	92
2	83.83	10.89	94.72
	87.96	12.34	100.30
	24	6	30
3	26.59	4.97	31.56
	23.68	5.15	28.83
	13	1	14
4	7.17	1.71	8.88
	4.78	1.43	6.21
	1	0	1
5+	1.73	1.19	2.92
	0.89	0.36	1.25
	604	35	639
TOTAL	606.52	32.48	639.00
	604.94	34.06	639.00

Estimates: $\hat{\lambda}_2 = 0.06322$, $\hat{\lambda}_1 = 0.80777$, $\hat{p} = 0.06352$

Model	χ^2	d.f.	Significance Level
Generalized	8.54	9	0.5
Leiter and Hamdan	19.11	10	0.04

TABLE 2: Observed and fitted distributions for the number of injury accidents (X) and the number of fatal accidents (Y) for 1969.

X	Y = 0	Y = 1+	TOTAL
	154		154
0	153.65		153.65
	144.81		144.81
	107	12	119
1	109.39	8.45	117.84
	118.25	9.13	127.38
	43	6	49
2	45.83	6.78	52.61
	48.28	7.74	56.02
	15	4	19
3	14.79	3.15	17.94
	13.14	3.29	16.43
	7	0	7
4	4.08	1.11	5.19
	2.68	0.93	3.61
	1	0	1
5+	1.01	0.75	1.76
	0.51	0.24	0.75
	327	22	349
TOTAL	328.75	20.24	349.0
	327.67	21.33	349.0

Estimates: $\hat{\lambda}_1 = 0.82039$, $\hat{\lambda}_2 = 0.06736$, $\hat{p} = 0.07166$

Model	χ^2	d.f.	Significance Level
Generalized	5.99	9	0.75
Leiter and Hamdan	12.53	10	0.25

TABLE 3: Observed and fitted distributions for the number of injury accidents (X) and the number of fatal accidents (Y) for 1970 (through October 31).

X	Y = 0	Y = 1+	TOTAL
0	132 131.22 125.02		132 131.22 125.02
1	91 93.00 99.59	6 5.23 5.61	97 98.23 105.20
2	39 38.00 39.66	4 4.09 4.59	43 42.10 44.25
3	9 11.78 10.53	2 1.82 1.88	11 13.60 12.41
4+	6 3.07 2.48	1 1.77 0.64	7 4.85 3.12
TOTAL	277 277.06 277.28	13 12.91 12.72	290 290.00 290.00

Estimates: $\hat{\lambda}_1 = 0.79294$, $\hat{\lambda}_2 = 0.05755$, $\hat{p} = 0.05327$

Model	χ^2	d.f.	Significance Level
Generalized	3.99	7	0.80
Leiter and Hamdan	6.66	8	0.55

TABLE 4: Observed and fitted distributions for the number of injury accidents (X) and the number of fatalities (Z) for the entire study.

X	Z = 0	Z = 1	Z = 2+	TOTAL
0	286 284.89 269.78			286 284.89 269.78
1	198 201.48 217.52	17 13.18 14.61	1 1.37 0.50	216 216.03 232.63
2	82 82.39 87.69	10 10.78 11.78	0 1.55 0.83	92 94.72 100.30
3	24 25.60 23.57	5 5.02 4.75	1 0.94 0.51	30 31.56 28.83
4	13 6.71 4.75	1 1.75 1.27	0 0.42 0.19	14 8.88 6.21
5+	1 2.06 0.88	0 0.67 0.31	0 0.19 0.06	1 2.92 1.25
TOTAL	604 603.13 604.19	33 31.34 32.72	2 4.47 2.09	639 639.0 639.0

Estimates: $\hat{\lambda}_1 = 0.80777$, $\hat{\lambda}_2 = 0.06322$, $\hat{\lambda}_3 = 0.06970$

Model	χ^2	d.f.	Significance level
Generalized	11.02	14	0.80
Leiter and Hamdan	20.51	15	0.15

TABLE 5: Observed and fitted distributions for the number of injury accidents (X) and the number of fatalities (Z) for 1969.

X	Z = 0	Z = 1	Z = 2+	TOTAL
0	154 153.65 144.81			154 153.65 144.81
1	107 109.22 118.19	12 7.75 8.85	0 0.86 0.57	119 117.84 127.38
2	43 45.19 48.23	6 6.41 7.23	0 1.00 0.57	49 52.61 56.03
3	15 14.28 13.12	3 3.04 2.95	1 0.61 0.36	19 17.94 16.43
4	7 3.83 2.68	0 1.08 0.80	0 0.27 0.13	7 5.19 3.61
5+	1 1.20 0.50	0 0.42 0.20	0 0.13 0.04	1 1.76 0.74
TOTAL	327 327.37 327.53	21 18.70 20.03	1 2.87 1.44	349 349.00 349.00

Estimates: $\hat{\lambda}_1 = 0.82039$, $\hat{\lambda}_2 = 0.06736$, $\hat{\lambda}_3 = 0.07594$

Model	χ^2	d.f.	Significance level
Generalized	9.21	14	0.90
Leiter and Hamdan	14.50	15	0.50

TABLE 6: Observed and fitted distributions for the number of injury accidents (X) and the number of fatalities (Z) for 1970 (through October 31).

X	Z = 0	Z = 1	Z = 2+	TOTAL
0	132 131.22 125.02			132 131.22 125.02
1	91 92.34 99.32	5 5.38 5.70	1 0.50 0.17	97 98.23 105.20
2	39 37.20 39.46	4 4.34 4.53	0 0.55 0.27	43 42.10 44.25
3	9 11.29 10.45	2 1.97 1.80	0 0.33 0.16	11 13.60 12.41
4+	6 3.78 2.45	1 0.88 0.59	0 0.18 0.98	7 4.85 3.12
TOTAL	277 275.83 276.70	12 12.57 12.62	1 1.56 0.68	290 290.00 290.00

Estimates: $\hat{\lambda}_1 = 0.79294$, $\hat{\lambda}_2 = 0.05755$, $\hat{\lambda}_3 = 0.06180$

Model	χ^2	d.f.	Significance level
Generalized	3.82	11	0.88
Leiter and Hamdan	11.55	12	0.51

REFERENCES

Consul, P. C. and Jain, G. C. (1973). On some interesting properties of the generalized Poisson distribution. *Biometrische Zeitschrift*, 15, 495-500.

Consul, P. C. (1975). On a characterization of Lagrangian Poisson and Quasi-Binomial distributions. *Communications in Statistics*, 4, 555-563.

Gupta, R. C. (1974). Modified power series distribution and some of its applications. *Sankhyā, Series B*, 35, 288-298.

Leiter, R. E. and Hamdan, M. A. (1973). Some bivariate probability models applicable to traffic accidents and fatalities. *International Statistical Review*, 41, 87-100.

Patil, G. P. (1963). Certain properties of the generalized power series distribution. *Annals of Institute of Statistical Mathematics*, 14, 179-182.

Riordan, J. (1968). *Combinatorial Identities*. Wiley, New York.

[*Received July* 1980. *Revised October* 1980]

ROLE AND USE OF STATISTICAL DISTRIBUTIONS IN INFORMATION THEORY AS APPLIED TO CHEMICAL ANALYSIS

V. ŠTĚPÁNEK

Environmental Research Center, Prague
Hradčanské nám. 8
110 00 Prague 1, Czechoslovakia

SUMMARY. Various procedures in both qualitative and quantitative chemical analysis can be evaluated and selected with the help of information theory. The quantification in information theory requires some kind of pre-information which is formulated in terms of probabilities. These probabilities enter the formulas for the information theoretic criteria together with statistical distributions of the results. A number of statistical models are employed explicitly in order to meet some practical situations.

KEY WORDS. uncertainty, entropy, information content, probability distributions, chemical analysis methods.

1. INTRODUCTION

According to a recent definition, Analytical Chemistry is the science of acquiring information on material systems and interpreting it with regard to its exploitation, employing the methods of natural science (Fresenius, 1977). Also the quantification of information in Analytical Chemistry has been the subject of several studies and papers, in which mainly terms from information and communication theories have been used although an analytical process cannot be straightforwardly translated into a communication process.

Applications of information theory to analytical problems concern the areas of both the qualitiative and the quantitative analysis. Separate attention has been paid to trace analyses and analytical quality control; new information quantities have been

C. Taillie et al. (eds.), Statistical Distributions in Scientific Work, Vol. 6, 105–110.

derived for purposes of optimizing analytical procedures (Eckschlager and Štěpánek, 1979).

2. QUALITATIVE ANALYSIS

In qualitative analysis the goal is to achieve an unambiguous assessment of the identity of the unknown element or pure compound and thus to gain a maximum decrease of uncertainty. Since reducing the uncertainty corresponds with gathering information, the decrease of uncertainty can be a measure for the amount of information obtained.

Prior to a qualitative analysis some pre-information is required to specify a set of possible identities together with their probabilities. Pre-information can be gathered from experience (records) or interpreted in a Bayesian way as subjective expectations or estimates. Often *a priori* probabilities are taken equal to $1/n$ for a set of n possible identities.

The output of an identification procedure is one of a set of possible signals accompanied with conditional probabilities $P(Y_k|X_i)$ of measuring a signal Y_k when X_i is the analyzed element (compound) ($i = 1,2,\cdots,n$; $k = 1,2,\cdots,m$). These probabilities can be obtained by calibration or from the reproducibility of the measuring process. If a signal Y_k has been measured, the *a posteriori* probabilities $P(X_i|Y_k)$ can be calculated from Bayes' Theorem as

$$P(X_i|Y_k) = \frac{P(X_i)\cdot P(Y_k|X_i)}{P(Y_k)}, \quad i = 1,2,\cdots,n, \tag{1}$$

where

$$P(Y_k) = \sum_{i=1}^{n} P(X_i)\cdot P(Y_k|X_i) \tag{2}$$

is the probability of measuring a signal Y_k.

If uncertainty is to be used as a criterion for judging the quality of the analytical result, Shannon's entropy is the most frequent function among those complying with reasonably defined desirable properties (Shannon, 1947). It has the form

$$H = -\sum_{i=1}^{n} p(i) \log p(i) \tag{3}$$

with the value 0 for $p(i) = 0$ and the units of uncertainty depend on the base of the logarithm.

3. QUANTITATIVE ANALYSIS

In quantitative analyses pre-information is expressed by an *a priori* continuous probability distribution which reflects our previous knowledge of the content of the component to be determined. Very often the probability density is considered to equal zero outside an interval and to be uniform within it. The result of a quantitative analysis is a realization of a random variable having a continuous probability distribution. This distribution is usually known from experience and is pertinent to the analytical method employed.

The amount of information obtained in this case can be measured as the information content (gain) achieved by replacing the *a priori* probability distribution by the *a posteriori* one (Eckschlager and Štěpánek, 1979), i.e.

$$I(q,p) = \int_{-\infty}^{\infty} q(x) \log [q(x)/p(x)] \, dx \tag{4}$$

where $p(x)$ and $q(x)$ are the probability densities before and after the analysis, respectively (P and Q are absolutely continuous distribution functions and Q is absolutely continuous with respect to P).

4. UNCERTAINTY AND INFORMATION

4.1 Discrete Signals. In qualitative analyses the amount of information obtained in case of an output signal Y_k can be defined as the decrease of uncertainty, i.e., the difference of uncertainties before and after analysis.

$$I(X\|Y_k) = -\sum_{i=1}^{n} P(X_i) \log P(X_i) + \sum_{i=1}^{n} P(X_i|Y_k) \log P(X_i|Y_k). \tag{5}$$

This formula leads sometimes to negative values of $I(X\|Y_k)$. However, in the important case of discrete uniform *a priori* probabilities $P(X_i) = 1/n$ it yields the same values as another information measure, analogous to (4),

$$I_1(X\|Y_k) = \sum_{i=1}^{n} P(X_i|Y_k) \log[P(X_i|Y_k)/P(X_i)]$$

which is always positive.

Since generally an analytical procedure is desirable for which $I(X \| Y_k)$ is close or equal to the entropy before analysis for every possible value of the signal, it is reasonable to describe the quality of the procedure by the average of the information

$$I = \sum_{k=1}^{m} P(Y_k) \cdot I(X \| Y_k)$$

(Kerridge, 1961; Cleij and Dijkstra, 1979). Introducing probabilities and using (1) and (2), I can be evaluated from the following relation

$$I = \sum_{i=1}^{n} \sum_{k=1}^{m} P(Y_k) \cdot P(X_i | Y_k) \log[P(X_i | Y_k)/P(X_i)]$$

and is always non-negative.

4.2 Continuous Signals. Besides discrete signals, in chemical analysis also continuously variable signals are measured. Then a conditional probability density $p(y|X_i)$ represents measuring a signal y when the measurement is performed with element (compound) X_i and probabilities $P(X_i|y)$ enter the formula for the uncertainty after measuring a signal of magnitude y and the formula for information obtained about the identity of the unknown element (compound)

$$I(X \| y) = - \sum_{i=1}^{n} P(X_i) \log P(X_i) + \sum_{i=1}^{n} P(X_i | y) \log P(X_i | y).$$

The average information content is defined by

$$I = \int_{-\infty}^{\infty} p(y) \cdot I(X \| y) \, dy$$

where $p(y)dy$ is the probability element of measuring a signal with magnitude y (prior to the anlaysis).

5. INFORMATION CONTENT WITH SPECIFIC STATISTICAL DISTRIBUTIONS

In quantitative analyses the most frequent case arises in such a way that the analyst does not know anything more about the content of the component to be determined that it lies within an interval (x_1, x_2). Thus the uniform (rectangular) distribution describes his knowledge of the true value whereas the results of the used method are distributed normally. Then the information content according to equation (4) takes on the value (Eckschlager, 1975)

$$I(q,p) = \log \frac{x_2 - x_1}{\sigma\sqrt{(2\pi e)}}$$

provided that the method yields unbiased results and that $\mu \in (x_1 + 3\sigma,\ x_2 - 3\sigma)$ where σ is the standard deviation of the method.

If the adopted analytical method is not unbiased and a systematic error $\delta = |X - \mu| > 0$ is present, the information content of the results reduces to

$$I(q,p) = \log \frac{x_2 - x_1}{\sigma\sqrt{(2\pi e)}} - \frac{1}{2}\left(\frac{\delta}{\sigma}\right)^2 .$$

In case that the method does not confirm the assumptions above with respect to the interval (x_1, x_2), the *a posteriori* normal distribution needs to be replaced by a truncated one. Then the information content is given by (Eckschlager and Štěpánek, 1979)

$$I(q,p) = \log \frac{x_2 - x_1}{[\phi(z_2) - \phi(z_1)]\sqrt{(2\pi e)}\sigma} + \frac{1}{2}\,\frac{z_2\varphi(z_2) - z_1\varphi(z_1)}{\Phi(z_2) - \Phi(z_1)}$$

where Φ and φ are the distribution and the frequency function of the standardized normal variable respectively and $z_j = (x_j - \mu)/\sigma$ for $j = 1,2,$.

Two normal distributions can be substituted in the formula for the information content if the *a priori* distribution is given by the probability distribution of the results of a preliminary (semiquantitative) determination. Setting $p(x)$ normal $N(\mu_0, \sigma_0^2)$ and $q(x)$ normal $N(\mu, \sigma^2)$ with $\sigma_0 \geqslant \sigma$ we get

$$I(q,p) = \frac{1}{2}(A^2 + B^2 - 1) - \log A$$

with $A = \sigma/\sigma_0 \leqslant 1$ and $B = (\mu - \mu_0)/\sigma_0$.

Another situation arises in trace analyses carried out with a method having a determination limit x_0 (Eckschlager and Štěpánek, 1978; Štěpánek and Eckschlager, 1979). If there is no response we can merely state that the content of the component $X \in (0, x_0)$. Then two uniform distributions yield a simple form of the information content $I(q,p) = \log (x_1/x_0)$ where x_1 is the highest assumed content of the trace component $(x_1 > x_0)$.

For detectable contents it is known that, in most cases, the results have a log-normal distribution. Substituting it in the form

$$q(x) = \frac{\exp\left\{-\frac{1}{2}\left[\frac{\log(x-x_0) - \log kx_0}{\sigma}\right]^2\right\}}{(x-x_0)\sigma\sqrt{(2\pi)}}, \quad x \in (x_0, x_1),$$

with the parameter of asymmetry k, we obtain for the information content an approximate value

$$I(q,p) = \log \frac{x_1}{x_0} \frac{\sqrt{n_p}}{k\sigma\sqrt{(2\pi e)}}$$

(here n_p is the number of parallel determinations).

REFERENCES

Cleij, P. and Dijkstra, A. (1979). Information theory applied to qualitative analysis. *Fresenius Zeitschrift für Analytische Chemie*, 298, 97-109.

Eckschlager, K. (1975). Informationsgehalt analytischer Ergebnisse. *Zeitschrift für Analytische Chemie*, 277, 1-8.

Eckschlager, K. and Štěpánek, V. (1978). Information content of trace analysis results. *Mikrochimica Acta*, I, 107-114.

Eckschlager, K. and Štěpánek, V. (1979). *Information Theory as Applied to Chemical Analysis*. Wiley, New York.

Fresenius, W. (1977). *Reviews on Analytical Chemistry*. Akad. Kiadó, Budapest.

Kerridge, D. P. (1961). Inaccuracy and inference. *Journal of the Royal Statistical Society, Series B*, 23, 184.

Shannon, E. and Weaver, W. (1947). *The Mathematical Theory of Information*. The University of Illinois Press, Urbana.

Štěpánek, V. and Eckschlager, K. (1979). Information content of chemical analysis results and methods. In *The Proceedings of the Sixth International CODATA Conference*. Pergamon Press, Oxford. Pages 239-241.

[*Received June* 1980. *Revised October* 1980]

MODELING THE DISTRIBUTION OF FINGERPRINT CHARACTERISTICS

STANLEY L. SCLOVE

Department of Industrial Engineering
and Management Sciences
The Technological Institute
Northwestern University
Evanston, Illinois 60201 USA

SUMMARY. Quantitative aspects of fingerprints are discussed. A study undertaken to develop methods for assigning probabilities to partial fingerprints is summarized, with emphasis on distributional aspects.

KEY WORDS. Fingerprints, two-way series, multinomial distribution, Markov process, Poisson process.

1. INTRODUCTION

This paper focuses on the distributional aspects of a study reported in three earlier articles (Osterburg, *et al.*, 1977; Sclove, 1979; Sclove, 1980) concerning the assingment of probabilities to partial fingerprints based on the numbers and locations of occurrences of the ten Galton characteristics. In the study a grid of cells was superimposed on the fingerprints. The number of characteristics in the cells is modeled as a multivariate two-way series (i.e., a multivariate stochastic process with two-dimensional indexing parameter). The statistical parameters were estimated from the data (fingerprints). Estimation of the probability of partial prints is illustrated. Some comparisons are made with the estimates provided by an assumption of independence between cells. Some analysis based on statistical results for infinitely divisible distributions (see Sclove, 1981) is discussed.

C. Taillie et al. (eds.), Statistical Distributions in Scientific Work, Vol. 6, 111–130.

2. BACKGROUND INFORMATION ON FINGERPRINTS

2.1 Types.

The three types. The bulb of each finger of the human hand contains ridge lines that form themselves into patterns, thus providing a basis for classification. Ridge-line patterns are of three major *types*: loops (ca. 65%), whoris (ca. 30%), and arches (ca. 5%). There is further subdivision within each major pattern. Arches are either plain or tented; loops are radial or ulnar; whorls are plain, central pocket loops, double loops, or accidentals. This further subdivision within each pattern allows a classification scheme to be organized so that for the ten fingers many categories of fingerprint-pattern combinations result. Within each category there are many fingerprints from different individuals which, to the untrained eye, appear to be the same. This process of separation through classification results in relatively small sets of fingerprints which are of manageable proportions for the purpose of search and comparison.

Search. Chernoff (1977) has treated the problem of selecting a subset of files such as fingerprint files for careful comparison with a target print to decide if the corresponding individual is represented in the files. It is assumed that much of the data in the files and on the target are subject to noise or random error. The (likelihood-ratio) solution depends upon the joint distribution of the filed data and the target data and their marginal distributions.

Computer assistance. Computer classification of single fingerprints into types (subdivisions of arch, loop, and whorl) by a syntactic approach has been achieved; see, e.g., Rao and Balck (1980).

Enhancement of latent fingerprints by numerical processing of the image has been treated; see e.g., Chiralo and Berdan (1978).

2.2 Ridge Counts. In the loop pattern there is a point where the three opposing ridge systems come together. (The outer and the lower ridge lines change concavity at that point.) This point is the triradius, or delta. If a straight line is drawn from the delta to the core, a certain number of ridge lines will be crossed. This number is the *ridge count.* Patterns with no triradii (simple arches) have no ridge count. In the case of patterns with two triradii (whorls and double loops) there are two counts; sometimes then one just works with the higher count. Sometimes the sum across fingers of the individual ridge counts, when defined, is considered; it is also called the "ridge count."

Holt (1951-52) has studied the correlation between numbers of crossings on different fingers.

2.3 The Galton Details. Fingerprints and dermatoglyphics in general have found use in medicine and genetics as correlates of genetic abnormalities; see, e.g., Holt (1968) and Priest, Tishler and Rosner (1976). The emphasis here, however, is on the use of fingerprints in identification, as in criminalistics. Partial prints such as those left at crime scenes do not always permit determination of the type or number of crossings. Even if they did, the individuality of a print would have to be based on the *details* of the print. The ridge-line details are termed *Galton characteristics* since Sir Francis Galton was among the first to study them systematically (Galton, 1892). He defined ten kinds of minutiae. One is a ridge ending, an abrupt ending to a ridge line; ridge endings are by far the most frequent characteristic. A ridge line may suddenly divide into two branches, much like a fork in a road; such a characteristic is termed a bifurcation (or fork). Similarly, eight other characteristics are defined. There is general agreement upon these ten types of ridge-line details. (See, e.g., Osterburg, *et al.*, 1977, for details and diagrams of the ten characteristics. See Appendix A for working definitions used for some of the characteristics.) The purpose of the study discussed here was to model the occurrence of the Galton characteristics, with a view toward the development of formulas for the calculation of probabilities of partial fingerprints.

The study was made as follows. A grid of one millimeter squares was placed over a fingerprint. Each fingerprint is considered as a *configuration* of the cells of the grid. For each cell of the grid there are several possibilities: one or more of the ten characteristics is there, or no characteristic is present. Thus a configuration is a grid of cells, where each cell may be thought of as either being empty or else being occupied by one or more words, the words representing the characteristics present. E.g., if a cell contains the words "dot, dot, ending ridge," it means that the area corresponding to that cell contained two dots and one ending ridge. Table 1 shows a configuration of 43 cells with 4 ridge endings and two forks.

A *match* between a suspect's full print and a partial print exists when there is a section of the full print that is the same as the partial print. Since we are working in terms of a grid of cells, for our purposes a match exists when a grid can be laid on the full print in such a way that the resulting configuration contains a section which is the same as the configuration corresponding to the partial print.

TABLE 1: Configuration of 43 *cells with* 4 *ending ridges and* 2 *forks.* 0 = *empty cell*, E = *ending ridge*, F = *fork.*

	1	2	3	4	5	6
a	0	0	0	0	0	0
b	E	0	0	0	E	0
c	0	0	F	0	0	0
d	0	0	0	0	0	0
e	0	0	0	E	0	0
f	0	0	0	0	F	0
g	0	E	0	0	0	0
h			0			

The fingerprints studied were enlarged to ten times actual size, making a full rolled print about 8" by 10". The cells of the grid were one centimeter square after enlargement. Members of the project staff coded the ten Galton characteristics, cell by cell. (See Appendix A for precise working definitions of the characteristics.) Thirty-nine prints were coded. (Osterburg had earlier examined 40 prints, from 40 different individuals, but one was missing, leaving 39 for re-examination.) There is no problem with representativeness of the sample. The Galton characteristics are "accidental." They are not genetic. With regard to these characteristics, two siblings, even two twins, are no more alike than two random persons. (On this point see e.g., Kingston (1964, p. 26), and the references given there.) Therefore, with respect to the Galton characteristics, each and every person is "representative."

3. DATA DESCRIPTION

By an *occurrence* we mean the occurrence of any one of the ten Galton characteristics. The 39 fingerprints used yielded a total of 8591 cells which could be coded. In all there were 2536 occurrences, or 0.295 per cell. Table 2 gives the distribution of the number of occurrences per cell, without regard to type.

TABLE 2: Distribution of number of occurrences.

Number of occurrences	0	1	2	3	4	5	Total
Number of cells	6584	1594	320	72	19	2	8591
Proportion of cells	.766	.185	.0372	.00838	.0022	.0023	1.00

The abbreviations used for the characteristics are as follows:

B:	bridge	L:	lake (eye)
D:	dot	O:	delta
E:	ending ridge	S:	spur
F:	fork (bifurcation)	T:	trifurcation
I:	island	Z:	double bifurcation

The symbol DE, for example, denotes the occurrence of one dot and one ending ridge in a cell; BEE would denote the occurrence of a bridge and two ending ridges in a cell; etc. Altogether 54 combinations occurred, including one DEEEE and one DDDDE. Table 3 gives the distribution of these *cell configurations*. (Of course, with a larger data set, many more cell configurations would occur.) Note in particular that 77% of the cells were empty; i.e., the probability that a cell is occupied is .23.

TABLE 3: Distributions of cell configurations (8591 cells).

	Frequency			Frequency	
Cell configuration	Number of cells	Number of cells	Cell configuration	Number of cells	Number of cells
Empty	6584	76.6%	ES	10	0.116
E	715	8.32	DDI	10	0.116
F	328	3.82	II	9	0.105
I	152	1.77	FI	9	0.105
D	130	1.51	BF	7	0.0815
EE	119	1.39	DEE	7	0.0815
B	105	1.22	FF	5	0.0582
S	64	0.745	T	5	0.0582
L	55	0.640	EEF	4	0.0466
EL	32	0.372	BEE	4	0.0466
DE	32	0.372	EII	4	0.0466
EEE	21	0.244	FL	3	0.0349
EI	21	0.244	BB	3	0.0349
O	17	0.198	FS	2	0.0233
DD	15	0.175	BD	2	0.0233
BE	13	0.151	DDE	2	0.0233
Z	12	0.140	LL		0.0233
DI	11	0.128	Other*	67	0.780
EEEE	10	0.116			

*19 other multiple occurrences.

3.1 Data Processing. One physical record corresponded to one cell and took the following form (the abbreviation "cc." means "card columns"):

cc. 1-2	cc.3	cc.4	cc.5-6	cc.7-8	cc.9-13
Fingerprint number, n	Hand	Finger	Row, i	Column, j	Alphabetic information giving cell contents
25	R	I	13	11	BEE

The line of data above signifies that there is a bridge (B) and two ridge endings (E) in the cell corresponding to row 13 and column 11 of fingerprint number 25, which is from the index finger (I in cc. 4) of somebody's right hand (R in cc. 3).

Actually these data were not card-punched but rather typed on a terminal and stored directly on disk, so "cc." is used only figuratively. Cc. 9-13 contain alphabetic information giving the contents of the cell. This shows the need either for programming in a language such as PL1 which allows alphabetic variables or for use of a text editor to convert the alphabetic data to numerical. The latter method was used, the field of cc. 9-13 being replaced by a field of ten columns (cc. 9-18) of the form X(1), X(2),···, X(10), where, for $v = 1,2,\cdots,10$, X(v) is the number of occurrences of the v*th* characteristic in the cell. E.g., BEE would be translated as (0,1,0,0,2,0,0,0,0,0) since, in the numbering used for the ten characteristics, X(2) = number of bridges and X(5) = number of ridge endings.

3.2 Notation. The process of occurrence of the Galton characteristics was modeled as a multivariate two-dimensional stochastic process, more specifically, a ten-variate process with two-dimensional indexing parameter. The index designates location (row and column) in the grid. The ten variates are the numbers of occurrences of the ten Galton characteristics. That is, the process is $\{\underset{\sim}{X}_{ij}: (i,j) \text{ in } G\}$, where G is the set of cells corresponding to the fingerprint impression. If the impression were rectangular, with I rows and J columns, then the grid G would be simply $\{(i,j): i = 1,2,\cdots,I;\ j = 1,2,\cdots,J\}$. Let the subscript n range over the 39 prints. Then the data set is

$$\{\underset{\sim}{x}_{nij}: n = 1,2,\cdots,39;\ (i,j) \text{ in } G_n\},$$

where G_n denotes the grid of usable cells in the n*th* print. The basic scalar datum is x_{vnij}, the number of occurrences of the v*th* Galton characteristic ($v=1,2,\ \ ,10$) in the (i,j)*th* cell of the n*th* print. Note that (i,j) is nested in n, in the sense that (i,j) has no absolute meaning; it is not the case that the core (center) of the print always has the same location.

4. THE MULTINOMIAL MODEL

Two aspects of the modeling process are modeling within cells and modeling between cells. Osterburg *et al.* (1977) used a multinomial model within cells and independence between cells. Sclove (1979) used the same multinomial model within cells but a Markov model between cells. Sclove (1980a) used the same Markov model between cells but a Poisson model within cells.

This section treats the multinomial model. The next section treats probabilities of various configurations under the multinomial model with independence. Section 6 summarizes the multinomial model with a Markov between-cells model. Section 7 summarizes the Poisson within-cells model in the context of the Markov between-cells model.

The multinomial within-cells model is as follows: For any cell there are 13 possibilities; either the cell is empty or one of the following twelve possibilities has occurred: B, D, E, F, I, L,), S, T, Z, EE (broken ridge), or other multiple occurrence. (By "multiple occurrence" we mean more than one occurrence in a cell.)

In regard to the selection of the multinomial categories priority was given to the ten standard Galton characteristics, occurring as singletons. The number of possible combinations (multiple occurrences) of these individual characteristics is enormous. Among the combinations, we selected the double ridge ending because it was the most frequent; also, it includes a broken ridge, which is different from a ridge coming to an end. A consequence of lumping rare multiple occurrences together into the single category "other multiple occurrence" is to give the benefit of the doubt to the suspect, in the sense of giving a conservative, i.e., large, probability estimate for the given configuration.

In terms of random variables the use of the multinomial model corresponds to using random vectors

$$\underset{\sim}{Y}_{nij} = (Y_{0nij}, Y_{1nij}, \cdots, Y_{12,nij}),$$

defined as follows:

Y_{0nij}, indicator of empty cell $= 1$ if $\underset{\sim}{X}_{nij} = (0,0,0,0,0,0,0,0,0,0)'$
$= 0$ otherwise

Y_{1nij}, indicator of island $= 1$ if $\underset{\sim}{X}_{nij} = (1,0,0,0,0,0,0,0,0,0)'$
$= 0$ otherwise

Y_{2nij}, indicator of bridge $= 1$ if $\underset{\sim}{X}_{nij} = (0,1,0,0,0,0,0,0,0,0)'$

$= 0$ otherwise

.
.
.

$Y_{10,nij}$, indicator of delta $= 1$ if $\underset{\sim}{X}_{nij} = (0,0,0,0,0,0,0,0,0,1)'$

$= 0$ otherwise

$Y_{11,nij}$, indicator of two ridge endings $= 1$ if $\underset{\sim}{X}_{nij} = (0,0,0,0,2,0,0,0,0,0,)'$

$= 0$ otherwise

$Y_{12,nij}$, indicator of multiple occurrence $= 1$ if $Y_{vnij} = 0$ for $v = 0,1,2,\cdots,11$

$= 0$ otherwise.

5. THE MULTINOMIAL MODEL WITH INDEPENDENCE

The model we employ in this section can be summarized as follows. First there are the two within-cells modeling assumptions developed in the preceding section.

(1) A fingerprint is characterized as a configuration of the cells of a grid.
(2) For any cell there are 13 possibilities; either the cell is empty, or one of the following twelve possibilities has occurred: B, D, E, F, I, L, O, S, T, Z, EE (broken ridge), or other multiple occurrence. (By "multiple occurrence" we mean more than one occurrence in a cell.)

Now, for between-cells modeling, we consider

(3) The cells are statistically independent.

Assumption (1) is used throughout the study; (2) is used in this and the next section but replaced in Section 7 by a Poisson assumption; (3) is used in this section and replaced in Section 6 by a Markov model.

The probability P of a given configuration is, under this model, given by the point multinomial distribution, as

$$\log P = k(0)\log P(0) + k(1)\log P(1) +\cdots+ k(12)\log P(12),$$

where the $k(i)$, $i = 0,1,2,\cdots,12$, are non-negative integers summing to t, the total number of cells in the print, and the P(i)'s are the probabilities of the 13 possibilities and hence sum to one. (For notational reasons and because the probabilities involved are small it is convenient to work in terms of logarithms.)

Estimates of probabilities. The parameters P(i) of the model were estimated from the data. See Table 4. The variance of the estimate of any one of the P(i) is P(i)[1-P(i)]/n, i=0,1,2,⋯,12, where n = 8591 cells. Table 4 gives estimates p(i) of P(i), i=0,1,2,⋯,12, and also estimates of the corresponding standard deviations. (We use upper case P for the parameter and lower case p for the estimate.)

TABLE 4. Estimates of probability parameters (Source: Table 3).

Probability parameter	Cell configuration	Frequency	Estimate of probability parameter	Estimated standard deviation of estimate
P(0)	Empty	6584	.766	.0045
P(1)	Island (I)	152	.0177	.0014
P(2)	Bridge (B)	105	.0122	.0012
P(3)	Spur (S)	64	.00745	.00093
P(4)	Dot(D)	130	.0151	.0013
P(5)	Ending ridge (E)	715	.0832	.0030
P(6)	Fork (F)	328	.0382	.0021
P(7)	Lake (L)	55	.00640	.00086
P(8)	Trifurcation (T)	5	.000582	.00024
P(9)	Double bifurcation (Z)	12	.00140	.00040
P(10)	Delta (D)	17	.00198	.00048
P(11)	Broken ridge (or EE)	118	.0139	.0013
P(12)	Other multiple occurrence	305	.0355	.0020

The probability P of a configuration of k(0) empty cells, k(1) cells containing islands, k(2) cells containing bridges, ⋯, k(10) cells containing deltas, K(11) cells containing two ending ridges, and k(12) cells containing other multiple occurrences is estimated by an estimate p given by

$$\log p = k(0)\log p(0) + k(1)\log p(1) + \cdots + k(12)\log p(12).$$

Let E, for *entropy (information)* be defined as E = -log P and e = -log p. We have

$$E = -k(0)\log P(0) - k(1)\log P(1) - \cdots - k(12)\log P(12).$$

Appendix B gives confidence bounds for E, based on the estimate e.

The study of inter-cell dependence discussed in the next section (from Sclove, 1979) indicates that the approximations of the present section should give results that are sufficiently accurate.

The preceding has dealt with the assignment of a probability P to the occurrence of a given configuration in a given set of cells. For inferential purposes is is necessary to estimate the probability that a person has this configuration anywhere on his fingers. A discussion of this aspect of the problem is given in Osterburg *et al.* (1977).

6. THE MULTINOMIAL MARKOV MODEL

The next analysis, relating to dependence among cells, shows that the probability that a cell is occupied increases monotonically with the number of neighbors occupied. Square blocks of 9 cells, 3 cells by 3 cells, were examined to determine the extent of inter-cell dependence. The data set of Osterburg *et al.* (1977) yielded 845 such blocks of cells. For i = 1,2,•••,845 blocks, let the variable y(i) = 1 or 0 according as the center cell of the *ith* block is occupied or not, and let x(i) be the number of adjacent cells which are occupied; x(i) is between 0 and 8. Table 5 gives the cross-tabulation of y and x and gives, for each value of x, the proportion of y's that are equal to 1, i.e., the proportion of center cells which are occupied.

TABLE 5: Cross-tabulation of occupancy of center cell and number of adjacent cells occupied.

	y		
x	0	1	Total
0	152(84.4)	28(15.6)	180(100)
1	170(79.15)	45(20.9)	215(100)
2	163(78.4)	45(21.6)	208(100)
3	97(77.0)	29(23.0)	126(100)
4	44(65.7)	23(34.3)	67(100)
5	23(63.9)	13(36.1)	36(100)
6	7(58.3)	5(41.7)	12(100)
7	0(---)	0(----)	0(---)
8	0(0.0)	1(100.0)	1(100)
	656	189	845 blocks of cells

y = 1 if given (center) cell is occupied; = 0 if it is empty
x = number of adjacencies (number of adjacent cells occupied)

The probability of occupancy increases monotonically with x. Such absolute consistency was not expected, firstly because it seems so rare in data analyses and secondly because it was thought that occurrences in most of the adjacent cells might crowd out occurrence in the center cell.

The value of the chi-square statistic for testing independence based on Table 5 is 18.77 (P<.005, 6 d.f., the categories x = 6, 7, and 8 having been pooled). The decomposition of this overall value based on the value .1404 of the correlation coefficient between x and y is given in Table 6. The value of chi-square due to correlation, 16.65, is the sample size (845 blocks) times the square of the correlation coefficient.

TABLE 6: Decomposition of chi-square according to correlation between x *and* y.

Source of Variation	d.f.	Value of Chi-square	
Overall	6	18.77	(P<.005)
Correlation	1	16.65	(P<.005)
Residual	5	2.12	(.80<P<.85)

In order to achieve independent trials for the chi-square test, separate blocks of 9 cells were used. This greatly reduces the effective sample size. The results here were clear, so it was not necessary to be more efficient. It should be noted, however, that such problems can be handled in a more efficient manner by Besag's (1974) "coding" scheme, used and discussed in later sections where necessary.

The above analysis demonstrated the necessity of developing a model which took account of inter-cell dependence. (It was subsequently found that the model based on inter-cell independence gave adequate results, but this determination could be made only in the context of a model incorporating dependence.) Accordingly, then, the model of Section 5 was extended to consider inter-cell dependence and the occurrence of the characteristics was modeled as a two-dimensional Markov-type process. This analysis is reported in Sclove (1979). The model can be described by saying that it is a nearest-neighbor, Markov-type model where the conditioning is on the sum and the allocation across types of characteristics is independent of the value of the sum.

Under this model, the estimated probability of the configuration of Table 1 is -12.0. Compare this with the figure of -11.4 given by the approximation based on an assumption of independence between cells. The difference in logarithms is 0.6; the ratio of the two estimates is thus 4:1. This difference is unimportant since we are interested only in order of magnitude. Note further that the estimate based on independence is a larger probability, i.e., it is conservative in this sense. (See Osterburg *et al.* (1977) for some discussion of the bearing of these probabilities on the guilt or innocence of a suspect. A large probability estimate is conservative in favor of a suspect, in the sense that it gives the suspect the benefit of the doubt). In general, independence gives too much weight (too low a probability) to configurations with a lot of clustering of occurrences. In the configuration of Table 1 there is some but not a great deal of clustering.

7. THE POISSON MARKOV MODEL

The categories defined in Assumption 2 are somewhat arbitrary. The ten categories corresponding to the occurrence of each of the ten characteristics as singletons are natural enough; it is the lumping together of multiple occurrences which warrants alternative treatment. In the preceding section the occurrence of the characteristics was modeled as a two-dimensional multinomial process, taking account of dependence among cells but not dealing differently with the problem of multiple occurrences. In Sclove (1980a) the occurence of the characteristics is modeled as a two-dimensional Poisson process, not only taking account of dependence among cells but also providing alternative treatment of multiple occurrences.

According to the between-cells data analysis discussed above, the probability that a cell is occupied increases monotonically with the number of neighbors occupied. Accordingly, we introduced an assumption that the expected number of occurrences in a cell depends upon the outcomes in neighboring cells only through the *number* of such cells that are occupied.

A within-cells data analysis is discussed in Appendix C. It was found that negative binomial distributions provided a good fit to the distribution of the number of characteristics per cell, and to the numbers of different characteristics. This is consistent with a model of a mixture of Poisson distributions, for a negative binomial distribution can be obtained as a mixture of Poisson distributions. Accordingly, we set out to test the hypothesis that the number of occurrences in a given cell is a Poisson random variable, at least conditionally.

These assumptions combined into an assumption that the number of occurrences in a cell is distributed according to a Poisson distribution with parameter M, say, which depends upon the random variable A, the number of adjacent cells occupied. In other words, the conditional distribution of the number of occurrences, given the number of adjacent cells that are occupied, i.e., given $A = a$, is Poisson with parameter $M(a)$, $a = 0, 1, 2, 3$, or 4.

This assumption was tested by fitting the number of occurrences of characteristics for each fixed number of adjacencies to a Poisson distribution and checking the goodness of fit. In making this test the dependence among cells had to be taken into account. The problem of dependence was treated by a method of "coding" discussed by Besag (1974); see the discussion by Bartlett (1975, p. 27). To understand the method, suppose the cells were labelled as in Table 7 with two symbols, y and o. This allows the values at the y-sites to be taken, conditional on the values at the o-sites, as independent. Table 8 gives the number of occurrences, by number of adjacencies, for the "y" cells in Besag's coding scheme. The results for the four orientations are given. The generalized likelihood ratio test was used to compare the Poisson fit with the empirical distribution. The "chi-square" values in Table 8 are values of $-2 \ln L$, where L is the generalized likelihood ratio. The Poisson fit appeared adequate. (The Pearson chi-square gave similar results.) Accordingly, a model was developed, based on these assumptions. Details are given in Sclove (1980a).

This Poisson Markov model gave for the configuration of Table 1 an estimated log probability of -12.0. Compare this with result of -11.4 given by the approximation based on independence and -11.8 given by the multinomial Markov model. We have $12.0 - 11.4 = 0.6$; the ratio of the two corresponding estimates is about 4:1.

8. THE INFINITELY DIVISIBLE MODEL

Alternative models considered include modeling the observed random vector giving the numbers of the ten characteristics per cell as an infinitely divisible random vector. (See Sclove, 1980b, for a discussion of multivariate infinitely divisible random vectors.)

A random variable X (which may be a scalar, vector or matrix) is infinitely divisible if there exists a triangular sequence $Y(1,1); Y(2,1), Y(2,2); \cdots; Y(n,1), Y(n,2), \cdots, Y(n,n); \cdots$, such that, for each $n = 1,2,\cdots$, the n random variables $Y(n,1), Y(n,2), \cdots, Y(n,n)$ are independent and identically distributed and the variables $X(1), X(2), \cdots, X(n), \cdots$, defined by

TABLE 7: Coding scheme for obtaining conditionally-indepedent trials in a second-order process.

```
o   o   o   o   o   o   o
o   y   o   y   o   y   o   y
o   o   o   o   o   o   o   o
o   y   o   y   o   y   o   y
```

TABLE 8: Distribution of number of occurrences, by number of adjacencies, with Besag's coding scheme, for test of goodness-of-fit of Poisson distribution.

Adja- cen- cies	Orien- tation	Number of occurrences 0	1	2	3	4	n	Mean	Vari- ance	Chi- square	d.f.	P
0	1	174	34	7	1	0	216	.24	.27	2.79	2	.25
	2	175	34	4	0	0	220	.20	.20	0.58	1	.45
	3	180	37	2	0	0	219	.19	.17	1.19	1	.28
	4	187	41	7	0	0	235	.23	.24	1.64	1	.20
1	1	161	44	5	1	0	211	.27	.27	0.55	2	.76
	2	177	32	8	0	3	220	.27	.44	26.9	3	<.001
	3	167	47	8	1	0	225	.33	.43	10.1	3	.02
	4	152	46	6	2	2	206	.31	.33	1.80	2	.41
2	1	78	33	7	1	1	120	.45	.52	2.54	3	.47
	2	76	25	6	3	0	110	.42	.52	4.48	2	.11
	3	70	28	8	0	0	106	.42	.40	2.48	1	.11
	4	77	1	7	3	0	98	.35	.56	17.3	2	<.001
3	1	24	12	5	2	1	44	.73	.99	2.88	3	.41
	2	25	10	3	0	1	39	.51	.73	5.06	3	.17
	3	21	8	4	2	0	35	.63	.83	3.27	2	.19
	4	23	15	2	1	0	41	.54	.50	1.48	2	.48
4	1	4	2	1	0	0	7	.57	.62	0.48	1	.48
	2	3	5	2	0	0	10	.90	.54	2.08	1	.15
	3	7	2	3	0	0	12	.67	.79	3.62	1	.06
	4	2	4	3	0	0	9	1.11	.61	2.96	1	.08

n's differ somewhat due to border effects; the grids are not perfect rectangles.

$$X(1) = Y(1,1)$$

$$X(2) = Y(2,1) + Y(2,2)$$

$$\vdots$$

$$X(n) = Y(n,1) + Y(n,2) + \cdots + Y(n,n)$$

$$\vdots$$

all have the same distributions as X.

An assumption that $\underset{\sim}{X} = (X_1, X_2, \cdots, X_{10})'$ in the fingerprint study is infinitely divisible can be supported on both physical and probabilistic grounds, as follows.

Speaking first from the physical point of view, it is not at all unreasonable to consider a point process to be infinitely divisible. For, the random variables count the numbers of occurrences in some specified area, such as the cells of the grid. One can conceive of using finer and finer grids. The Y's in the decomposition necessary for infinite divisibility correspond to the cells of these finer partitions. The Galton characteristics may be considered as occurring at dimensionless points, a fork occurring at the point of bifurcation, a spur at the point of separation, and so on. Thus the assumption of infinite divisibility seems reasonable.

Arguing from probabilistic grounds, the assumption of infinite divisibility, at least under a hypothesis of independence of the variates, seems justified, on the grounds that negative binomial and Poisson distributions, which fit the marginal distributions, are infinitely divisible.

As discussed in Sclove (1981), Pierre (1971) defines the measure of dependence (h here, pi in his notation)

$$h(X,Y) = \mathrm{Cov}[X-EX)^2, (Y-EY)^2] - 2[\mathrm{Cov}(X,Y)]^2$$

for random variables X,Y in an infinitely divisible random vector with no Gaussian component. He further shows that $[h(X,Y)/[h(X,X)h(Y,Y)]^{1/2}$ is between zero and one and hence is a normalized measure of dependence analogous to a correlation coefficient. (The parameter h(X,Y) is the cumulant of order (2,2) of (X,Y); the corresponding k-statistic estimate can be used.) Estimates of h were used to estimate the normalized measure of the 10 x 9/2 = 45 pairs of characteristics. The values are small; in fact, the largest was only .018. (The square root of this is still only

.13.) Thus an assumption of independence of the ten variates is further supported by this analysis.

ACKNOWLEDGEMENTS

The initial study, reported in Osterburg, Parthasarathy, Raghavan, and Sclove (1977), was supported under a contract with the Center for Research in Criminal Justice, University of Illinois at Chicago Circle. Work on Sclove (1979) and Sclove (1980) was supported under Grant AFOSR 77-3454 from the Air Force Office of Scientific Research. Computations were performed using the facilities of the Computer Center of the University of Illinois at Chicago Circle. The author's current work on Markov models for two-way series is supported by the Office of Naval Research as Contract N00014-80-C-0408 under Task NR 042-443. These sources of support are gratefully acknowledged.

REFERENCES

Bartlett, M. S. (1975). *The Statistical Analysis of Spatial Pattern*. Chapman and Hall, London: Halsted Press, New York.

Besag, J. E. (1974). Spatial interaction and the statistical analysis of lattice systems. *Journal of the Royal Statistical Society, Series B*, 36, 192-236.

Bowman, K. O., Hutcheson, K., Odum, E. P., and Shenton, L. R. (1971). Comments on the distribution of indices of diversity. In *Statistical Ecology, Vol. 3*, G. P. Patil, E. C. Pielou, and W. E. Waters, eds. Pennsylvania State University Press, University Park, Pennsylvania.

Chernoff, H. (1977). Some applications of a method of identifying an element of a large multidimensional population. In *Multivariate Analysis-IV*, P. R. Krishnaiah, ed. North-Holland Publishing Company, Amsterdam.

Chiralo, R. P., and Berdan, L. L. (1978). Adaptive digital enhancement of latent fingerprints. *Proceedings of the Society of Photo-Optical Instrumentation Engineers*, 149, 118-125.

Cooke, T. D. (1974). Personal communication to Osterburg.

Cox, D. R., and Miller, W. D. (1965). *The Theory of Stochastic Processes*. Wiley, New York.

Galton, F. (1892). *Finger Prints*. Macmillan, London; republication (1965), DeCapo Press, New York.

Greenwood, M., and Yule, G. U. (1920). An inquiry into the nature of frequency-distributions representative of multiple happenings with particular reference to the occurrence of multiple attacks of disease or of repeated accidents. *Journal of the Royal Statistical Society*, 83, 255.

Holt, S. (1951-52). The correlation between ridge-counts on different fingers. *Annals of Eugenics*, 16, 287-297.

Holt, S. (1968). *The Genetics of Dermal Ridges*. Charles C. Thomas, Springfield, Illinois.

Kingston, C. R. (1964). *Probabilistic analysis of partial fingerprint patterns*. Doctoral dissertation, University of California, Berkeley.

Kingston, C. R. (1965a). Applications of probability theory in criminalistics--I. *Journal of the American Statistical Association*, 60, 70-80.

Kingston, C. R. (1965b). Applications of probability theory in criminalistics-II. *Journal of the American Statistical Association*, 60, 1028-1034.

Osterburg, J. W., Parthasarathy, T., Raghavan, T. E. S., and Sclove, S. L. (1977). Development of a mathematical formula for the calculation of fingerprint probabilities based on individual characteristics. *Journal of the American Statistical Association*, 72, 772-778.

Parzen, E. (1962). *Stochastic Processes*. Holden-Day, San Francisco.

Pierre, P. A. (1971). Ifinitely divisible distributions, conditions for independence, and central limit theorems. *Journal of Mathematical Analysis and Applications*, 33, 341-354.

Priest, J. H., Tishler, P. V., and Rosner, B. (1976). Dermatoglyphics in mosaic Down's syndrome. *Clinical Genetics*, 9, 417-426.

Rao, K., and Balck, K. (1980). Type classification of fingerprints: a syntactic approach. *IEEE Transactions on Pattern Analysis and Machine Intelligence*, 2, 223-231.

Sclove, S. L. (1979). The occurrence of fingerprint characteristics as a two-dimensional process. *Journal of the American Statistical Association*, 74, 588-595.

Sclove, S. L. (1980). The occurrence of fingerprint characteristics as a two-dimensional Poisson process. *Communications in Statistics*, A9, 675-695.

Sclove, S. L. (1981). Some recent statistical results for infinitely divisible distributions. In *Statistical Distributions in Scientific Work*, C. Taillie, G. P. Patil, and B. Baldessari, eds. Reidel, Dordrecht-Holland. Vol. 6, 267-280.

APPENDIX A: THE GALTON CHARACTERISTICS

Definitions of some of the Galton characteristics were refined by means of precise working definitions, necessary to accomplish the coding. A *bridge* was defined as less than two centimeters in length in the enlarged photograph (i.e., two millimeters in actuality); otherwise, it would be coded as a fork. A *dot* was defined as being large enough to encompass one pore. Smaller "dots" were not counted; larger "dots" were coded as short ridges. Distinct breaks in ridges were coded as two separate ending ridges to distinguish such breaks from ridges simply coming to an end. A *spur* was defined as being less than two centimeters in length in the enlargement (i.e., two millimeters in actuality); otherwise, it was coded as a fork. A spur was counted only once: the end of a spur was not counted as a ridge ending.

The sizes used are of an order suggested by T. Dickerson Cooke of the Institute of Applied Science, Chicago, Illinois (Cooke, 1974) and are consistent with recommendations of the Committee on Standardization of the International Association for Identification.

APPENDIX B: DERIVATION OF CONFIDENCE BOUNDS FOR THE ENTROPY

The negative log probabilities considered in the model based on independence are in terms of logs base 10 and are given by the expression

$$E = -\log P = -[k(0)\log P(0) + k(1)\log P(1) +\cdots+ k(10)\log P(10)],$$

where $P(0)=1-P(1)-P(2)-\cdots-P(12)$ and $k(0)=t-k(1)=k(2)-\cdots-k(12)$, t being the total number of cells in the print. For the estimate e of E we have $e = -\log p = c \ln p = cH$, where $h = \ln p$ and the constant c is the log base 10 of the base "e" of the natural logs (about 0.434). Thus $\mathrm{Var}(e) = c\, \mathrm{Var}(H)$. The asymptotic variance of H is given by Bowman *et al.* (1971); this gives

$$\mathrm{Var}(e) \simeq (1/n)c^2 \sum_{i=1}^{12} [k(i)]^2/P(i) - t^2].$$

This variance is estimated by substituting the estimates $p(i)$ for the $P(i)$. E.g., for 12 ending ridges and no other characteristics in a print of area $t = 72$ cells, we have 60 empty cells. Hence $k(0) = 60$, $k(5) = .0832$. Thus $e = -12 \log .0832 - 60 \log .766 = 19.9$, and $\mathrm{Var}(e) = (1/8591)\ 0.434^2\ [(60^2/.766) + (12^2/.0832 - 72^2] = 0.0273$. The corresponding standard deviation, the square root of this, is 0.165. Thus a 95% confidence interval is obtained from the point estimate by adding and subtracting 1.96(0.165) = 0.3.

APPENDIX C: MARGINAL DISTRIBUTION OF THE CHARACTERISTICS

The distribution of the number of occurrences per cell is given in Table 2.

For testing goodness-of-fit, it was necessary to use Besag's coding scheme to achieve independent trials. This gave the distribution of Table 9. A Poisson distribution is inadequate [-2 ln L = 9.69, 3 d.f., P = .021; the Pearson chi-square gave a similar result: chi-square statistic = 8.63, 3 d.f. (pooling categories), P = .03]. The distribution is well fit by a negative binomial distribution; in fact, the special case of a geometric distribution provides an adequate fit (Pearson chi-square = 3.11, 3 d.f., P = .38).

TABLE 9: Distribution of number of occurrences for subsample of independent cells.

	Number of occurrences							
	0	1	2	3	4	5	n	Mean
Number of cells	441	125	25	5	2	0	598	0.331
Proportion of cells	.737	.209	.042	.008	.003	.000	1.000	

An interpretation of the fit by the negative binomial family is that what is involved is a gamma-type mixture of Poisson distributions (see, e.g., Parzen, 1962, p. 57), resulting in a negative binomial distribution, as in the classical accident studies of Greenwood and Yule (1920). Empirical support for the assumption is demonstrated by the plausibility of the following assumptions [the usual axioms for a Poisson process (see, e.g. Parzen, 1962, p. 119), generalized to two dimensions]. Given any set S in the (x,y)-plane, let N(S) be the number of occurrences in S and let a(S) be the area of S. Given any point (x,y), let {S(n)} be a sequence of sets tending to (x,y) as n tends to infinity. Then the following assumptions are plausible. There is a positive number M(x,y) such that, as n tends to infinity,

$$\{1 - \Pr[N(S(n))=0]\}/a(S(n)) \text{ tends to } M(x,y),$$

$$\Pr[N(S(n))=1]/a(S(n)) \text{ tends to } M(x,y),$$

$$\Pr[N(S(n))\geq 2]/a(S(n)) \text{ tends to } 0.$$

The intensity parameter varies with position in the sense that occurrences are more probable in the pattern than the non-pattern area and the intensity may be a decreasing function of the distance from the core of the pattern. The two-dimensional non-homogeneous Poisson process may be termed a Poisson random *field*, since the intensity parameter varies with position. See, e.g., Cox and Miller (1965) for a discussion of such Poisson processes over general spaces. The monograph by Bartlett (1975) provides a discussion of general models and methods for analysis of two-dimensional processes; see the Appendix in Sclove (1979) for a synopsis.

[*Received June* 1980. *Revised September* 1980]

STOCHASTIC MODELING IN POLITICAL SCIENCE RESEARCH

MANUS I. MIDLARSKY

Department of Political Science
University of Colorado at Boulder
Boulder, Colorado 80309 USA

SUMMARY. Applications of stochastic models in political science research are reviewed in (a) the occurrence of international and national conflict behavior, (b) international alliance behavior, (c) exponential models of governmental turnover, attrition and inequality and (d) the establishment of formal identities. Major efforts have been directed at the detection of contagion, diffusion, and reinforcement effects as well as the examination of other forms of statistical interdependence. Departures from stochasticity provide an increasingly common and richly varied basis for the examination of hypotheses of interest. Formal identities among different forms of conflict behavior have been established, while the use of bivariate negative binomial distributions based on compound Poisson and reinforcement assumptions are now being used in the analysis of political behavior.

KEY WORDS. Poisson, negative binomial, stochastic modeling, conflict, alliance, war, interdependence, contagion, diffusion, bivariate distributions.

1. INTRODUCTION

This paper is intended as a review of the use of stochastic models in political science research. Its purpose is to outline the major areas of political science research where probability distributions have been used, with appropriate illustrations, and simulataneously to suggest additional areas for future applications. Most important, the use of probability distributions in connection with either the development or testing of theory is emphasized. As a consequence, simulations, regressions,

C. Taillie et al. (eds.), Statistical Distributions in Scientific Work, Vol. 6, 131–146.

discriminant analyses, the casting of deterministic theory in probabilistic format, or any other statistical methods which do not explicitly incorporate probability distributions as intrinsic components of the model are omitted. One exception will be made in the instance of Markov chain analyses because of their use in the testing of important substantive theory and their connection with implied distributions.

Substantively, virtually all of the applications of stochastic models in political science research occur in weakly or non-regulated systems. These are systems without central direction in which events can occur independently of each other. It is this quality of a mutual, or at times, a conditional independence which allows the use of stochastic models in certain areas of political science research. Here, the occurrence of violence between nations and cooperative efforts in the form of international alliances have, for the most part, satisfied certain independence assumptions, thus allowing the use of stochastic models. This is not to exclude certain conflict behaviors internal to the nation-state such as coups d'état and urban disorders, for as we shall see, these too can be modeled stochastically. Governmental turnover, representative tenure, and the growth of bureaus over time constitute another class of behaviors not centrally directed and, therefore, susceptible to stochastic modeling.

Theoretically, in the construction of the models, questions of independence or mutual dependence have occupied a major share of attention. Whether events are mutually independent or display qualities of dependence or contagion can be assessed by means of stochastic models and related analyses.

2. INTERNATIONAL AND NATIONAL CONFLICT BEHAVIOR

2.1 The Poisson Process and the Negative Binomial Distribution. The first known application of stochastic models to political science concerns is that of Richardson (1960, p. 128; first published in 1945). Prompted by the discovery that radioactive particles were emitted according to a Poisson law, Richardson applied this distribution to the frequency of the outbreaks of war in the range of approximately 3,000 to 30,000 war dead during the period 1820 to 1929. The fit between observed and Poisson predicted frequencies was excellent and this finding later was validated for another data set by Singer and Small (1972, p. 205). In this later application, the intervals between wars were found to be exponentially distributed in the period 1816-1965, thus establishing a clear connection with the Poisson. In a series of other applications based on variants of the binomial distribution, Richardson (1960, pp. 247-314) found a randomness in dyadic confrontations between nations, and concluded that "chaos" modified by infectiousness and the geographic opportunity for war was the most accurate representation of the data. These and other probability models of war used by Richardson are reviewed extensively in Zinnes (1976, pp. 246-299).

The application of the Poisson, of course, depends on the independence of the conflict events. The possibility of the diffusion of violence also has occupied a considerable share of attention. The diffusion of political instability in the form of the coup d'état was examined by means of a comparison between the Poisson and a modified Poisson which was altered to reflect the increased probability of a coup occurring after the occurrence of a prior coup (Midlarsky, 1970). The modified Poisson gave the better fit by far, especially for the period 1935-1949 in Latin America. The modified Poisson is of the form (Coleman, 1964, p. 300)

$$p_i = \frac{\alpha(\alpha+\beta)\cdots(\alpha+[i-1]\beta)e^{-\alpha t}(1-e^{-\beta t})^i}{i!\beta^i} \qquad (1)$$

where β is a measure of the extent of diffusion given by

$$\beta = \ln(\sigma^2/\mu) \quad \text{and} \quad \alpha = \mu^2\beta/(\sigma^2-\mu)$$

and μ and σ^2 are respectively the mean and variance of the observed distribution. If we set $\alpha/\beta = \lambda p$ and $p = 1/(e^{-\beta t} - 1)$ then the distribution (1) turns out to be identical to the negative binomial as a limiting form of the Polya distribution given by (Feller, 1968, p. 143)

$$p_i = \binom{\lambda p + i - 1}{i} \left(\frac{p}{1+p}\right)^{\lambda p} \left(\frac{1}{1+p}\right)^i .$$

The diffusion of instability in Latin America also was found by Li and Thompson (1975) in comparing the Poisson and modified Poisson, but, in addition, using autocorrelation functions.

The diffusion of other forms of violence also has been examined. Davis *et al.* (1978) found an interdependence of the instances of dyadic warfare in the period 1816-1965 by comparing the Poisson, negative binomial and autoregressive functions. Their finding of such an interdependence is consistent with the Singer and Small (1972, pp. 209-212) conclusion as the result of a spectral analysis that there exist periodicities of between 20 and 40 years in the amount of war *underway* in the period 1816-1965, although no such periodicities were found in the amount of war *begun* in this period (cf. the Richardson Poisson finding).

Most and Starr (1980) also used the Poisson and modified Poisson models of the form (1) to assess the extent of diffusion of war in the period 1946-1965. They concluded that the occurrence of a new war participation will increase the probability that both the same nation and other nations will experience subsequent new war participations. The effect of borders with other nations on the probability of war was explored here, as it was in Starr and Most (1978), Richardson (1960, p. 177) and Midlarsky (1975, pp. 68-71).

2.2 *Distinguishing Among the Sources of Interdependence.* A major issue in all of these analyses is that of distinguishing among reinforcement, diffusion and heterogeneity. (Contagion as the violation of stochastic independence will be treated shortly.) The modified Poisson, equation (1), and its mathematical twin, the negative binomial, can be derived by assuming any one of several sources for a differing propensity towards experiencing the same phenomenon by the units in question. Any of several conditions, namely, reinforcement as the tendency of the same unit (country or person) to have a higher probability of experiencing the phenomenon after an earlier incident, or diffusion as the increased probability after another unit has experienced it, or simply heterogeneity as an intrinsically differing probability across units, can lead to the applicability of the negative binomial.

Solutions to this problem have included the use of turnover tables or transition matrices by Most and Starr (1980) and by Midlarsky *et al.* (1980). As shown in the matrix below, a pure diffusion or contagion can be seen in the row 0 to $1, 2, \cdots, n$ incidents where x_0 incidents are experienced by a unit (country) during the time interval δ_0 and x_1 incidents are experienced during the adjacent interval δ_1. If countries in that row of the matrix increase over time, from δ_0 to δ_1, then such an inference can be made. Other ways of examining this matrix also exist.

		x_1				
		0	1	2	3 ···	n
	0	a	b	c	······	
	1	·	·	·		
x_0	2	·	·	·		
	3	·	·	·		
	·	·	·	·		
	m	·	·	·		

Currently, bivariate models are being used to distinguish among the various possibilities. Specifically, one can posit complete independence of the units in question over space and over time, and in this instance, a bivariate Poisson is appropriate (Feller, 1968, p. 172). Transition matrices of the form shown above can be used to assess the applicability of this model, or alternatively, that of a bivariate compound Poisson which assumes heterogeneity in its derivation (Arbous and Kerrich, 1951, p. 415) and is of the form

$$p(x_0,x_1) = \frac{k^k \alpha^x}{(k+\alpha\delta)^{k+x}} \; \frac{\Gamma(k+x)\, \delta_0^{x_0} \delta_1^{x_1}}{\Gamma(k)\, x_0! x_1!} \tag{2}$$

where x_0 and x_1 are defined as before, x is the number of incidents experienced during the entire interval δ, α and k are constants and Γ is the gamma function.

This bivariate distribution can be compared with another resulting from the reinforcement assumption given by (Arbous and Kerrich, 1951 p. 421)

$$p(x_0,x_1) = e^{-\beta\delta} \frac{\Gamma(\beta/\gamma + x)}{\Gamma(\beta/\gamma) x_0! x_1!} \left(e^{-\gamma\delta_1} - e^{-\gamma\delta}\right)^{x_0} \left(1 - e^{-\gamma\delta_1}\right)^{x_1} \tag{3}$$

where β and γ are constants. Bates and Neyman (1952, p. 266) derive a related function based on similar assumptions.

These models can be used with any set of conflict or other societal data, and in conjunction with the examination of cells of the transition matrix itself, can be used to differentiate among the various possible sources for the applicability of the negative binomial. Note that both (2) and (3) become the univariate negative binomial when applied to a single time period, δ.

An important distinction to consider is that between diffusion as a modal response to a unique precipating event and contagion as a direct imitation effect, not requiring an independent precipitant (Midlarsky, 1978). This differentiation can be accomplished by the use of diffusion models which incorporate statistical independence in the initiation of events but which yield outcomes which are dependent on the occurrence of prior events. The lognormal distribution as a solution to a variant of the Kolmogorov equation (Bharucha-Reid, 1960, pp. 154-158, 189-192; Bartlett, 1966, pp. 89-90; and Tintner and Sengupta, 1972, pp. 65-66) incorporates a proportionate effect model for the growth of a process in which the state of the process at time t_1 is dependent on the state of that process at time t_0, but each event contributing to the growth of the random variable is independent of others (Cramér, 1946, pp. 218-221). The number of arrests in American urban disorders of 1966-67 was found to obey the lognormal distribution with the independent precipitants being the interaction between police and black city residents in large cities. However, the intensity and visibility of the Newark and Detroit disorders led to the hierarchial contagion of the violence from large to small cities, as evidenced by the inapplicability of the lognormal distribution during this segment of the overall time period (Midlarsky, 1977, 1978). Another use of the lognormal distribuion is seen in Russett's (1968) finding of national population size as lognormally distributed under the assumption of a proportionate growth rate which is independent of a nation's size. There was no evidence, therefore, of an increased concentration of power over time in the largest countries, using population as one measure of power.

A distinction between diffusion and contagion also was found in the analysis of the occurrence of international terrorism in the period 1968-74, using Poisson and negative binomial distributions (Midlarsky *et al.*, 1980). The Polya distribution, which has the negative binomial as a limiting form, was a useful model of diffusion. The urn model of withdrawal and replacement wherein each withdrawal of a ball at random leads to the replacement of s additional balls of the same color is an accurate representation of a diffusion condition in which the probabilities of later events are conditioned by earlier ones (increased number of balls of a given color) and the event in question such as an act of terrorism is triggered randomly, say by the repressive policies of an authoritarian government. The inapplicability of the negative binomial for one segment of the time period along with the theory of hierarchies led to the inference of contagion as a direct imitation process for this time interval. Yamamoto and Bremer (1980) also used the Polya distribution as a model for the increased probability of major power entry into an ongoing war in the period 1816-1965. This distribution was found to be superior to a one-way choice model developed by Rutherford (1954).

2.3 Markovian Analyses. Several studies of international conflict have employed Markov chain analyses to good effect. These studies generally are of international crises, linkages between domestic and foreign conflict behavior, or the duration of wars. In the former category, Zinnes *et al.* (1972) studied the 1914 crisis in detail using Markov type transition matrices and found that decision-makers in their expressions of hostility after perceiving it, reacted in proportion to the stimuli received, despite the demands and time constraints of an international crisis. In two separate studies, Zinnes and Wilkenfeld (1971) and

Wilkenfeld and Zinnes (1973) used Markov analyses of factor scores from studies of domestic and foreign conflict behavior to examine the relationships between these two conflict forms as well as the effect of one conflict form (say internal) on later occurrences of the same type. The latter hypothesis was confirmed as a first order Markov chain.

The Sino-Indian crisis of 1959-64 also was studied by means of Markov chains (Duncan and Siverson, 1975). An open-open model of communication was contrasted with open-closed and closed-closed models in which openness refers to receptivity to the other party's communications and closedness implies a receptivity only to one's own past behavior. The closed-closed model was found to be the best descriptor of the data. The results of simulation studies of international crises such as that of the Taiwan Straits crisis of 1957 also have been analyzed using Markov chains (Leavitt, 1972).

Both one-dimensional and two-dimensional Markov processes were employed by Weiss (1963) in his analyses of the duration and magnitude of wars. Here, the analysis is explicitly tied to a particular probability distribution. Under the Markovian assumption that the end of a war after x deaths and before $x + 1$ depends only on x, Weiss analyzed a negative power function distribution which effectively represents the Richardson data on wars. This, in fact, is the Pareto distribution which Simon (1955) derived along with the Yule distribution as one among many skew distributions which can be derived as a consequence of a first-order Markovian assumption. (See Ijiri and Simon, 1977, for further applications.) Weiss then incorporated time in his two-dimensional Markov process and arrived at a somewhat more complex functional relationship between the duration and magnitude of wars.

3. INTERNATIONAL ALLIANCE BEHAVIOR

The issue of statistical independence and various forms of dependence, or alternatively, heterogeneity, also appears in the study of alliances. Following Kaplan's (1957) analysis of the nineteenth century balance of power system as one based on random conjunctions or meetings of countries which lead to alliance formation, McGowan and Rood (1975) found that alliance formation was Poisson distributed over time. Although Lawson (1976) criticized this application saying that the randomness in time of alliance formation was irrelevant to balance of power theory, the finding that alliance formation is Poisson distributed over time emerges directly from an equilibrium assumption that there exists an average equality in the formation and dissolution of alliances over time (Midlarsky, 1981). This equilibrium assumption emerges from the treatment of increases in power within a nineteenth century balance of power system as dependent solely on the accession of new allies. With the absence of citizen armies in the nineteenth century and a relatively unchanging weapons technology, at least until the end of the century, power could not be generated internal to the nation-state. Further, each ally which is lost would have to be replaced at some arbitrary time in the future in order to maintain the existing balance or power equilibrium. Given equation (4)

$$C_i = \langle D_i \rangle_{av} = \sum_{n=i}^{\infty} P(n) D_i \qquad (4)$$

where D_i and C_i are respectively the probabilities of alliance dissolution and creation and $P(n)$ is the probability that n alliances *exist* in the time interval t, and the alliances are assumed to be randomly distributed, then the Poisson distribution for alliance *formation* then follows. Equations for the transition probabilities from one state of the system to the next are derived and, in fact, are very similar in form to those used by Smoluchowski (1916) in his analysis of Brownian motion. A version of Smoluchowski's derivation can be found in Chandrasekhar (1943). The transition probabilities $P(n, n+j)$ and $P(n, n-j)$ are given by

$$P(n, n+j) = e^{-\mu p} \sum_{i=0}^{n} \frac{n!}{i!(n-i)!} p^i (1-p)^{n-i} \times (\mu p)^{i+j}/(i+j)! \qquad (5)$$

$$P(n, n-j) = e^{-\mu p} \sum_{i=j}^{n} \frac{n!}{i!(n-i)!} p^i (1-p)^{n-i} \times (\mu p)^{i-j}/(i-j)! \qquad (6)$$

where p is the probability of a single alliance disappearing in t, and μ is the mean of the observed distribution.

In this analysis, the nineteenth century also was found to be very heterogeneous, using Fisher's logarithmic series (Fisher, *et al.*, 1943; Quenouille, 1949), which suggests the importance of small powers in maintaining the structure of the nineteenth century system (Midlarsky, 1981). Yamamoto (1974) also applied Fisher's log series to nineteenth century peacetime alliance activity following Horvath and Foster's (1963) suggestion that wartime alliances likely were Yule distributed, whereas peacetime alliances either were distributed according to the truncated Poisson or Fisher's log series.

Stochastic treatments of alliance behavior include those by Job (1976) and Siverson and Duncan (1976). Both studies differentiated between alliance behavior in the nineteenth and twentieth centuries and found that the modified Poisson of equation (1) provided a better fit to the data on alliance formation than did the Poisson itself for most time periods. This suggests either a heterogeneity or diffusion for countries, but not for alliances as discrete organizations. These studies conclude that the nineteenth century experienced random alliance formation, but the periods prior to 1914 and after World War II were subject to diffusion effects. The interwar period likely witnessed a heterogeneity of alliance behaviors. Li and Thompson (1978) used autocorrelation functions to investigate alliance formation and found a randomness in the nineteenth century and a diffusion in post World War II alliance formation, but also a randomness for the interwar period. It is possible that apparent inconsistencies between the findings of several studies, particularly in regard to heterogeneity, diffusion, or randomness, could be resolved by the use of bivariate distributions of the form (2) or (3).

4. EXPONENTIAL MODELS OF GROWTH, ATTRITION, AND INEQUALITY

Another area of application for probability distributions is that of governmental growth and attrition. The model of exponential growth and decline has been applied to governmental bureaus in the United States and to the survival rate of members of a governmental body or organization. This was shown, for example, in the instance of the Canadian House of Commons (Casstevens and Denham, 1970), the Soviet Central Committee (Casstevens and Ozinga, 1974) and for U. S. government bureaus (Casstevens, 1980). The assumption here, of course, is that the probabilities of reelection of a given candidate, or some other form of political survival, or, indeed, the probability of existing beyond a certain number of years, are unrelated to those of any other candidate or government agency.

The exponential distribution also has been used as a basis for the analysis of inequality under conditions of scarce resources. It can be shown (Midlarsky, 1980) that under conditions of scarcity, an exponential distribution, or even more accurately, a geometric distribution of the scarce goods can be expected. The geometric distribution emerges from the fact that the frequency of male family members of the n*th* generation follows a geometric distribution (Feller, 1968, p. 294) and in the absence of primogeniture and the inability to migrate, even an initially equal set of holdings will eventually yield a geometric distribution. This theory has been empirically confirmed for island nations with their delimited land areas (Midlarsky, 1980) and for farm land in the U. S. and elsewhere (Dovring, 1973; Boxley, 1971; Clark, 1972). Such distributions imply that in a situation of increasing scarcity, inequality along lines of the exponential model must follow. Another use of the exponential model was derived from a conceptualization of the exercise of power as uncertainty reduction (Midlarsky, 1974), and led to the finding that international crises likely were to be found at logarithmically increasing distances from home for great powers (Midlarsky, 1976).

5. DEPARTURES FROM STOCHASTICITY

Departures from statistical independence provide a fruitful and increasingly common mode of inquiry. Hayes (1973) showed that data on international events can be compared with theoretical Poisson expected frequencies to ascertain the extent of departure from independence, or as a measure of change over time. Chan (1978) then applied this method to the Vietnam War. He was able to delineate periods of escalation and deescalation in U. S. behavior towards both North Vietnam and the National Liberation Front, differences that had not been suspected until the completion of this analysis.

Another effective usage of departures from prediction is that by Casstevens and Ozinga (1974) in the estimation of purges in the Central Committee of the Communist Party of the U.S.S.R. The authors used the exponential model of personnel turnover to estimate the predicted membership size between 1957 and 1961 -- the period of the Khrushchev purge -- and then compared it with the actual size. They estimated that ten full members and six candidates were purged in this period.

Other applications of departures from randomness can be found in the finding of the contagion of urban disorders as a departure from lognormality during the Newark and Detroit disorders in 1967 (Midlarsky, 1978) or departures from the negative binomial as a limiting condition of the Polya distribution, as an indication of the contagion of international terrorism in 1973 (Midlarsky *et al.*, 1980). As interesting early application of departures from stochasticity is found in Stokes and Iversen's (1962) analysis of political party competition in the U.S. They demonstrated that party vote does not conform to a generalized random walk, thus implying the existence of equilibrium forces in American politics which restore party competition.

Galton's problem as a difficulty in statistical inference (Naroll, 1973) also has received a share of attention in political science research. Here, the problem is not some departure from stochasticity as the result of central decision-making in Vietnam or the widespread visibility of large urban disorders, but rather the absence of strict independence as a consequence of diffusion among statistical units (e.g., countries) in correlation or other inferential analyses. Gillespie (1970) brought this problem to the attention of political scientists and Ross and Homer (1976) demonstrated its impact on values of the correlation coefficient in cross-national studies. Klingman (1980) has now formulated an overall expression for the analysis of social change as the result of diffusion effects, as well as intra-systemic sources of change.

6. FORMAL IDENTITIES

A potentially fruitful, although thus far little used function of stochastic models is the discovery of formal identities among various processes, especially those pertaining to conflict behavior. Wesley's (1962) model of geographical opportunity for war reproduced Richardson's relationship between the frequency and magnitude of war, which in turn was found to bear strong similarity to that found by Richardson (1960, p. 149) for individual murders, based on his argument that the opportunity for murder varied positively with the size of towns as geographic entities. More directly, Horvath (1968) found a strong similarity between the duration of wars and strikes. Under the assumption that each war or strike is a stochastic process in which there are many impediments to solution, Horvath used the theory of extreme values to arrive at the Weibull distribution as an accurate descriptor of the duration of wars between 1820 and 1949, and strikes in the U.S. in the years 1960 and 1961. These findings raise the possibility of the existence of other formal identities along similar lines.

7. CONCLUSION

The past decade has witnessed an ever-increasing use of stochastic models in political science research. Not only have an increasing number of phenomena ranging from wars to coups, alliances, and governmental bureaus been studied successfully using probability models, but an increasing number of theoretically important questions are being answered by the use of departures from stochastic formulations. Ultimately, societal polarizations in both national and international societies can be cast in the form of departures from the appropriate stochastic model. Problems of collective decision-making also may be pursued successfully as in the pioneering treatments in Niemi and Weisberg (1972). Additional applications to political science concerns could emerge from the treatments of diffusion processes by Bartholomew (1973) and Hamblin *et al.*, (1973), Cohen's (1971) stochastic analyses of group size, and the applications of fuzzy sets to decision-processes (Zadeh *et al.*, 1975).

Thus far, the applications have been mostly to conflict events or forms of cooperative behavior such as alliances which have no central organizing focus. The majority of the treatments also have been univariate. These phenomena may continue to be analyzed in this fashion, but with the additional impetus now of bivariate and multivariate models detailed above. In this way, not only may we answer some of the still-nagging questions of distinguishing adequately among various forms of statistical interdependence, but also may open new avenues for the exploration of additional processes not yet considered.

REFERENCES

Arbous, A. G. and Kerrich, J. E. (1951). Accident statistics and the concept of accident proneness. *Biometrics*, 7, 340-432.

Bartholomew, D. J. (1973). *Stochastic Models for Social Processes*, 2nd Ed. Wiley, New York

Bartlett, M. S. (1966). *An Introduction to Stochastic Processes: With Special Reference to Methods and Applications*. Cambridge University Press, Cambridge, England.

Bates, G. E. and Neyman, J. (1952). Contributions to the theory of accident proneness. *University of California Publications in Statistics*, 1, 215-275.

Bharucha-Reid, A. T. (1960). *Elements of the Theory of Markov Processes and their Applications*. McGraw-Hill, New York.

Boxley, R. F. (1971). Farm size and the distribution of farm numbers. *Agricultural Economics Research*, 23, 87-94.

Casstevens, T. W. (1980). Birth and death processes of governmental bureaus in the United States. *Behavioral Science*, 25, 161-165.

Casstevens, T. W. and Denham, W. A. III (1970). Turnover and tenure in the Canadian House of Commons. *Canadian Journal of Political Science*, 3, 655-661.

Casstevens, T. W. and Ozinga, J. R. (1974). The Soviet Central Committee since Stalin: A longitudinal view. *American Journal of Political Science*, 18, 559-568.

Chan, S. (1978). Temporal delineation of international conflicts: Poisson results from the Vietnam War, 1963-1965. *International Studies Quarterly*, 22, 237-265.

Chandrasekhar, S. (1943). Stochastic problems in physics and astronomy. *Reviews of Modern Physics*, 15, 1-89.

Clark, C. (1972). The extent of hunger in India. *Economic and Political Weekly*, 7, 2019-2027.

Cohen, J. E. (1971). *Casual Groups of Monkeys and Men: Stochastic Models of Elemental Social Systems*. Harvard University Press, Cambridge, Mass.

Coleman, J. S. (1964). *Introduction to Mathematical Sociology*. The Free Press, New York.

Cramér, H. (1946). *Mathematical Methods of Statistics*. Princeton University Press, Princeton.

Davis, W. W., Duncan, G. T. and Siverson, R. M. (1978). The dynamics of warfare: 1816-1965. *American Journal of Political Science*, 22, 772-792.

Dovring, F. (1973). Distribution of farm size and income: Analysis by exponential functions. *Land Economics*, 49, 133-147.

Duncan, G. T. and Siverson, R. M. (1975). Markov chain models for conflict analysis: Results from Sino-Indian relations, 1959-1964. *International Studies Quarterly*, 19, 344-374.

Feller, W. (1968). *An Introduction to Probability Theory and its Applications, Vol. I*, 3rd Ed. Wiley, New York.

Fisher, R. A., Corbet, A. S. and Williams, C. B. (1943). The relation between the number of species and the number of individuals in a random sample of an animal population. *The Journal of Animal Ecology*, 12, 42-58.

Gillespie, J. V. (1970). Galton's problem and parameter estimation error in comparative political analysis. Paper presented at the Annual Meeting of the Midwest Political Science Association.

Hamblin, R. L., Jacobsen, R. B. and Miller, J. L. (1973). *A Mathematical Theory of Social Change*. Wiley, New York.

Hayes, R. E. (1973). Identifying and measuring changes in the frequency of event data. *International Studies Quarterly*, 17, 471-493.

Horvath, W. J. (1968). A statistical model for the duration of wars and strikes. *Behavioral Science*, 13, 18-28.

Horvath, W. J. and Foster, C. C. (1963). Stochastic models of war alliances. *Journal of Conflict Resolution*, 7, 110-116.

Ijiri, Y. and Simon, H. A. (1977). *Skew Distributions and the Sizes of Business Firms*. North-Holland, Amsterdam.

Job, B. L. (1976). Membership in inter-nation alliances, 1815-1965: An exploration utilizing mathematical probability models. In *Mathematical Models in International Relations*, D. A. Zinnes and J. V. Gillespie, eds. Praeger, New York. 74-109.

Kaplan, M. A. (1957). *System and Process in International Politics*. Wiley, New York.

Klingman, D. (1980). Temporal and spatial diffusion in the comparative analysis of social change. *American Political Science Review*, 74, 123-137.

Lawson, F. H. (1976). Communication: Alliance behavior in nineteenth century Europe. *American Political Science Review*, 70, 932-934.

Leavitt, M. R. (1972). Markov processes in international crises: An analytical addendum to an event-based simulation of the Taiwan Straits crises. In *Experimentation and Simulation in Political Science*, J. A. Laponce and P. Smoker, eds. University of Toronto Press, Toronto. Pages 280-290.

Li, R. P. Y. and Thompson, W. R. (1975). The "coup contagion" hypothesis. *Journal of Conflict Resolution*, 19, 63-88.

Li, R. P. Y. and Thompson, W. R. (1978). The stochastic process of alliance formation behavior. *American Political Science Review*, 72, 1288-1303.

McGowan, P. J. and Rood, R. M. (1975). Alliance behavior in balance of power systems: Applying a Poisson model to nineteenth century Europe. *American Political Science Review*, 69, 859-870.

Midlarsky, M. I. (1970). Mathematical models of instability and a theory of diffusion. *International Studies Quarterly*, 14, 60-84.

Midlarsky, M. I. (1974). Power, uncertainty, and the onset of international violence. *Journal of Conflict Resolution*, 18, 395-431.

Midlarsky, M. I. (1975). *On War: Political Violence in the International System*. The Free Press, New York.

Midlarsky, M. I. (1976). Power and distance in international conflict behavior. In *Mathematical Models in International Relations*, D. A. Zinnes and J. V. Gillespie, eds. Praeger, New York. Pages 132-155.

Midlarsky, M. I. (1977). Size effects and the diffusion of violence in American cities. *Papers, Peace Science Society (International)*, 27, 39-47.

Midlarsky, M. I. (1978). Analyzing diffusion and contagion effects: The urban disorders of the 1960s. *American Political Science Review*, 72, 996-1008.

Midlarsky, M. I. (1980). The revolutionary transformation of foreign policy: Agrarianism and its international impact. In *The Political Economy of Foreign Policy Behavior*. (Sage International Yearbook of Foreign Policy Studies), C. Kegley and P. McGowan, eds. Sage, Beverly Hills, California (to appear).

Midlarsky, M.I. (1981). Equilibria in the nineteenth century balance of power system. *American Journal of Political Science*, 25 (to appear).

Midlarsky, M.I., Crenshaw, M. & Yoshida, F. (1980). Why violence spreads: The contagion of international terrorism. *International Studies Quarterly*, 24, 262-298.

Most, B. A. and Starr, H. (1980). Diffusion, reinforcement, geopolitics and the spread of war. *American Political Science Review*, 74 (to appear).

Naroll, R. (1973). Galton's problem. In *A Handbook of Method in Cultural Anthropology*, R. Naroll and R. Cohen, eds. Columbia University Press, New York. Pages 974-989.

Niemi, R. G. and Weisberg, H. F., eds. (1972). *Probability Models of Collective Decision Making*. Merrill, Columbus, Ohio.

Quenouille, M. H. (1949). A relation between the logarithmic, Poisson, and negative binomial series. *Biometrics*, 5, 162-164.

Richardson, L. F. (1960). *Statistics of Deadly Quarrels*. Boxwood, Pittsburgh.

Ross, M. H. and Homer, E. L. (1976). Galton's problem in cross-national research. *World Politics*, 29, 1-28.

Russett, B. M. (1968). Is there a long-run trend toward concentration in the international system? *Comparative Political Studies*, 1, 103-122.

Rutherford, R. S. G. (1954). On a contagious distribution. *Annals of Mathematical Statistics*, 25, 703-713.

Simon, H. A. (1955). On a class of skew distribution functions. *Biometrika*, 42, 425-440.

Singer, J. D. and Small, M. (1972). *The Wages of War, 1816-1965: A Statistical Handbook*. Wiley, New York.

Siverson, R. M. and Duncan, G. T. (1976). Stochastic models of international alliance initiation, 1885-1965. In *Mathematical Models in International Relations*, D. A. Zinnes and J. V. Gillespie, eds. Praeger, New York. Pages 110-131.

Smoluchowski, M. v. (1916). Drei vorträge über diffusion, brownsche molekularbewegung and koagulation von kolloidteilchen. *Physikalische Zeitschrift*, 17, 557-571.

Starr, H. and Most, B. A. (1978). A return journey: Richardson, "frontiers" and wars in the 1946-1965 era. *Journal of Conflict Resolution*, 22, 441-467.

Stokes, D. E. and Iversen, G. R. (1962). On the existence of forces restoring party competition. *Public Opinion Quarterly*, 26, 159-171.

Tintner, G. and Sengupta, J. K. (1972). *Stochastic Economics: Stochastic Processes, Control, and Programming*. Academic Press, New York.

Weiss, H. K. (1963). Stochastic models for the duration and magnitude of a "deadly quarrel." *Operations Research*, 11, 101-121.

Wesley, J. P. (1962). Frequency of wars and geographical opportunity. *Journal of Conflict Resolution*, 6, 387-389.

Wilkenfeld, J. and Zinnes D. A. (1973). A linkage model of domestic conflict behavior. In *Conflict Behavior and Linkage Politics*, J. Wilkenfeld, ed. David McKay, New York. Pages 325-356.

Yamamoto, Y. (1974). *Probability models of war expansion and peacetime alliance formation*. Ph.D. dissertation. The University of Michigan.

Yamamoto, Y. and Bremer, S. (1980). Wider wars and restless nights: Major power intervention in ongoing war. In *The Correlates of War: II, Testing Some Realpolitik Models*. J. D. Singer, ed. The Free Press, New York. Pages 199-229.

Zadeh, L. A., Fu, K., Tanaka, K. and Shimura, M. (1975). *Fuzzy Sets and their Applications to Cognitive and Decision Processes*. Academic Press, New York.

Zinnes, D. A. (1976). *Contemporary Research in International Relations: A Perspective and a Critical Appraisal*. The Free Press, New York.

Zinnes, D. A. and Wilkenfeld, J. (1971). An analysis of foreign conflict behavior of nations. In *Comparative Foreign Policy: Theoretical Essays*, W. F. Hanrieder, ed. David McKay, New York. Pages 167-213.

Zinnes, D. A., Zinnes, J. L. and McClure, R. D. (1972). Hostility in diplomatic communication: A study of the 1914 crisis. In *International Crises: Insights from Behavioral Research*, C. F. Hermann, ed. The Free Press, New York. Pages 139-162.

[*Received October* 1980]

STATISTICAL DISTRIBUTION MODELS IN THE BEHAVIORAL SCIENCES: A REVIEW OF THEORY AND APPLICATIONS

P. R. MORGAN

Center for Social Organization of Schools
The Johns Hopkins University
Baltimore, Maryland 21218 USA

SUMMARY. In this paper a classification is suggested for discrete distributions that have proved useful in education, political science, psychology and sociology. The importance of making careful inferences from these models in policy-related research is emphasized. Examples are given of how complex models may be applied in research on collective violence.

KEY WORDS. Simple random processes, complex random processes, propensity function, Gurland distribution.

1. INTRODUCTION

Roughly three decades have elapsed since the first applications of the simple Poisson distribution to the present day applications of complex chance mechanisms in behavioral science research. Lack of contact between behavioral disciplines and theoretical statistics has slowed the enrichment of methodology in the former and stemmed the flow of substantive stimuli to the latter, however. On the other hand, interdisciplinary contact within the behavioral sciences has been enhanced by a shared substantive interest in collective behavior. This paper will therefore attempt to review and extend discrete distribution models in the behavioral sciences with urban collective violence as the primary focus.

Distribution models of episodic processes are discrete probability distributions that provide a theoretical count of objects or events. In the context of issues and problems relevant

C. Taillie et al. (eds.), Statistical Distributions in Scientific Work, Vol. 6, 147–159.

to research in education, psychology, political science and sociology, it is convenient to classify episodic processes into two broad categories. In the first category, *simple random processes,* chance mechanisms are assumed to operate singly to produce observed distributions. What this implies to behavioral scientists is that the processes that produce the observed data are inscrutably random and exhibit no underlying or secondary mechanisms that operate together with any simple, primary mechanism.

Complex random processes, on the other hand, provide a stimulus to causal theory building and testing in the behavioral sciences. These mechanisms are assumed to involve underlying secondary mechanisms that may provide a theoretical explanation for an observed count of events. It is possible, indeed, for these complex mechanisms to involve more than two processes in operation simultaneously (Gurland, 1957).

2. SIMPLE RANDOM PROCESSES

Simple random processes are generic chance mechanisms that may be identified with simple discrete distributions such as the Poisson, binomial and negative binomial distributions. These theoretical distributions suggest a classification of simple random processes into three major types: (1) regular (binomial), (2) irregular (Poisson), and (3) aggregated (negative binomial). The three simple random mechanisms may be considered to represent the varieties of uncomplicated randomness, i.e. randomness that is free of underlying mechanisms (Pielou, 1960; Boswell *et al.*, 1979). The simple random distributions share a common identity as birth-and-death process models. Although these distributions do not exhaust the possibilities of the general birth-and-death process model (Feller, 1968), they appear to be the major alternative expressions of simple randomness in episodic behavioral processes (Pielou, 1960; Rogers, 1969; Boswell *et al.*, 1979). In this context, moreover, it is convenient to derive the three simple mechanisms in terms of models of collective violence such as political coups or urban riots.

Let the probability that a single city has experienced k riots by the time t be given by $p_k(t)$ and the probability of one riot occurring in the period succeeding time t (the $k+1$ riot) be given by $f_k(t)$. We also assume that no more than one riot may occur in a single time interval and that no two riots may occur simultaneously in a single city.

The probability of the system being in state k (the proportion of cities with k riots) is therefore given by the differential equation

$$\frac{dp_k(t)}{dt} = p_{k-1}(t)f_{k-1}(t) - p_k(t)f_k(t).$$

The two conditions described suggest the probability generating functions (p.g.f.'s),

$$F(s) = \sum_{k=0}^{\infty} p_k(t)s^k \quad \text{and}$$

$$H(s) = \sum_{k=0}^{\infty} f_k(t)p_k(t)s^k, \quad \text{respectively, so that}$$

$$\frac{d}{dt}F(s) = (s-1)(\sum_{k=0}^{\infty} f_k(t)p_k(t)s^k) = (s-1)H_k(s) .$$

$F(s)$ is therefore the p.g.f. for simple random mechanisms whether regular, irregular or aggregated (Rogers, 1969, pp. 19-20; Feller, 1968, pp. 248-261). The specific form of $F(s)$ and therefore the determination of a particular mechanism and its distribution depends on the form taken by the function $f_k(t)$.

2.1 The Regular Random Process. In terms of outbreaks of urban collective violence, the regular random distribution of riots is such that conditions tend to reduce the accumulation of outbreaks. On the basis of this assumption, where k riots having already occurred in a city diminishes the probability of another riot occurring, we let $f_k(t)$ be expressed by a diminishing linear function such as $f_k(t) = c - bk$, where $bk < c$ and $c > 0$. Then

$$H_k(s) = \sum_{k=0}^{\infty} (c - bk)p_k(t)s^k$$

and $F(s) = (1 - p + ps)^r$, the p.g.f. for the binomial distribution, where $p = e^{bt} - 1$ and $r = c/b$.

A regular random distribution of riot events, represented by a binomial distribution, may be produced either by chance or by design. In planned rioting, the intent may have been simply to space events evenly. This case is exemplified in rioting theories that postulate a "master plan" or "manipulation by outside agitators." Alternatively, riots may escalate so far as to paralyze the national urban system as a result of exhaustion and saturation. We may hypothesize that the nearness to which a binomial series represents a rioting pattern, the closer an urban system has approached a condition of planned rioting or saturation rioting.

2.2 The Irregular Random Process. The mechanism of irregular randomness, when viewed in the context of urban rioting, is one that assumes complete independence of disturbances within cities or levels of disturbances (q) within a national urban system. This probabilistic independence takes form in the expression for $H_k(s)$ by letting $f_k(t) = f(t)$. Consequently, $H_k(s) = f(t)F(s)$ and $F(s) = e^{-qa(1-s)}$, the p.g.f. of the Poisson distribution (where $qa = \int_c^t f(t)dt$).

In a stochastic theory of collective violence, it may be important to distinguish between "binomial rioting" resulting from random saturation and "Poisson rioting" resulting from a widely shared ideology of defensive action (Tomlinson, 1968). In the former case, the probability of rioting may be raised to a very high level by escalation in events and the rioting may be very frequent, though evenly spaced in time or distance. In the latter case, the probability of rioting is uniformly low and actual outbursts will be infrequent, despite the shared beliefs in collective violence as a viable defense against oppressive actions of the government or majority.

2.3 The Aggregated Random Process. In an aggregated random process of urban disorders, conditions tend to increase the likelihood of continued rioting beyond initial outbreaks, suggesting that

$$f_k(t) = c + bk, \quad c > 0, \quad b > 0$$

and

$$H_k(s) = \sum_{k=0}^{\infty} (c + bk)p_k(t)s^k.$$

In this case, $F(s) = (1 + p - ps)^{-r}$, the p.g.f. of the negative binomial distribution, where $p = e^{bt} - 1$ and $r = c/b$. Rioting that exhibits a process of simple aggregated randomness is one in which no casual or underlying process can be found to account for a grouping effect in the pattern of outbursts. Obviously, complex processes may underlie grouped random patterns, but that is not *necessarily* the case (Patil and Stiteler, 1974). This point is crucial to theorizing about the course of civil outbursts: a pattern that fits the NBn may be one that owes its origins to no underlying or causal process at all. In this case, the grouping of events may occur completely at random with the tendency of outbursts to become less dispersed rather than more so.

To summarize, simple random processes occur in three forms that permit theorizing about behavioral processes: regular, irregular and aggregated. Each of these simple processes may be viewed as completely random and therefore as giving no evidence of causal or other non-random mechanisms.

3. COMPLEX RANDOM PROCESSES

Complex random processes involve the combination of at least two chance mechanisms. The primary mechanism is that which counts events or objects on the highest ecological level. For example, in a complex process that counts insect larvae in a vegetable field, a partition of the field into small areas, called quadrats, will provide counts of insect larvae per quadrat as well as a count of larvae-bearing quadrats. The count of quadrats on the higher ecological level is determined by the primary mechanism. The count of larvae within quadrats, on the other hand, is determined by the secondary mechanism. Secondary mechanisms may themselves be complex in type.

In research on human behavior which may conceivably be intervened upon and altered, the factors that are allowed to influence the conceptualization of episodic processes as simple or, alternatively, complex processes deserve careful scrutiny. Complex models of behavioral and social processes invite theorizing about the nature of underlying mechanisms that may not be evident in the available data. For example, a pattern of riot data that resembles a random binomial process may have been produced by more than one chance mechanism. Such a result may have been produced by two or more mechanisms operating in some combined form. In this section we will attempt to review and develop the major complex episodic processes that have become the object of interest to behavioral scientists. These complex mechanisms appear to take three basic forms: (1) contagious, (2) heterogeneous and (3) clustered.

3.1 Contagion. Although the complex process of contagion has been described in numerous ways in the statistics literature, for the purposes of behavioral science research it is convenient to classify this process into three general types: (1) reinforcement, (2) equilibrium and (3) diffusion (Morgan, 1978). These three types of contagion will be referred to respectively as type I, type II and type III. A contagious process is one that cumulatively alters the prevalence of a characteristic in a system. In type I contagion (reinforcement), the propensity for a certain characteristic to increase in a system (e.g., rate of victimage, consumption of certain goods, etc.) is continually changed as a function of the propensity at any given state of the system. In type II contagion (equilibrium), elements of the system are considered to be simultaneously increasing and decreasing in a certain attribute. Type III contagion (diffusion), involves the transfer of some attribute throughout a system by means of its elements. This process may engage a simple random process (e.g., Poisson) as a primary mechanism and a function of simple propensity as a secondary mechanism. The general form of the contagious process is referred to in the statistics literature as the "birth-death model" (Feller, 1957; Boswell *et al.*, 1979) and may be given a shorthand expression such as:

$$\sum_{\phi=k-1}^{k+1} p_{\phi}\lambda(\phi) \quad \text{(Kolmogorov)} ,$$

where ϕ denotes the frequencies for three consecutive states of the system and $\lambda(\phi)$ denotes the *propensity function* that differentiates between the three major types of contagion and the various sub-types of contagion.

A. Type I contagion provides a model of social processes in which escalation in the occurrence of a certain event (e.g., rioting) depends directly upon the current state of the system. This model is sometimes referred to as the linear reinforcement model in the literature of psychology and sociology (Coleman, 1964). The linear reinforcement model involves a primary Poisson process and a propensity function $\lambda(\phi)$ of the form

$$\lambda_{\phi=k} = \alpha + k\beta$$

as the secondary process. The resultant distribution is the negative binomial.

Making inferences from models such as the type I contagious Poisson can be difficult for the social scientist whose formalizations might have some bearing on public policy. For example, does the statistical model of linear reinforcement when in agreement with civil disruption data taken for a large system of cities indicate that the individual cities experience "auto"-reinforcement? Does some diffusion process give the entire system the attribute of reinforced escalation in the aggregate rate of disturbances? Answers to such questions may only be provided with the aid of longitudinal surveys.

B. Type II contagion, referred to in the literature as "equilibrium," (Coleman, 1964, pp. 315-380) is a condition wherein attributes of the system accumulate and diminish simultaneously. The propensity function, λ_{ϕ}, for type II contagion involves parameters for each direction of a two-way process. In the behavioral sciences this model has been utilized in the study of reciprocal action between elements of a system, e.g., studies of reward structures and allocation of effort in group activities. This model requires propensity functions for each direction of the process in the form

$$\lambda_{\phi=k} = \alpha + k\gamma; \lambda_{\phi=k} = \beta.$$

C. Type III contagion involves the spread of an attribute among the elements of a system. This eventually leads to a condition of systemic exhaustion in which all elements have acquired the attribute. The propensity function, λ_{ϕ}, for the contagious Poisson type III takes the following form:

$$\lambda_{\phi=k} = \alpha(N-k).$$

Continuous distribution functions have proved valuable in behavioral science research on diffusion processes. Some type III processes that involve discrete counts of events may approximate continuous distributions such as the logistic (Dodd, 1955; Bailey, 1957; Atkinson, 1972), and it is sometimes useful to derive discrete distributional equivalents of processes for which continuous distribution models have already been developed (Diekmann, 1979).

3.2 Heterogeneity. The complex process frequently referred to as heterogeneity has been identified with compound distributions and labelled "apparent contagion" (Greenwood and Woods, 1919), although the term "compound distribution" has sometimes described the *clustered* complex mechanism (Feller, 1968). The heterogeneous process is one in which propensities for transition are distributed as well as the events. In other words, both the propensity and frequency functions are statistical distributions.

Using Gurland's notation (1957), a heterogeneous variate, X_3 is equivalent to the compounding of the variate X_1 by the variate X_2: $X_3 \sim X_1 \wedge X_2$. Where X_1 denotes the primary variate of the counting distribution,

$$p_{x_1} = \int_{-\infty}^{\infty} F_1(x_1|x_2)F_2(x_2)dx_2.$$

In terms of the p.g.f., we have

$$G_3(s) = \sum_{-\infty}^{\infty} G_1(s)F_2(x_2) \quad \text{where} \quad X_2 \quad \text{is discrete}$$

and

$$G_3(s) = \int_{-\infty}^{\infty} G_1(s|x_2)F_2(x_2)dx_2, \quad \text{where} \quad X_2 \quad \text{is continuous.}$$

As an example of the heterogeneous process in the context of urban riots, let us assume that an urban area at risk is placed under a spatial or temporal grid (quadrat). The variate representing the number of riots in each quadrat is denoted by k_1. The quadrats, units in which observations are made, may be temporal (years, months, weeks, etc.) as well as spatial. The measure of riots per quadrat is given by the distribution function $f_1(k_1)$. In heterogeneity, where more than one parameter of the probability distribution is subject to random variation, we may take the second randomly distributed parameter (usually the one representing propensity to riot) of the distribution, $f_1(k_1)$, to be k_2; this parameter, k_2, in turn becomes the variate with distribution f_2. The propensity to riot varies randomly over quadrats in an urban area according to the secondary distribution. In this case we find that the primary distribution, f_1, counts the disturbances whereas the secondary distribution does not *count* propensities to rioting but provides a distribution of those propensities for the entire urban system. This distribution of propensities is the tendency of the urban system to increase its aggregate level of rioting determined by the proportion of cities at each level of disorder (i.e., having experienced k riots). In Spilerman's study of the 1960's urban riots in the U.S.A. (Spilerman, 1970), the heterogeneous model consists of a primary Poisson model and a secondary *gamma* distribution, resulting in the negative binomial: $P \wedge Ga \sim NBn$. Where k is the counting variate and λ is the propensity variate,

$$P_k = \int_{-\infty}^{\infty} f_1(k|\lambda)\ f_2(\lambda)\ d\lambda.$$

The fact that the secondary process may be continuous indicates that the secondary process does not reflect *kinetic* attributes of the system such as the riot propensity of *each* city in an urban system. It is important to recognize that the secondary heterogeneous process, whether discrete or continuous, only reflects the distribution of values in an attribute of the systemic, primary variable.

When the secondary function of a heterogeneous process is discrete in type, it is important to consider that, although values of the attribute such as riot propensity are countable, one cannot safely say that the secondary distribution assigns a riot propensity value to each riot-prone city in an urban system.

3.3 Clustered Mechanism. The complex mechanism sometimes described as the "generalized" distribution is referred to as the clustered mechanism. In the urban riots example, the clustered mechanism provides a count of individual cities as well as individual events taking place within those cities. This process involves a primary and secondary distribution, the latter of which must be discrete in type (Feller, 1943), notwithstanding Gurland's oversight to the contrary (Gurland, 1957). This process has a very simple form which follows, for the counting variate, X_2, and where F_1 and F_2 are the primary and secondary distributions, respectively:

$$p_{x_2} = \sum_{x_1=0}^{\infty} F_2(x_2|x_1)F_1(x_1),$$

where X_1 is the primary variate. This model may represent the contribution to the national level of urban rioting due to individual cities. If the Poisson distribution represents a stochastic count of "rioting cities per year," then a discrete secondary distribution such as the logarithmic series can give us a stochastic count of "riots per city" (Quenouille, 1949; Patil and Boswell, 1975; Eaton and Fortin, 1978). Clustered distributions reflect *a single dimension* of the episodic process on two distinct ecological levels. In other words, the clustered model may reflect frequencies on two distinct ecological levels (e.g., "rioting cities per year," "riots per city"). The resultant distribution gives the combined count provided by the primary and secondary distributions, such as "riots per year." The clustered distribution produced by a Poisson primary and log series secondary distribution is negative binomial:

$$\text{NBn} = \text{P} \vee \text{LS}.$$

The clustered model of aggregation in episodic processes may be generally characterized as follows in terms of the p.g.f.:

$$G_3(s) = G_1(G_2(s)).$$

The clustering mechanism is distinguished by the secondary process: it can only be a *discrete* variable (Feller, 1943; Gurland, 1957; Morgan, 1979). This characteristic of the clustering mechanism insures that the secondary process counts a frequency, but on the kinetic level (distinguished from the systemic level in which the primary frequency is counted). The fact that a continuous variate can not *practically* have a counting distribution deserves more attention in the literature (Kotz, 1980).

Applications of the clustered mechanism are surprisingly infrequent in the behavioral sciences. In terms of the urban riots model, riots-per-city may be clustered within riot-cities per unit time.

Similar to the other forms of aggregation, we find there are several models of clustering involving various pairs of probability distributions. Each of these may be used to substantively describe distinct patterns of activity, whether social or biological. A time-based model of clustered urban rioting would involve the following assumptions: (1) riots occur in clusters dependent upon the city of incidence; (2) the number of clusters, or cities, in a single quadrat of time, say one year, is given by the primary distribution, $f_1(X_1)$ of the random counting variable X_1; (3) the number of riots-per-cluster for any given cluster (city) is distributed by a secondary distribution in the same way for all clusters, $f_2(X_2)$; and (4) the clustered distribution, f_2, is discrete.

3.4 Mechanism Identification and Gurland's Theorem. The problem of distinguishing between underlying mechanisms that generate distributions such as the negative binomial has attracted the attention of biostatisticians and other mathematical ecologists as well as behavioral scientists (Kemp, 1970).

In the case of urban rioting, three distinct theories of rioting that call for three different types of probability distributions all give the identical estimates to data in terms of the negative binomial distribution. If we assume that this negative binomial data has evolved through a heterogeneous process, we will suggest a systemic theory of riot escalation that calls for variation in the propensity to escalate according to a gamma process if the overall pattern of riot frequencies follows a

Poisson process. In other words, we can postulate that the systemic proportions for rioting follows a Poisson distribution whereas the variation in the tendency for the level of rioting to increase follows a gamma distribution. There are, however, other complex processes also equivalent to the negative binomial. One of these involves a primary binomial and a secondary negative binomial process. It is not difficult for a sociologist to generate an appropriate theory of riot causation and pattern on the basis of such a process. Both models just described are heterogeneous and would therefore support an identical behavioral organization of causality. We have found that the negative binomial distribution may result from (1) a linear reinforcement effect on a Poisson process, (2) a linear reinforcement effect on a Poisson process in equilibrium, (3) a linear reinforcement effect on a Poisson diffusion process, and (4) in the clustering category, Poisson clusters of riots distributed as a logarithmic series; to name some of its generating mechanisms (but not all of them). Faced with such an array of generating mechanisms, how can the behavioral scientist choose a theory to explain data that is estimated by the negative binomial distribution considering that the choice may influence public policy? Results have been reported in the literature to resolve this problem (Arbous and Kerrich, 1951; Bates and Neyman, 1952; Allison, 1978; Feller, 1943). Equivalence analysis of episodic processes has become a focus of research in a stochastic models of collective behavior since very different theories of occurrences of events and, consequently, radically different policy implications, may be precipitated by the numerous alternatives presented by the data. Theoretically, the identification of underlying mechanisms in complex processes is aided by the fact that certain primary distributions (*Gurland distributions*) have a property that determines equivalence of complex mechanisms. A Gurland distribution, G, has the following property: if a parameter, λ, of G is the variate in a mixing distribution, F, and also occurs as an exponent in the p.g.f. of G, $\{h(s)\}^{\lambda}$, then

$$G \wedge F \equiv F \vee G \quad \text{(Gurland 1957).}$$

4. CONCLUSION

Over the past three decades, behavioral scientists have begun to show interest in statistical distribution models of behavioral processes. Most of the research activity, primarily in sociology and political science, indicates that much has to be done to promote interdisciplinary research involving distributional approaches. Although the vast majority of discrete distributions reported in the behavioral science literature are of the simple random process type (usually Poisson), studies of collective behavior have encouraged the application of complex and multivariate distributions as well as other distributional statistics.

ACKNOWLEDGEMENTS

Part of the research was conducted under NIE-G-80-0113, a research grant from the Department of Health Education and Welfare.

REFERENCES

Allison, P. (1978). Estimating and testing for a model of reinforcement. Department of Sociology, Cornell University (Mimeographed).

Arbous, A. G. and Kerrich, J. E. (1951). Accident statistics and the concept of accident-proneness. *Biometrics*, 7, 340-432.

Atkinson, A. C. (1972). A test of the linear-logistic and Bradley-Terry models. *Biometrika*, 59, 37-42.

Bailey, N. T. J. (1957). *The Mathematical Theory of Epidemics*. Griffin, London.

Bates, G. E. and Neyman, J. (1952). Contributions to the theory of accident-proneness. II: True or false contagion. *University of California Publications in Statistics*, 1, 255-276.

Boswell, M. T. and Patil, G. P. (1971). Chance mechanisms generating the logarithmic series distribution used in the analysis of number of species and individuals. In *Statistical Ecology, Vol. I*, G. P. Patil, E. C. Pielou, and W. E. Waters, eds. The Pennsylvania State University Press, University Park. pages 99-130.

Boswell, M. T., Ord, J. K. and Patil, G. P. (1979). Chance mechanisms underlying univariate distributions. In *Statistical Distributions in Ecological Work*, J. K. Ord, G. P. Patil, and C. Taillie, International Co-operative Publishing House, Fairland Maryland.

Coleman, J. S. (1964). *Introduction to Mathematical Sociology*. The Free Press, New York.

Diekmann, A. (1979). A dynamic stochastic version of the Pitcher-Hamblin-Miller model of collective violence. *Journal of Mathematical Sociology*, 6, 277-282.

Dodd, D. (1955). Diffusion is predictable: Testing probability models for laws of interaction. *American Sociological Review*, 20, 392-402.

Eaton, W. W. and Fortin, A. (1978). A third interpretation for the generating process of the negative binomial distribution. *American Sociological Review*, 43, 264-267.

Feller, W. (1943). On a general class of 'contagious' distributions. *Annals of Mathematical Statistics*, 14, 389-400.

Feller, W. (1968). *An Introduction to Probability Theory and Its Applications, Vol. I*, 3rd Ed. Wiley, New York.

Greenwood, M. and Woods, H. M. (1919). A report on the incidence of industrial accidents upon individuals with special reference to multiple accidents. H. M. Stationery Office, London.

Gurland, J. (1957). Some interrelations among compound and generalized distributions. *Biometrika*, 44, 265-268.

Kemp, C. D. (1970). "Accident proneness" and discrete distribution theory. In *Random Counts in Scientific Work, Vol. II*, G. P. Patil, ed. The Pennsylvania State University Press, University Park. Pages 41-66.

Kotz, S. (1980). Personal communication on the clustering mechanism.

Lieberson, S. and Silverman, A. (1965). The precipitants and underlying conditions of race riots. *American Sociological Review*, 30, 887-898.

Morgan, P. R. (1978). *Formal Models of Collective Violence.* Ph.D. dissertation, University of Colorado.

Morgan, P. R. (1979). The epidemiology of collective violence. Center for Social Organization of Schools, The Johns Hopkins University.

Patil, G. P. and Stiteler, W. M. (1974). Concepts of aggregation and their quantification: A critical review with some new results and applications. *Researches on Population Ecology*, 15, 238-254.

Patil, G. P. and Boswell, M. T. (1975). Chance mechanisms for discrete distributions in scientific modeling. In *Statistical Distributions in Scientific Work, Vol. II*, G. P. Patil, S. Kotz, and J. K. Ord, eds. Reidel, Dordrecht-Holland. Pages 11-24.

Pielou, E. C. (1960). A single mechanism to account for regular, random and aggregated populations. *Journal of Ecology*, 48, 575-584.

Quenouille, M. H. (1949). A relation between the logarithmic, Poisson and negative binomial series. *Biometrics*, 5, 162-164.

Rogers, A. (1969). Quadrat Analysis of Urban Dispersion. Working Paper No. 93, Center for Planning and Development Research, University of California, Berkeley.

Spilerman, S. (1970). The causes of racial disturbances: A comparison of alternative explanations. *American Sociological Review*, 35, 627-649.

Tomlinson, T. M. (1968). The development of a riot ideology among urban (blacks). *American Behavioral Scientist*, 2, 27-31.

[*Received July* 1980. *Revised November* 1980]

SOME ISSUES ASSOCIATED WITH THE MEASUREMENT OF INCOME INEQUALITY

JAMES B. McDonald

Department of Economics
Brigham Young University
Provo, Utah 84601 USA

SUMMARY. The paper deals with problems encountered in measuring income inequality and the most commonly used measures are discussed. Parametric and nonparametric methods of estimation are considered for data both in a grouped format as well as individual observations. Commonly adopted distribution functions are reviewed along with derived expressions relating some measures of inequality to the distribution parameters.

The question of the impact of measurement errors upon the distribution of measured income and associated measures of inequality is considered for some leading cases. The associated distribution functions are seen to involve such special functions as Whittaker functions and generalized hypergeometric series.

KEY WORDS. Measures of income inequality, coefficient of variation, Pietra index, Gini coefficient, Theil entropy measure, estimation of measures of income inequality, measurement error.

1. INTRODUCTION

It is well known that alternative economic policies can have diverse impact upon the level of economic activity. An equally important consideration of such policies is the impact upon the distribution of income as well as on the average level of income. This is important both because of possible economic externalities as well as possible social preferences for one distribution over another, say on the basis of egalitarian consideration. Numerous measures of income inequality have been seriously considered. Each of the measures is perfectly valid for descriptive purposes;

C. Taillie et al. (eds.), Statistical Distributions in Scientific Work, Vol. 6, 161–179.

however, if a particular measure is used to evaluate the desirability (or rank) alternative distributions of income, then an implicit assumption is being made about the form of a social welfare function. The basic problem is that a single characteristic is being used to compare two density functions. Such comparisons and some measures of income inequality can frequently be interpreted in terms of the Lorenz Curve which depicts the relationship between the percent of income received by different percentages of the population.

Let $f(y)$ and $F(y)$, respectively, denote the density and distribution functions of income (y). The Lorenz Curve can be defined in terms of $f(y)$ by

$$L(p) = \mu^{-1} \int_0^p F^{-1}(t)dt \qquad 0 \leqslant p \leqslant 1$$

where $L(p)$ denotes the fraction of total income that holders of the lowest p*th* fraction of income possess and $\mu = E(Y)$ (Gastwirth, 1971).

Sen (1973) reviews several of the more commonly used measures of income inequality and their properties. These include the coefficient of variation (CV), the Pietra index (or Relative mean deviation, P), standard deviation of logarithms (H), Gini Coefficient (G), and Theil's entropy measure (T). Measures of inequality are frequently evaluated on the basis of such characteristics as their sensitiviy to changes in the units of measurement or to transfers from one group to another. The satisfaction of the Pigou-Dalton Principle has generally been accepted as being desirable, i.e., a transfer from a richer person to a poorer person always reduces "inequality."

The *coefficient of variation* is defined by

$$CV = \sigma/\mu \tag{1}$$

where σ^2 denotes the variance of Y. CV is invariant with respect to scale, attaches equal weights to income transfers at different levels and is sensitive to all income transfers and satisfies the Pigou-Dalton Principle.

The *Pietra index* is defined by

$$P = E(|Y-\mu|)/2\mu. \tag{2}$$

P is invariant with respect to scale, but is not sensitive to income transfers from rich to poor if they are on the same side of the mean.

The *standard deviation of logarithms* is defined by

$$H = \sqrt{E(\ln(y/\mu))^2} \,. \tag{3}$$

H is invariant with respect to scale, attaches greater weight to income transfers at low income levels, but doesn't satisfy the Pigou-Dalton Principle.

Perhaps the most widely used measure of income inequality is the *Gini coefficient* which is defined by

$$G = \frac{1}{2\mu} \int_0^\infty \int_0^\infty |x-y|\ f(x)f(y)\,dx\,dy = 1 - \frac{\int_0^\infty (1-F(y))^2 dy}{\int_0^\infty (1-F(y))\,dy} \tag{4}$$

(cf. Dorfman, 1979). The Gini coefficient is unaffected by a change in scale and satisfies the Pigou-Dalton Principle, but attaches more weight to income transfers affecting the middle class.

Theil's entropy (expected information) measure of inequality is defined by

$$T = \int_0^\infty \left(\frac{y}{\mu}\right) \ln \left(\frac{y}{\mu}\right) f(y)\ dy \tag{5}$$

T is scale invariant and satisfies the Pigou-Dalton Principle.

If the Lorenz Curves for two different income distributions do not intersect, then most measures of inequality will yield the same ranking. In fact, Atkinson (1970) has shown that if the Lorenz Curve of one distribution $(\vec{x})$ lies strictly inside the Lorenz curve of another distribution $(\vec{y})$ then $W(\vec{x}) > W(\vec{y})$ where $W(\vec{x})$ denotes an additive welfare function,

$W(x_1,\cdots,x_n) = \sum_{i=1}^{n} U(x_i)$, and $U(x_i)$ is strictly concave in x_i.

However, if the Lorenz curves intersect, then a ranking based upon say a Gini coefficient may differ from rankings based upon another measure of inequality or upon a social welfare function. Newberry (1970) proved that there does not exist an additive social welfare function which will rank income distributions the same as the Gini coefficient. However, Sheshinski (1972) demonstrated that the welfare function $W(\vec{y})=\sum_{i=1}^{n} y_i(1-G)$ will yield the same ranking as that obtained from the use of the Gini coefficient where G denotes the Gini coefficient associated with $\tilde{y}$. Atkinson defines a measure of income inequality which depends upon the form of the utility function. In spite of these limitations, measures of inequality are frequently employed to compare different income distributions.

There are two general approaches to estimation: nonparametric and parametric depending on whether an underlying density function is assumed. These approaches can be applied to data sets which report individual observations as well as observations reported in a grouped format.

In Section 2, we consider nonparametric estimators of some of the various measures of inequality for both types of data sets. In Section 3, we consider measures of inequality which are based upon an assumed functional form for income and consider alternative methods of estimation. In Section 4, we consider the impact of measurement error upon the distribution of measured income and measures of income inequality.

2. NONPARAMETRIC METHODS

The measures of income inequality defined in Section 1 lead to natural estimators of the measures which do not depend upon the particular form of the underlying density function. In a few cases, the associated asymptotic distributions have been determined. In empirical work the data is generally reported in one of two ways: (1) Individual observations are presented or (2) the data is summarized and reported in a grouped format with the relative frequency associated with each income group being reported. With the grouped format, the group means are not always reported.

2.1 Individual Observations. Let $(y_1, y_2, \cdots, y_n)$ denote a random sample drawn from the same population and let $\bar{Y}$ and s be the sample mean and standard deviation, respectively. The following estimators of the previously discussed inequality measures have been widely adopted.

$$\widehat{CV} = s/\bar{Y} \tag{6}$$

$$\hat{P} = \sum_{i=1}^{n} |y_i - \bar{Y}|/2n\bar{Y} \tag{7}$$

$$\hat{H} = \sqrt{\sum_{i=1}^{n} (\ln(y_i/\bar{Y}))^2/n} \qquad (8)$$

$$\hat{G} = \sum_{i,j} |y_i - y_j|/2n^2\,\bar{Y} \qquad (9)$$

$$\hat{T} = \sum_{i=1}^{n} (y_i/\bar{Y})\ln(y_i/\bar{Y})/n. \qquad (10)$$

It should be remembered that the estimators have statistical distributions and a comparison of two different populations requires that this be considered and appropriate techniques of statistical inference adopted. In other words, a comparison of two populations based upon a particular inequality measure requires more than a comparison of the estimated measures. Two estimators will assume different values with a probability of one and it is the statistical significance of the difference which is of interest.

In the absence of specifying an underlying density function for income, we must appeal to asymptotic distribution theory. Given certain regularity conditions, the asymptotic distributions of $\hat{CV}$, $\hat{P}$ and $\hat{G}$ are given by

$$\sqrt{n}(\hat{CV} - CV) \sim N[0,\sigma^2_{\hat{CV}}] \qquad (11)$$

$$\sqrt{n}(\hat{P} - P) \sim N[0, \sigma^2_{\hat{P}}] \qquad (12)$$

$$\sqrt{n}(\hat{G} - G) \sim N[0, \sigma^2_{\hat{G}}] \qquad (13)$$

Formulas for the standard deviations of these estimators can be found in Kendall and Stuart (1961), Gastwirth (1974) and Hoeffding (1948), respectively. An important question related to an application of (11), (12), and (13) is how large n must be for different underlying income distributions. The author is unaware of any reported results for $\hat{H}$ and $\hat{T}$.

2.2 Grouped Observations. Assume that the interval $[0, \infty)$ is partitioned by $0 = a_0 < a_1 < a_2 < \cdots < a_{g-1} < a_g = \infty$ with the *ith* income group being defined by $I_i = [a_{i-1}, a_i)$. One of the most common formats for income data is that in which the relative frequency (n_i/n) associated with each income group is reported.

The sample means $(\bar{y}_i)$ for each income group are not always reported. The problem then becomes how to estimate the inequality measure of interest. A common method of estimation assumes that there is no intragroup variation and that all observations in each interval lie at a particular point and proceed from there. If the group means are available, then all observations are typically assumed to lie at the sample mean, if not then the observations are frequently assumed to lie at the "midpoint." The "midpoint" of the last group is not defined and this case needs to be treated separately. One approach is to fit a Pareto to the upper tail of the income distribution.

Estimators of CV, P, H, G and T which might be considered in this case are

$$CV^* = s^*/\bar{Y}^* \tag{14}$$

$$\text{where } s^* = \sqrt{\sum_{i=1}^{g} n_i(\mu_i - \bar{Y}^*)^2/(n-1)}$$

$$\bar{Y}^* = \sum_{i=1}^{g} n_i\mu_i/n$$

$$P^* = \sum_{i=1}^{g} n_i|\mu_i - \bar{Y}^*|/2n\bar{Y}^* \tag{15}$$

$$H^* = \sqrt{\sum_{i=1}^{g} n_i(\ln(\mu_i/\bar{Y}^*))^2/n} \tag{16}$$

$$G^* = \sum_{i,j=1}^{g} \left(\frac{n_i}{n}\right)\left(\frac{n_j}{n}\right)|\mu_i - \mu_j|/2\bar{Y}^* \tag{17}$$

$$T^* = \sum_{i=1}^{g} n_i(\mu_i/\bar{Y}^*)\ln(\mu_i/\bar{Y})/n. \tag{18}$$

The μ_i in (14)-(18) can be interpreted as $\bar{y}_i$ if the sample means are reported, as "midpoints" if sample means are not given, or possibly some other estimates, see Needleman (1978). The distributions of (14)-(18) are subject to the usual sources of sampling variation, but are also sensitive to the implicit assumption of no intragroup variation.

As a case in point Gastwirth (1972) considered the estimator G^* when group means are given and noted that it would consistently underestimate $\hat{G}$ (equation 9) and for large "n" G (equation 4) would also be underestimated by G^*. Gastwirth didn't report the distribution of G^* (μ_i replaced by $\bar{y}_i$), but proposed estimators of upper and lower bounds defined in terms of

$$GL = \sum_{i,j=1}^{g} p_i p_j |\mu_i - \mu_j| / 2\mu \tag{19}$$

and

$$GU = GL + \sum_{i=1}^{g} p_i^2 \frac{(\mu_i - a_{i-1})(a_i - \mu_i)}{(a_i - a_{i-1})\,\mu} \tag{20}$$

where p_i and μ_i, respectively, denote the fraction of the population and mean income associated with the *ith* group. Estimates of GL and GU, $\hat{GL}$ and $\hat{GU}$, can be obtained by replacing p_i and μ_i in (19) and (20) by (n_i/n) and $\bar{y}_i$. It then follows from the construction of GL and GU that

$$GL < G < GU \tag{21}$$

and

$$\hat{GL} < \hat{G} < \hat{GU}. \tag{22}$$

However, it doesn't follow that

$$\hat{GL} < G < \hat{GU} \tag{23}$$

is always satisfied. Smith (1972) obtained the asymptotic distribution of $\hat{GL}$ and $\hat{GU}$ corresponding to underlying density functions with a single shape parameter. In a Monte Carlo study, McDonald and Ransom (1979a) found that as the number of groups (g) increases (GU - GL) decreases, but the relative frequency with which (23) is violated increases. In fact for a sample of 1000 (5000), (23) was violated approximately 50%(20%) when $g = 10$ (equal probability groups) were used in the study. This highlights the importance of questions of statistical inference in issues dealing with income inequality. The author is not aware of any results dealing with the distribution of CV*, P*, H*, or T*.

Murray (1978) developed a method of obtaining bounds on $\hat{G}$ for the case in which group means are unknown. He didn't consider their distribution and found the bounds to be quite far apart which supports the case for including some information about intragroup behavior.

3. PARAMETRIC APPROACH

Much of the empirical work in the area of the distribution of income is based upon the assumption of a particular functional form. The parameters defining the function are estimated and then corresponding estimates of measures of inequality can be obtained directly from the relationship between the parameters and the measure of inequality or through an indirect approach such as being inferred from the associated Lorenz Curve.

3.1 Functional Forms. Numerous functional forms have been considered as models for the distribution of income. Perhaps the most commonly used distributions are the Pareto, lognormal and gamma distributions. These distributions while being relatively simple do not provide a uniformly good fit over the entire income range e.g., the Pareto provides a good fit for higher income levels but not for middle or lower income levels. Salem and Mount (1974) found that the gamma provided a better fit to U. S. family income than the lognormal. The Pareto, gamma, and lognormal distributions involve two parameters.

On approach has been to adopt more general distributions which depend upon more than two parameters. In an interesting paper by Singh and Maddala (1976) the notion of hazard-rate was used to obtain a three parameter distribution function which is a generalization of the Pareto, Weibull and Sech^2 distribution considered by Fisk (1961). Singh and Maddala estimated the parameters of the "Singh-Maddala" function and compared the results with Salem and Mount's results and concluded that the Singh-Maddala function provided a better fit to U. S. family income data than the gamma. Singh and Maddala used a different estimation technique than Salem and Mount and hence the comparisons are not strictly appropriate.

Thurow (1970) used a beta distribution to analyze macroeconomic explanatory factors associated with the distribution of income. McDonald and Ransom (1979c) compared the gamma, lognormal, Singh-Maddala and beta distributions using several different estimation techniques and U. S. family income data. A comparison of functional forms for a given estimation technique was possible as well as the impact of different estimation techniques for a given functional form. The lognormal was found to provide the worst fit, with the gamma providing a much improved fit and the best fit corresponding to the beta and Singh-Maddala distributions. Table 1 provides definitions of these density functions and associated characteristics.

We note that the entries in Table 1 facilitate estimation of measures of inequality once parameter estimates have been obtained. The results for the beta can be shown to include those of the gamma as a limiting case ($q \to \infty$ for a fixed mean) and hence the beta must provide as a good a fit as the gamma.

Kloek and Van Dijk (1978) have considered, such three parameter distributions as the log t distribution, the generalized gamma and Champernowne distribution and estimated them for Dutch income-earnings data. The authors note that the log t includes a "heavy-tailed" log Cauchy distribution and a "light-tailed" lognormal distribution as limiting cases. The generalized gamma includes the gamma distribution as a special case. Based upon the data used, the authors concluded that the lognormal and gamma distributions didn't provide an accurate approximation to the distribution of income. They found that considerable improvements in fit were obtained by going from two parameter distributions (lognormal and gamma) to three or more parameter distributions. To some extent this would be expected since some of the three parameter distributions include the lognormal and gamma distributions as special cases.

Esteban and Pastor (1977) examine a class of distribution functions which behave like the Pareto for large income levels. These functions are characterized in terms of an elasticity,

$$\eta(y) = -yf'(y)/f(y). \tag{24}$$

They consider the family of distributions which corresponds to

$$\eta(y) = 1 + \alpha - \beta x^{-\varepsilon} . \tag{25}$$

The associated density function is the generalized gamma

$$f(y) = \frac{|\xi| \, (\frac{\beta}{\varepsilon})^{\frac{\alpha}{\varepsilon}} \, y^{-(1+\alpha)}}{\Gamma(\frac{\alpha}{\varepsilon})} \, e^{-\frac{\beta}{\varepsilon} y^{-\varepsilon}} \tag{26}$$

and includes the Pareto, gamma, normal, a special case of Pearson type V, Weibull and lognormal as special or limiting cases. The Gini coefficient associated with (26) can be shown to be

$$G = \frac{\Gamma(\rho) \quad 2^{-\rho}}{\Gamma(\gamma)\Gamma(\delta)} \left\{ \frac{1}{\gamma} \, {}_2F_1 \begin{bmatrix} 1, \ \rho; \ 1/2 \\ \gamma + 1; \end{bmatrix} - \frac{1}{\delta} \, {}_2F_1 \begin{bmatrix} 1, \ \rho; \ 1/2 \\ \delta + 1; \end{bmatrix} \right\}$$

where $\rho = (2\alpha - 1)/\varepsilon$, $\gamma = \alpha/\varepsilon$, $\delta = (\alpha - 1)/\varepsilon$.

3.2 Parameter Estimation. If individual observations are available then such standard estimation procedures as maximum likelihood and methods of moments can be used to estimate the unknown parameters. If only grouped data is available, then these

TABLE 1: Density functions and associated characteristics.*

Random Variable	Density	Mean(u)	CV	P	G	T
Pareto Johnson and Kotz (1970) Thiel (1967)	$\frac{ak^a}{y^{a+1}}$ $y \geq k$	$\frac{ak}{a-1}$	$\frac{1}{\sqrt{a(a-2)}}$	$\frac{(a-1)^{a-1}}{a^a}$	$\frac{1}{2a-1}$	$\frac{1}{a-1} - \ln(\frac{a}{a-1})$
Lognormal* Atkinson and Brown (1970)	$\frac{\exp\{-(\ln y-\mu)^2/2\sigma^2\}}{y\sigma\sqrt{2\pi}}$	$\exp(\mu+\sigma^2/2)$	$\exp(\sigma^2)-1$	$2N[\sigma/2\|0,1]-1$	$2N[\sigma/\sqrt{2}\|0,1]-1$	$\sigma^2/2$
Gamma McDonald and Jensen (1979) Salem and Mount (1974)	$\frac{\beta^\alpha y^{\alpha-1} e^{-y\beta}}{\Gamma(\alpha)}$	α/β	$1/\sqrt{\alpha}$	$\frac{\alpha^\alpha e^{-\alpha}}{\Gamma(\alpha+2)}\, {}_1F_1\left[\begin{matrix}2; \\ \alpha+2;\end{matrix}\ \alpha\right]$	$\frac{\Gamma(\alpha+1/2)}{\sqrt{\pi}\,\Gamma(\alpha+1)}$	$\frac{1}{\alpha} + \Psi(\alpha) - \ln(\alpha)$
Beta McDonald (1979)	$\frac{y^{p-1}(b-y)^{q-1}}{B(p,q)b^{p+q-1}}$ $0 \leq y \leq b$	${}^{bp}/_{(p+q)}$	$\sqrt{\frac{q}{p(p+q+1)}}$	$\frac{(\frac{p}{p+q})^p \cdot {}_2F_1}{p(p+1)B(p,q)}\left[\begin{matrix}p,1-q; \\ p+2;\end{matrix}\ \frac{p}{p+q}\right]$	$\frac{\Gamma(p+q)\Gamma(p+\frac{1}{2})\Gamma(q+\frac{1}{2})}{\Gamma(p+q+\frac{1}{2})\Gamma(p+1)\Gamma(q)\sqrt{\pi}}$	$\Psi(p+1)-\Psi(p+q+1)$ $-\ln(p/(p+q))$
Singh-Maddala* Singh and Maddala (1976) Cronin (1979) McDonald and Ransom (1979c)	$\frac{a_1a_2a_3y^{a_2-1}}{(1+a_1y^{a_2})^{a_3+1}}$	$\frac{\Gamma(\frac{1}{a_2}+1)\Gamma(a_3-\frac{1}{a_2})}{(a_1^{1/a_2})\Gamma(a_3)}$	CV_{sm} (see foot-note)	P_{sm} (see footnote)	$1-\frac{\Gamma(2a_3-\frac{1}{a_2})\Gamma(a_3)}{\Gamma(2a_3)\Gamma(a_3-\frac{1}{a_2})}$	T_{sm} (see footnote)

*Footnote to Table 1.

$N[x|0,1]$ denotes the cumulative distribution for a standard normal. ${}_1F_1[a; b; x]$ and ${}_2F_1[a, c; b; x]$ respectively denote the confluent and hypergeometric series. $\Psi(x)$ is the digamma function defined by $d\ln(x)/dx$.

The references for known results or derivations of new results are available from the author upon request. Expressions for CV_{sm}, P_{sm} and T_{sm} can be shown to be

$$CV_{sm} = \sqrt{\frac{\Gamma(a_3)\Gamma(1+2/a_2)\Gamma(a_3-2/a_2)}{\Gamma^2(1+1/a_2)\Gamma^2(a_3-1/a_2)} - 1}$$

$$P_{sm} = \frac{a_1 a_2 a_3 \mu^{a_2}}{(1+a_1\mu^{a_2})} \left\{ \frac{1}{a_2} \; {}_2F_1\left[\begin{matrix}1, 1-a_3; \\ 2;\end{matrix} \; z\right] - \frac{{}_2F_1\left[\begin{matrix}1+1/a_2, 1-a_3+1/a_2; \; z \\ 2+1/a_2;\end{matrix}\right]}{(1+a_2)(1+a_1\mu^{a_2})^{1/a_2}} \right\}$$

$$\text{where } z = \frac{a_1\mu^{a_2}}{1+a_1\mu^{a_2}}$$

$$T_{sm} = \frac{1}{a_2}\left(2\Psi(1/a_2) - \Psi(a_3+1) - \Psi(a_3 - 1/a_2)\right) - \ln a_3 - \ln B(1 + 1/a_2, a_3 - 1/a_2).$$

techniques can't be directly applied. The distribution of income literature contains examples of many estimation techniques which range from very simplistic approaches to those which could be expected to be associated with desirable statistical properties.

In order to facilitate the discussion of estimation associated with grouped data let $p_i(\theta)$ denote the predicted fraction of the population in the *ith* interval, I_i, i.e.,

$$p_i(\theta) = \int_{I_i} f(y;\theta)dy \tag{27}$$

where $f(y;\theta)$ denotes the assumed density function and θ denotes the unknown parameters. The estimation problem can be viewed as selecting estimates of θ so as to obtain the "best fit" between the observed frequencies n_i and the predicted frequencies $(np_i(\theta))$ where $n = \sum_{i=1}^{g} n_i$. This is equivalent to trying to obtain the "best fit" between the *observed relative frequencies* (n_i/n) and the predicted probabilities $p_i(\theta)$.

An important question is how to define the "best fit." We might consider parameter estimators defined by

$$\underset{\theta}{\text{minimize}} \sum_{i=1}^{g} ({}^{n}i/n - p_i(\theta))^2 \tag{28}$$

or

$$\underset{\theta}{\text{minimize}} \sum_{i=1}^{g} |{}^{n}i/n - p_i(\theta)|. \tag{29}$$

These estimators are referred to as least squares and least lines respectively. Aigner and Goldberger (1970) considered the use of (28) for the Pareto distribution. (29) has not been seriously considered in the context of estimating the parameters of density functions. Least squares and least lines estimators attribute an equal weight to each interval and for cases in which there are unequal observations in the intervals more efficient estimation techniques are available (cf. Cox and Hinkley, 1974).

The n_i can be viewed as having been generated from a multinomial distribution; hence, maximum likelihood estimators of the parameters can be obtained from the solution of

$$\underset{\theta}{\text{Maximize}}\ n! \prod_{i=1}^{g} \left(\frac{p^{n_i}}{n_i!}\right). \tag{30}$$

This method is referred to as the method of scoring. Two other methods which are asymptotically equivalent to the method of scoring estimators are based upon a "Chi-square goodness-of-fit." The Pearson and Neyman-Wald estimators, respectively, are defined by

$$\underset{\theta}{\text{Minimize}}\ n \sum_{i=1}^{g} \frac{((n_i/n) - p_i(\theta))^2}{p_i(\theta)} \quad \text{and} \tag{31}$$

$$\underset{\theta}{\text{Minimize}}\ n \sum_{i=1}^{g} \frac{(n_i/n - p_i(\theta))^2}{(n_i/n)} \quad . \tag{32}$$

Note that if the *ith* term in (28) is "weighted by n/p_i or n^2/n_i the Pearson or Neyman-Wald estimators are obtained and unequal interval frequencies are taken into account. Also note that if all $n_i = n/g$ or $p_i = 1/g$, the Pearson or Neyman-Wald estimators, respectively, yield the same results as obtained by least squares. The estimators obtained by methods of scoring, Pearson, and Neyman-Wald are asymptotically equivalent and are asymptotically efficient.

3.3 An Example. In this section several of the estimation techniques discussed in the previous section are used to estimate the parameters defining the lognormal, gamma, beta, and Singh-Maddala density functions and calculate the associated means and values for the Gini coefficient. The program GQOPT (Princeton University) was used for all calculations and a convergence criterion of 10^{-8} was specified. The results of this estimation are given in Table 2. The data used is 1975 income data for families. The rows labeled SSE and χ^2 are obtained by evaluating $\sum_{i=1}^{g} (n_i/n-p_i)^2$ and $n \sum_{i=1}^{g} ((n_i/n) - p_i)^2/p_i$ corresponding to the estimated values of the population parameters.

We note that the estimated Gini coefficient and mean associated with the lognormal density function are higher than obtained from either the gamma, beta, or Singh-Maddala functions. Based upon either a SSE or χ^2 criterion the lognormal provides the poorest fit, followed by the gamma, with the beta being associated with the best fit. There appears to be greater variation in the estimated means and Gini coefficients across functional forms than across estimation techniques. For a given functional form, the efficient estimation techniques generally yield

TABLE 2: Parameter estimates (1975 family income).

		Scoring	Pearson	Least Squares
Lognormal	μ	2.5261	2.5204	2.6337
	σ	.7562	.7767	.6174
	Gini	.4079	.4179	.3385
	Mean	16.642	16.811	16.849
	SSE	.009423	.010550	.004904
	χ^2	2590.33	2518.63	21034.40
Gamma	α	2.3558	2.3534	2.6245
	β	.1535	.1534	.1677
	Gini	.3487	.3489	.3322
	Mean	15.345	15.342	15.646
	SSE	.001395	.001403	.000967
	χ^2	238.38	283.36	589.44
Beta	p	1.9454	1.9449	1.8993
	q	6.0053	6.0328	5.3371
	b	61.3536	61.5484	57.1665
	Gini	.3282	.3285	.3274
	Mean	15.012	15.005	15.004
	SSE	.000122	.000124	.000112
	χ^2	59.93	59.88	62.90
Singh-Madalla	$a_1 \cdot 10^2$	.0278	.0198	.0197
	a_2	1.7496	1.7389	1.7389
	a_3	26.0700	37.5510	37.5510
	Gini	.3330	.3328	.3328
	Mean	15.131	15.102	15.173
	SSE	.000159	.000151	.000144
	χ^2	89.32	88.85	90.64

TABLE 3: Estimated income characteristics.*

	Estimates Reported by Singh and Maddala			Census		New Estimates		
	Implied Mean	Implied Median	OBJ	Mean	Median	Implied Mean	Implied Median	OBJ
1960	11.505	9.875	10.459		5.620	6.464	5.416	.007
1967	11.579	9.865	2.047	9.019	7.974	9.084	7.779	.015
1972	10.956	9.414	.476	12.625	11.625	12.478	10.991	.001

*OBJ is equal to the objective function evaluated at the estimated parameter values. The median is equal to $((2^{1/a_3}-1)/a_1)^{1/a_2}$.

estimated means and Gini coefficients which are quite close; whereas, least squares estimators may differ substantially from these results. It appears that both the functional form and the estimation techniques are important determinants of the values of estimated population characteristics, with the functional form probably being the most important of the two.

Another important factor is the accuracy of the algorithm adopted. As a case in point Singh and Maddala estimated a_1, a_2 and a_3 by minimizing

$$\Sigma[\ln(1-F) + a_3 \ln(1 + a_1 y^{a_2})]^2. \qquad (33)$$

The implied values for the mean and median income appeared unreasonable and so new parameter estimates were obtained by minimizing (33) using GQOPT, cf. McDonald and Ransom (1979c). Some typical results are shown in Table 3. The new estimates are seen to agree quite closely with the corresponding census estimates. These results underscore the importance of numerical accuracy for a given functional form and estimation technique.

4. MEASUREMENT ERROR

The previous discussion has assumed that the income data is accurate. In many situations one may reasonably expect that the reported magnitudes may in fact be understatements of the true values of the variables. The existence of measurement error will clearly have an impact on the accuracy of estimated income characteristics. The errors in this case may be onesided as opposed to the usual errors in variables models.

Hartley and Revankar (1974) consider this problem and assume that the true distribution is a Pareto distribution and the fraction of income reported (u) is distributed independently of the level of density $g(u) = pu^{p-1}$, $0 \leqslant u \leqslant 1$. They consider maximum likelihood estimation of the parameters of the Pareto density function.

Cramer (1980) notes that a strict application of a Chi Square goodness of fit test in the work of Kloek and Van Dijk (1978) and McDonald and Ransom (1979c) results in the rejection of virtually all of the models considered. He investigates whether this might be due to the presence of an error in the observed income series. He adopts method of moments estimation and concludes that "neither the Pareto nor gamma are redeemed by the introduction of an error term," but that "the verdict on the lognormal must be postponed; as a by-product of the analysis we find that it does better than was suggested some years ago by Salem and Mount."

In order to explicitly consider the impact of measurement errors on characteristics of measured income let $y = uy^*$ where y, y^*, u denote measured income, true income and the fraction reported. It is easily shown that

$$CV = \sqrt{CV_u^2\ (CV^{*2} + 1) + CV^{*2}}$$

where CV, CV_u and CV^* denote the coefficient of variation for y, u and y^*, cf. Israelsen, McDonald and Newey (1980). The presence of measurement error increases income inequality as indicated by the coefficient of variation, i.e., $CV > CV^*$ for $CV_u > 0$. CV is independent of $E(u)$ for $Var(u) = 0$, $CV_u = 0$.

Let the density functions of y^*, u and y be denoted $f(y^*; \theta_1)$, $g(u; \theta_2)$, and $h(y; \theta)$ where θ_1, θ_2 and θ denote vectors of parameters. A knowledge of any two of $f(\ )$, $g(\ )$, and $h(\ )$ is sufficient to deduce the remaining density function, e.g.,

$$h(y; \theta) = \int f(y/u)\ g(u)\ (du/u). \tag{34}$$

Israelsen, McDonald and Newey (1980) assume the density of measurement errors to be

$$g(u; \alpha,\beta,\gamma) = \frac{u^{\alpha-1}(\gamma-u)^{-1}}{\gamma^{\alpha+\beta-1} B(\alpha,\beta)} \qquad 0 \leqslant u \leqslant \gamma.$$

This formulation includes that of Hartley and Revankar (1974) as a special case ($\beta = \gamma = 1$) and is extremely flexible, ranging from "$\cup$" or "$\cap$" shaped to "J" shaped depending upon the values of α and β. If $\gamma > 1$, overreporting as well as underreporting is admitted. The authors consider different densities for $f(\ )$ or $h(\ )$ and then deduce the other. As an example, if the density of y^* is

$$f(y^*; p,\lambda) = \frac{y^{*p-1} e^{-y^*/\lambda}}{\lambda^p \Gamma(p)}$$

then the corresponding density of y is

$$h(y; p,\lambda,\alpha,\beta,\gamma) = \frac{\Gamma(\beta) y^{\frac{\alpha+p-3}{2}} \exp(\frac{-y}{2\gamma\lambda})}{\Gamma(p)B(\alpha,\beta)(\lambda\gamma)^{\frac{\alpha+p-1}{2}}} \tag{35}$$

$$\cdot\ W_{\frac{p-\alpha-2\beta+1}{2},\ \frac{\alpha-\beta}{2}}\left(\frac{y}{\lambda\gamma}\right)$$

where $W_{a,b}(z)$ denotes the Whittaker function

For a given functional form for y^* maximum likelihood estimators of θ_1 can be obtained by (1) obtaining maximum likelihood estimators of θ, and (2) solving for the corresponding estimates of θ_1. The second step involves what is referred to as the *identification* problem in econometrics. From (34), θ can be seen to be a function of θ_1 and θ_2. Since we have observations on y and not y^*, the characterization of the density of y^* requires that we be able to infer θ_1 from a knowledge (estimate) of θ. (35) is associated with an identification problem since separate estimation of λ and γ is not possible. For the problem under consideration γ will be specified to be one and then MLE of $p, \lambda, \alpha, \beta$ can be obtained directly (numerical optimization) if individual data are available or using the method of scoring if grouped data is used. Maximum likelihood estimators of such characteristics of income as the Gini coefficient; coefficient of variation or mean income can then be obtained by using the appropriate formulas from Table 1. These estimators are clearly dependent upon the assumptions made about the functional forms for y, y^* and u.

5. SUMMARY

Measurement of income inequality is an important area and has received considerable attention in the literature. The discussion has ranged from the comparative merits of alternative measures to associated problems of estimation, functional forms and statistical inference. The nature of the problems is sensitive to the type of data, grouped or individual observations. Income data may be subject to onesided measurement errors - underreporting. Peterson (1979) has demonstrated that grouped data may be associated with special problems in the presence of inflation. The analysis of these problems has provided the motivation for developing new distribution functions as well as a more thorough analysis of existing distributions. Much progress has been made, but much remains to be done.

REFERENCES

Aigner, D. and Goldberger, A. (1970). Estimation of Pareto's law from grouped observations. *Journal of the American Statistical Association*, 65, 712-23.

Aitchison, J. and Brown, J. A. C. (1969). *The Lognormal Distribution*. Cambridge University Press.

Atkinson, A. B. (1970). On the measurement of inequality. *Journal of Economic Theory*, 2, 244-263.

Cox, D. R. and Hinkley, D. V.(1974). *Theoretical Statistics*. Chapman and Hall, London.

Cramer, J. S. (1980). Errors and disturbances and the performance of some common income distribution functions. University of Amsterdam: Mimeographed Manuscript.

Dorfman, R. (1979). A formula for the Gini coefficient. *Review of Economics and Statistics*, LXI, 147-149.

Easteban, J. and Pastor, A. (1977). A general density function for the distribution of income. University of Barcelona: Mimeographed Manuscript.

Fisk, P. R. (1961). The graduation of income distribution. *Econometrica*, 29, 171-85.

Gastwirth, J. L. (1971). A general definition of the Lorenz Curve. *Econometrica*, 39, 1037-1039.

Gastwirth, J. (1972). The estimation of the Lorenz Curve and Gini index. *Review of Economics and Statistics*, 54, 306-316.

Gastwirth, J. (1974). Large sample theory of some measures of income equality. *Econometrica*, 42, 191-196.

Hartley, M. J. and Revankar, N. S. (1974). On the estimation of the Pareto law from under-reported data. *Journal of Econometrics*, 2, 327-341.

Hoeffding, W. (1948). A class of statistics with asymptotically normal distribution. *Annuals of Mathematical Statistics*, 19, 293-325.

Israelsen, D. L., McDonald, J. B. and Newey, W. K. (1980). The impact of under-reporting on measures of income inequality. Brigham Young University: Mimeographed Manuscript.

Johnson, N. L. and Kotz, S. (1970). *Continuous Univariate Distributions*. Wiley, New York.

Kendall, M. G. and Stuart, A. (1961). *The Advanced Theory of Statistics, Vol. 1*, 2nd edition. Hafner, New York.

Kloek, T. and van Dijk, H. K. (1978). Efficient estimation of income distribution parameters. *Journal of Econometrics*, 8, 61-74.

McDonald, J. B. (1978). The beta as a model for the distribution income. Brigham Young University: Mimeographed Manuscript.

McDonald, J. B. and Jensen, B. C. (1979). An analysis of estimators of alternative measures of income inequality associated with the gamma distribution function. *Journal of the American Statistical Association*, 74, 856-860.

McDonald, J. B. and Ransom, M. (1979a). A note on the bounds for the Gini coefficient: is closer better. Brigham Young University: Mimeographed Manuscript.

McDonald, J. B. and Ransom, M. (1979b). Alternative parameter estimators based upon grouped data. *Communications in Statistics*, A8, 899-917.

McDonald, J. B. and Ransom, M. (1979c). Functional forms, estimation techniques and the distribution of income. *Econometrica*, 47, 1513-1526.

Murray, D. (1978). Extreme values for Gini coefficients calculated from grouped data. *Economic Letters*, 1, 389-392.

Needleman, L. (1978). On the approximation of the Gini coefficient of concentration. *The Manchester School*, 46, 105-122.

Newberry, D. (1970). A theorem on the measurement of inequality. *Journal of Economic Theory*, 2, 264-266.

Peterson, H. (1979). Effects of growing incomes on classified income distributions, the derived Lorenz Curves, and Gini indices. *Econometrica*, 47, 183-198.

Salem, A. B. Z. and Mount, T. D. (1974). A convenient descriptive model of income distribution; the gamma density. *Econometrica*, 42, 1115-1127.

Sen, A. (1973). *On Economic Inequality*. Oxford University Press, Oxford.

Sheshinski, E. (1972). Relation between a social welfare function and the Gini index of income inequality. *Journal of Economic Theory*, 4, 98-100.

Singh, S.K. and Maddala, G. S .(1976). A function for size distribution of incomes. *Econometrica*, 44, 963-970.

Smith, John Terry. (1972). *Some statistical methods for data grouped by quantiles*. Ph.D. Dissertation. Johns Hopkins University.

Theil, H. (1967). *Economics and Information Theory*. Rand McNally, Chicago.

Thurow, L. C. (1970). Analyzing the American income distribution. *American Economic Review*, 60, 216-269.

United States Bureau of the Census: Current Population Reports, Series P-60 and P-20. (1976). U. S. Government Printing Office, Washington D.C.

[*Received May* 1980. *Revised October* 1980]

LORENZ ORDERING WITHIN THE GENERALIZED GAMMA FAMILY OF INCOME DISTRIBUTIONS

CHARLEȘ TAILLIE

Department of Statistics
The Pennsylvania State University
University Park, PA 16802 USA

SUMMARY. Most of the distributional families that have been employed for graduating income data have the property that any two members of the family are comparable with respect to the Lorenz ordering. This includes the gamma, the Pareto, and the lognormal family. The present paper investigates the Lorenz ordering for the generalized gamma family of distributions. The ordering is characterized for this family and is shown to be not linear. As an application, the variance of the logarithm is found to be isotonic with respect to the Lorenz ordering on the generalized gamma family.

KEY WORDS. Lorenz curve, inequality measures, star-shaped ordering, Young's inequality.

1. INTRODUCTION

A variety of distributions have been employed for the purpose of graduating income data. In particular, the Pareto, gamma, and lognormal families have received wide application. Each of these families has the feature that Lorenz curves corresponding to different members of the family do not intersect. In other words, the Lorenz ordering is linear when restricted to any one of these families. This is perhaps not surprising since each of the families in question has only a single parameter (aside from the scale parameter).

With real data, however, it is not uncommon for Lorenz curves to intersect (Atkinson, 1970). In these circumstances, use of any of the above-mentioned families would impose upon the

C. Taillie et al. (eds.), Statistical Distributions in Scientific Work, Vol. 6, 181–192.

fitted curves structural and comparative features not present in the data. This difficulty can be most easily avoided by using a multi-parameter family of distributions. The primary purpose of this paper is to characterize the Lorenz ordering within a particular multi-parameter family known as the generalized gamma family.

The paper opens with a discussion of the Lorenz and related orderings at a fairly abstract level. Readers who are interested primarily in the generalized gamma family may proceed directly to Section 4. Properties of the generalized gamma distribution are briefly reviewed in Section 5 and the main result is given in Section 6. Some applications are given in the concluding section. For example, it is pointed out that the Lorenz curve of any lognormal distribution intersects the Lorenz curve of any gamma or Pareto distribution. Also, the variance of the logarithm is shown to preserve the Lorenz ordering within the generalized gamma family.

2. YOUNG'S INEQUALITY

In 1912, W. H. Young proved a variant of the following result which has come to be known as Young's inequality:

> Let $F(x)$, $x \geqslant 0$, be a continuous and strictly increasing function, vanishing at the origin, and let F^{-1} denote its inverse function. Then for $a, \; b \geqslant 0$,
>
> $$ab \leqslant \int_0^a F(x)\,dx + \int_0^b F^{-1}(y)\,dy \quad , \tag{1}$$
>
> where equality holds if and only if $a = F^{-1}(b)$.

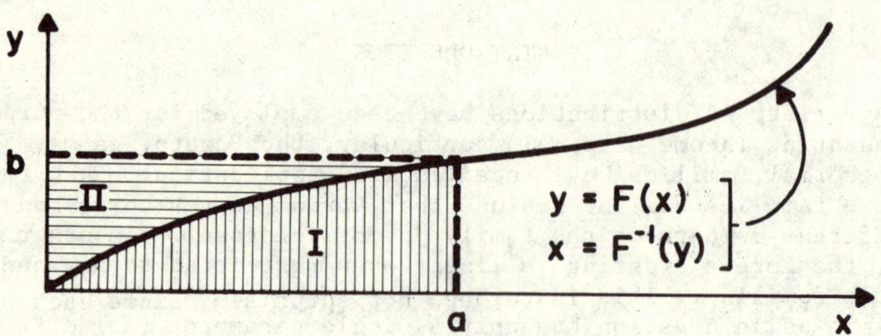

FIG. 1: Geometry of Young's inequality.

Geometric motivation for the inequality is displayed in Figure 1. The two shaded regions labeled I and II have areas equal to the respective integrals appearing on the right hand side of (1); the inequality asserts that these regions combine to give an area at least as large as that of the rectangle ab.

Despite its seeming simplicity, Young's inequality has a number of applications (see Krall, 1973), and some effort has been expended to produce a rigorous proof. The article by Diaz and Metcalf (1970) contains an interesting historical account of these efforts.

For the application that we have in mind, $F(x)$ is the cumulative distribution function of a nonnegative random variable and the above assumptions (continuous, *strictly* increasing, $F(0) = 0$) are unnecessarily restrictive. It is the purpose of this section to give a rigorous, but geometrically appealing, proof of Young's inequality assuming only that $F(x)$ is monotone increasing.

Let $F:[0, \infty) \to [0,\infty]$ be an increasing function, possibly taking infinite values, and define the extended inverse $F^{-1}:[0, \infty) \to [0, \infty]$ by

$$F^{-1}(y) = \sup\{x \geqslant 0 \mid F(x) \leqslant y\}, \qquad 0 \leqslant y \leqslant \infty,$$

where the supremum of the empty set is taken to be zero. The function F^{-1} is increasing and right continuous; its graph is obtained from that of F by reflecting across the 45° line and right continuizing. It is immediate from the definition that: (i) $(F^{-1})^{-1}$ is the right continuous version of F; (ii) all versions (right continuous, left continuous, etc.) of F have the same extended inverse; and (iii) $x \leqslant F^{-1}(y)$ if and only if $F(x-) \leqslant y$, $0 \leqslant x$, $y < \infty$, where $x-$ indicates a left hand limit and $F(0-) \equiv 0$.

Theorem 1. Young's inequality (1) holds for all increasing functions F and is an equality when $a = F^{-1}(b)$.

Proof. In the coordinate plane, consider the two measurable sets

$$\begin{aligned} A_1 &= \{(x, y) \mid 0 \leqslant x \leqslant a, \quad 0 \leqslant y < F(x-)\} \\ &= \{(x, y) \mid 0 \leqslant x \leqslant a, \quad 0 \leqslant y, \ F^{-1}(y) < x\}, \\ A_2 &= \{(x, y) \mid 0 \leqslant y \leqslant b, \quad 0 \leqslant x, \ F(x-) \leqslant y\} \\ &= \{(x, y) \mid 0 \leqslant y \leqslant b, \quad 0 \leqslant x \leqslant F^{-1}(y)\}, \end{aligned}$$

where we have used properties (ii) and (iii). Note that A_1 and

A_2 consist of points below and above the graph of F, respectively. Now A_1 and A_2 are disjoint and their union contains the rectangle $A = [0,a] \times [0,b]$. Using the Tonelli-Fubini theorem to replace double with iterated integrals, there results

$$ab = \iint_A dx\,dy \leq \iint_{A_1} dx\,dy + \iint_{A_2} dx\,dy$$

$$= \int_0^a \{\int_0^{F(x-)} dy\}\,dx + \int_0^b \{\int_0^{F^{-1}(y)} dx\}\,dy$$

$$= \int_0^a F(x-)\,dx + \int_0^b F^{-1}(y)\,dy\,.$$

Young's inequality now follows upon substitution of $F(x)$ for $F(x-)$ in the first integral of the last line. It is easily verified that $A = A_1 \cup A_2$ when $a = F^{-1}(b)$ so that equality holds in this case.

Remark. While we have no need of it, one may show that equality holds if and only if $F^{-1}(b-) \leq a \leq F^{-1}(b)$. From properties (i) and (iii), this last condition is equivalent to $F(a-) \leq b \leq F(a+)$.

Theorem 2. Let F, $G:[0,\infty) \to [0,\infty]$ be increasing functions. The following are equivalent:

A) $\int_0^a F(x)\,dx \leq \int_0^a G(x)\,dx$ for all nonnegative a,

B) $\int_0^b F^{-1}(y)\,dy \geq \int_0^b G^{-1}(y)\,dy$ for all nonnegative b.

Proof. By property (i), we need only show that A implies B. Let $0 \leq b$ and first suppose that $G^{-1}(b)$ is finite. By Young's inequality with $a = G^{-1}(b)$, we have

$$ab \leq \int_0^a F(x)\,dx + \int_0^b F^{-1}(y)\,dy$$

and
$$ab = \int_0^a G(x)\,dx + \int_0^b G^{-1}(y)\,dy < \infty.$$

Statement B follows directly. Next suppose that G^{-1} takes infinite values and put b_0 to $\inf\{b|G^{-1}(b) = \infty\}$. By limiting argument, statement B is true for $b = b_0$. Now $\int_0^b G^{-1}(y)\,dy = \infty$ for $b > b_0$, so that statement B can fail for $b > b_0$ only if $\int_0^b F^{-1}(y)\,dy$ is finite. But this implies that $F^{-1}(b')$ is finite for $b_0 < b' < b$, whence $G(x) \leq b_0 < b' \leq F(x)$ when $F^{-1}(b') < x$. But his contradicts statement A.

3. LORENZ ORDERING

Consider a nonnegative random variable X with distribution function $F_X(x)$, taken to be right continuous. When $E[X] = m$ is finite and nonzero, we write $\tilde{X}$ for the *unitized* variable X/m. Note that $F_{\tilde{X}}(x) = F_X(mx)$ and $F_{\tilde{X}}^{-1}(\cdot) = m^{-1}F_X^{-1}(\cdot)$.

The *Lorenz curve* of X is the graph of the function

$$\begin{aligned} L_X(s) &= m^{-1}\int_0^s F_X^{-1}(v)\,dv \\ &= \int_0^s F_{\tilde{X}}^{-1}(v)\,dv\,, \qquad 0 \leqslant s \leqslant 1. \qquad (2) \end{aligned}$$

The Lorenz curve is undefined when m is infinite. When F is continuous, the change of variable $s = F_X(x)$ yields the usual parametric representation of the Lorenz curve. On the other hand, when X is discrete, say $P(X=x_i) = p_i$ for $0 \leqslant x_1 < x_2 < \cdots < x_n$, the Lorenz curve is the polygonal path joining the points

$$\left(\sum_{j=1}^{i} p_j,\ \sum_{j=1}^{i} p_j x_j \Big/ \sum_{j=1}^{n} p_j x_j\right), \qquad i = 0,1,\cdots,n.$$

We also make the observation that the Lorenz curve is the graph of the cumulative distribution function corresponding to the probability density function $F_{\tilde{X}}^{-1}(v)$, $0 \leqslant v \leqslant 1$.

Definition. Given two nonnegative random variables X and Y with finite nonzero expectations, Y is said to have greater inequality than X (written $X <_L Y$) if the Lorenz curve of Y is everywhere below that of X, that is $L_Y(s) \leqslant L_X(s)$ for all $0 \leqslant s \leqslant 1$.

Theorem 3. The Lorenz order is a partial order: $X <_L Y <_L Z$ implies that $X <_L Z$. Also X and Y have the same Lorenz curve if and only if they have the same distribution apart from a change of scale.

Proof. The first assertion is obvious and the second follows from properties (i) and (ii) of Section 2.

The Lorenz ordering is but one of many partial orderings of probability distributions. Some others are:

Stochastic Ordering (Lehman, 1959).

$$X <_{St} Y \text{ if } 1 - F_X(x) \leqslant 1 - F_Y(x) \text{ for all } x.$$

Second Order Stochastic Ordering (Rolski, 1976).

$$X <_{St2} Y \text{ if } \int_x^\infty [1-F_X(t)]dt \leqslant \int_x^\infty [1-F_Y(t)]dt < \infty \text{ for all } x.$$

Our next theorem relates these various orderings. The theorem is not new; the equivalence of a) and e) was demonstrated by Atkinson (1970), at least for absolutely continuous distributions with the same expected value, while the equivalence of e) and f) has been noted by many authors, usually with unnecessary assumptions (see Atkinson, 1970, for references to the economics literature and Rolski, 1976, for references to the statistics literature).

Theorem 4. Let X and Y be nonnegative random variables with finite nonzero expectations. The following are equivalent:

a) $X <_L Y$. b) $\tilde{X} <_L \tilde{Y}$.

c) The distribution with density function $F_{\tilde{X}}^{-1}(v)$, $0 \leqslant v \leqslant 1$, is stochastically less than the distribution with density function $F_{\tilde{Y}}^{-1}(v)$, $0 \leqslant v \leqslant 1$.

d) The distribution with density function $1 - F_{\tilde{X}}$ is stochastically less than the distribution with density function $1 - F_{\tilde{Y}}$.

e) $\tilde{X} <_{St2} \tilde{Y}$.

f) $E[\Psi(\tilde{X})] \leqslant E[\Psi(\tilde{Y})]$ whenever $\Psi(x)$, $0 \leqslant x \leqslant \infty$, is a convex function for which the two expectations exist and are finite.

Proof. The equivalence of a), b) and c) is immediate from (2) while the equivalence of c) and d) follows from Theorem 2. Part 3) is merely a restatement of d). Next we show that d) implies f). There are two complications since the convex function Ψ need not be continuous at the origin and $\Psi(0+)$ may be infinite. First we suppose that $\Psi(0+)$ is finite. For positive x, we may write

$$\Psi(x) = \int_1^x \psi(t)\, dt + \text{constant},$$

where $\Psi(t)$, $0 < t < \infty$, is increasing. The additive constant may be ignored without loss of generality. Let $0 < \varepsilon < 1$. Since ψ is bounded from below on the interval (ε, ∞), an interchange in the order of integration is justified, giving

$$\int_{\varepsilon+}^{\infty} \Psi(x)\ dF_{\tilde{X}}(x) = \int_{\varepsilon+}^{\infty} \int_{1}^{x} \psi(t)\ dt\ dF_{\tilde{X}}(x)$$

$$= -\int_{\varepsilon}^{1} \psi(t) \int_{\varepsilon+}^{t} dF_{\tilde{X}}(x)\ dt + \int_{1}^{\infty} \psi(t) \int_{t}^{\infty} dF_{\tilde{X}}(x)\ dt$$

$$= -\int_{\varepsilon}^{1} \psi(t)[F_{\tilde{X}}(t) - F_{\tilde{X}}(\varepsilon)]dt + \int_{1}^{\infty} \psi(t)[1 - F_{\tilde{X}}(t)]dt$$

$$= -\int_{\varepsilon}^{1} \psi(t)[1 - F_{\tilde{X}}(\varepsilon)]dt + \int_{\varepsilon}^{\infty} \psi(t)[1 - F_{\tilde{X}}(t)]dt$$

$$= \Psi(\varepsilon)[1 - F_{\tilde{X}}(\varepsilon)] + \int_{\varepsilon}^{\infty} \psi(t)[1 - F_{\tilde{X}}(t)]dt.$$

Adding $\Psi(0)F_{\tilde{X}}(0) = \Psi(0)P(\tilde{X} = 0)$ to both sides and letting $\varepsilon \to 0$, we obtain

$$E[\Psi(\tilde{X})] = \Psi(0+) + [\Psi(0) - \Psi(0+)]F_{\tilde{X}}(0) + \int_{0}^{\infty} \psi(t)[1 - F_{\tilde{X}}(t)]dt. \tag{3}$$

This expression is to be compared with the corresponding expression for $E[\Psi(\tilde{Y})]$. The second term is greater for $\tilde{Y}$ since $F_{\tilde{X}}(0) \leqslant F_{\tilde{Y}}(0)$ and $\Psi(0) - \Psi(0+) \geqslant 0$ (Ψ is convex). Also ψ is monotone increasing and, by a well-known property of stochastic ordering (Lehman, 1959, p. 73), this implies that the last term is greater for $\tilde{Y}$. Next suppose $\Psi(0+) = \infty$. (Note that $\Psi(0+) = -\infty$ is impossible by convexity.) Let $0 \leqslant \varepsilon \leqslant 1$ and define

$$\Psi_{\varepsilon}(x) = \begin{cases} \Psi(\varepsilon) + M_{\varepsilon}(x-\varepsilon) & \text{if } 0 \leqslant x \leqslant \varepsilon \\ \Psi(x) & \text{if } \varepsilon \leqslant x < \infty, \end{cases}$$

where M_{ε} is the right hand derivative of $\Psi(x)$ at $x = \varepsilon$. The function Ψ_{ε} is convex with a finite value at the origin, so

$$E[\Psi_{\varepsilon}(\tilde{X})] \leqslant E[\Psi_{\varepsilon}(\tilde{Y})]. \tag{4}$$

When ε goes to zero, Ψ_{ε} converges pointwise to Ψ and, by the Legesgue dominated convergence theorem, the limit may be passed under the expectation in (4) to give the desired conclusion. Finally, we show that f) implies e). In fact this follows by inserting the convex function $\Psi(x) = (x-a)^{+}$ into (3).

Corresponding to any convex function Ψ we may define a measure of inequality H_{Ψ} by $H_{\psi}(X) = E[\Psi(\tilde{X})] = E[\Psi(X/m)]$. These measures are scale-free (a basic property of inequality measures) and, according to Theorem 4, they also preserve the Lorenz ordering: $X <_L Y$ implies that $H_\Psi(X) \leq H_\Psi(Y)$. Gastwirth (1975) has also attempted to associate inequality measures with convex functions. In general, however, his measures are not scale-free and do not preserve the Lorenz ordering.

Although not strictly necessary, it is usually convenient to take $\Psi(1) = 0$ so that degenerate (completely equal) distributions have inequality measure zero.

Corresponding to the convex functions $\Psi(x) = (x^{t+1} - 1)/[t(t+1)]$, $-\infty < t < \infty$, are the indices

$$H_t(X) = [E(\tilde{X}^{t+1}) -1]/[t(t+1)].$$

This rich family of indices will be used in Section 4. As special cases, we mention that H_1 is one-half the squared coefficient of variation (Herfindal index), $H_0 = E[\tilde{X} \log \tilde{X}]$ is the Theil index, while $H_{-1} = \log(AM/GM)$ and $H_{-2} = (1/2)\ AM/HM$, where AM, GM, HM are the arithmetic, geometric, and harmonic means, respectively.

4. STAR-SHAPED ORDERING

In practice, Theorem 4 is often difficult to apply because of the requirement that the variables first be unitized. In this section we consider another ordering of nonnegative random variables, one that is analytically more flexible than the Lorenz ordering.

Definition. Let X and Y be nonnegative random variables, not identically zero. Then Y has a star-shaped distribution with respect to X (written $X <_* Y$) if $F_Y^{-1}(v)/F_X^{-1}(v)$ is an increasing function of v for $0 < v < 1$.

It is immediate that the star-shaped ordering is a partial order and that $X <_* Y <_* X$ if and only if X and Y have the same distribution apart from a change of scale. Note that the star-shaped ordering is meaningful even when the variables have infinite expectations.

Theorem 5. Let X and Y have finite nonzero expectations. Then

$$X <_* Y \text{ implies } X <_L Y.$$

Proof. Assume that $X <_* Y$ so that $\tilde{X} <_* \tilde{Y}$ and $F_{\tilde{Y}}^{-1}(v)/F_{\tilde{X}}^{-1}(v)$ is increasing for $0 < v < 1$. Thus the density function $F_{\tilde{Y}}^{-1}$ has monotone likelihood ratio with respect to the density function $F_{\tilde{X}}^{-1}$ and this is known (Lehmann, 1959, p. 74) to imply statement c) of Theorem 4.

The converse of Theorem 5 is false. For a counter-example let X take the values 2, 3, 5 with equal probability and let Y take the values 1, 4, 5 with equal probability. By the principle of transfers, $X <_L Y$. On the other hand, F_Y^{-1}/F_X^{-1} takes the successive values 1/2, 4/3, 5/5 so that X and Y are not comparable in the star-shaped sense.

Remark. The star-shaped ordering is discussed in Barlow and Proschan (1966, 1975) in the context of reliability theory. They considered only continuous positive random variables and defined $X <_* Y$ if $F_Y^{-1}[F_X(t)]/t$ is increasing for $t > 0$. Making the substitution $v = F_X(t)$, this reduces to our definition when F_X is continuous (but not otherwise).

The following gives an interesting property of the star-shaped ordering. In general this property does not hold true for the Lorenz ordering.

Theorem 6. If $X <_* Y$ then $1/X <_* 1/Y$.

5. GENERALIZED GAMMA FAMILY

Consider a random variable U having a gamma distribution with pdf

$$u^{k-1}\exp(-u)/\Gamma(k), \quad u > 0.$$

Let γ be a non-zero real number. Then $X=U^{1/\gamma}$ is said to have a generalized gamma distribution with parameters k and γ. The pdf is

$$|\gamma|\ x^{k\gamma-1}\exp(x^{\gamma})/\Gamma(k), \quad x > 0,$$

where $k > 0$ and $\gamma \neq 0$. The generalized gamma family has been discussed by Stacey (1962) and, indirectly, by Ferguson (1962).

Included as special cases are the gamma distribution (γ=1), the reciprocal gamma distribution (γ=-1), and the Weibull and reciprocal Weibull families (k=1). Some limiting forms are:

Power function: $k \to 0$, $\gamma \to \infty$, $k\gamma \to$ const > 0.

Pareto: $k \to 0$, $\gamma \to -\infty$, $k\gamma \to$ const < 0.

Lognormal: $k \to \infty$ $\gamma \to 0$, $k\gamma^2 \to 1/\sigma^2$.

The generalized gamma distribution has its moments given by $E[X^t] = \Gamma(k+t/\gamma)/\Gamma(k)$, $k + t/\gamma > 0$. In particular,

$$1 + t(t+1)H_t(X) = \Gamma(k+(t+1)/\gamma)\Gamma(k)^t/\Gamma(k+1/\gamma)^{t+1}, \quad k+(t+1)/\gamma>0,$$

where the inequality measure $H_t(X)$ is defined in Section 3.

6. MAIN RESULT

The following theorem characterizes the Lorenz ordering within the generalized gamma family when $\gamma > 0$. The ordering is displayed in Figure 2.

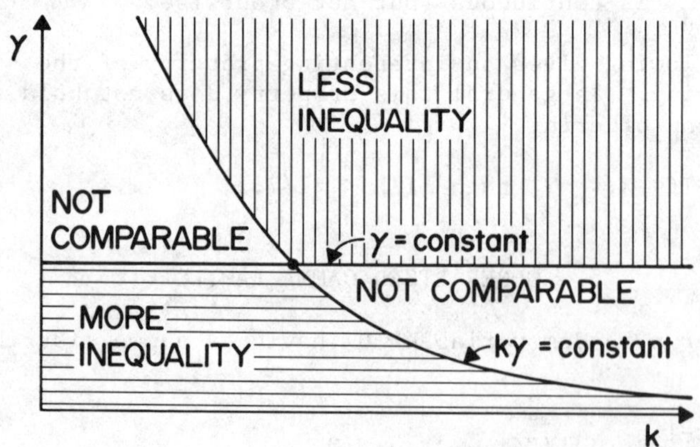

FIG. 2: Lorenz ordering for the generalized gamma family with positive γ.

Theorem 7. Let X and Y have generalized gamma distributions with respective parameters (k,γ) and $(\bar{k},\bar{\gamma})$ where γ and $\bar{\gamma}$ are both positive. The following are equivalent: a) $X <_* Y$; b) $X <_L Y$; c) $\gamma \geq \bar{\gamma}$ and $k\gamma \geq \overline{k\gamma}$.

Proof. It follows from Theorem 5 that a) implies b). Next we show that b) implies c). Suppose then that $X <_L Y$ so that $H_t(X) \leq H_t(Y)$ for all t. But $H_t(X)$ becomes infinite at

$t = -k\gamma-1$ while $H_t(Y)$ becomes infinite at $t = -\bar{k}\bar{\gamma}-1$. Thus $-k\gamma-1 \leqslant -\bar{k}\bar{\gamma}-1$ and $k\gamma \geqslant \bar{k}\bar{\gamma}$. In order to show that $\gamma \geqslant \bar{\gamma}$, use the fact that $H_t(Y)/H_t(X) \geqslant 1$ and let $t \to \infty$ while applying Stirling's formula. We do not give the details of the proof that c) implies a). The idea is to show that $F_Y^{-1}F_X$ is convex employing painstaking arguments similar to those of Van Zwet (1964).

At present we do not know how to compare two generalized gamma distributions whose γ parameters have opposite signs. But when both are negative, Theorem 6 may be applied to conclude that Theorem 7 remains valid provided γ and $\bar{\gamma}$ are replaced by their absolute values in part c).

7. APPLICATIONS

The main result allows one to quickly compare any two generalized gamma distributions and their limiting forms. As an illustration we prove the following:

Theorem 8. Let X have a lognormal distribution with $\mathrm{var}(\log X) = \sigma^2$ and let Y have a generalized gamma distribution with parameters (k,γ) where $\gamma \neq 0$. Then X and Y are not comparable in the star-shaped (and hence not in the Lorenz) sense.

Proof. For simplicity take $\gamma > 0$. Now X can be approximated as closely as we like by a generalized gamma distribution with parameters $(\bar{k},\bar{\gamma})$ where $\bar{k}$ is indefinitely large, $\bar{\gamma}$ is indefinitely small, and $\bar{k}\bar{\gamma}^2 = \sigma^2$. In particular, we may take $\bar{\gamma} < \gamma$. Since $\bar{\gamma}$ is indefinitely small, $\bar{k}\bar{\gamma}$ is indefinitely large so that $\bar{k}\bar{\gamma} > k\gamma$. Thus, X and Y are not comparable.

It is known that the variance of the logarithm does not preserve the Lorenz ordering for distributions in general (e.g., Hart, 1975). Our final result asserts that it is order-preserving within the generalized gamma family.

Theorem 9. Let X and Y have generalized gamma distributions with respective parameters (k,γ) and $(\bar{k},\bar{\gamma})$ where γ and $\bar{\gamma}$ have the same sign. If $X <_* Y$ (or $X <_L Y$) then $\mathrm{var}(\log X) \leqslant \mathrm{var}(\log Y)$.

Proof. For the generalized gamma family, the variance of the logarithm is given by $\mathrm{var}(\log X) = \psi'(k)/\gamma^2$, where ψ' is the trigamma function (Ferguson, 1962). For simplicity take γ and $\bar{\gamma}$ both positive. We proceed from (k,γ) to $(\bar{k},\bar{\gamma})$ in two steps, first holding $k\gamma$ constant while increasing γ and then holding

γ constant while decreasing k (refer to Figure 2). For the first step

$$\psi'(k)/\gamma^2 = k^2\psi'(k)/k^2\gamma^2.$$

Here the denominator is constant while the numerator increases (cf. formula 6.4.10 of Abramowitz and Stegun, 1972). For the second step, $\psi'(k)/\gamma^2$ increases since ψ' is a decreasing function.

REFERENCES

Abramowitz, M. and Stegun, I. (1972). *Handbook of Matehmatical Functions*. Dover, New York.

Atkinson, A. B. (1970). On the measurement of inequality. *Journal of Economic Theory*, 2, 244-263.

Barlow, R. E. and Proschan, F. (1966). Inequalities for linear combinations of order statistics from restricted families. *Annals of Mathematical Statistics*, 37, 1574-1592.

Barlow, R. E. and Proschan, F. (1975). *Statistical Theory of Reliability and Life Testing*. Holt, Rinehart, and Winston, New York.

Diaz, J. B. and Metcalf, F. T. (1970). An analytic proof of Young's inequality. *American Mathematical Monthly*, 70, 603-609.

Ferguson, T. S. (1962). Location and scale parameters in exponential families of distributions. *Annals of Mathematical Statistics*, 33, 986-1001.

Gastwirth, J. L. (1975). The estimation of a family of measures of economic inequality. *Journal of Economic Theory*, 3, 61-70.

Hart, P. E. (1975). Moment distributions in economics: An exposition. *Journal of the Royal Statistical Society, Series A*, 138, 423-434.

Krall, A. M. (1973). *Linear Methods of Applied Analysis*. Addison-Wesley, Reading, Massachusetts.

Lehmann, E. L. (1959). *Testing Statistical Hypotheses*. Wiley, New York.

Rolski, T. (1976). Order relations in the set of probability distribution functions. *Dissertationes Mathematicae*, no. 132.

Stacey, E. W. (1962). A generalization of the gamma distribution. *Annals of Mathematical Statistics*, 33, 1187-1192.

Van Zwet, W. R. (1964). *Convex Transformations of Random Variables*. Mathematical Center Tracts, Amsterdam.

Young, W. H. (1912). On classes of summable functions and their Fourier series. *Proceedings of the Royal Society, Series A*, 87, 225-229.

[*Received June* 1980. *Revised September* 1980]

THE CHOICE OF A DISTRIBUTION TO DESCRIBE PERSONAL INCOMES

J. K. ORD*

University of Warwick, England

G. P. PATIL and C. TAILLIE

Department of Statistics
The Pennsylvania State University
University Park, Pennsylvania 16802 USA

SUMMARY. If the average level of income and the extent of income inequality are specified then an entropy-maximizing distribution can be determined subject to these constraints. This provides a natural link between particular distributions and inequality measures used to describe incomes data. The links between the present inequality measures and previous proposals are indicated.

KEY WORDS. entropy, entropy maximizing distribution, Pareto distribution, inverse Gaussian distribution, income inequality, Lorenz ordering, reciprocal gamma distribution.

1. INTRODUCTION

A variety of positively skewed distributions have been used to describe personal incomes; these include the lognormal (Aitchison and Brown, 1957), Pareto (1897), Champernowne (1973), and, more recently, the gamma (Salem and Mount, 1974). The general shape of these distributions is often very similar in the central region, making it difficult to discriminate between them using published data, which is grouped into very wide classes in the upper tail. As a result, several authors have proposed stochastic

* Present Address: Department of Management Science, The Pennsylvania State University, University Park, Pennsylvania 16802 USA.

C. Taillie et al. (eds.), Statistical Distributions in Scientific Work, Vol. 6, 193–201.

models which give rise to one or other of the distributions mentioned (see Ord, 1975, and Steindl, 1965, for further details). The assumptions underlying these models are relatively simple and while this enables the familiar distributions to be generated, the lack of realism has been criticized by other writers (Stiglitz, 1969). In this paper we look at entropy-maximizing distributions, where the entropy is maximized subject to stipulated levels of income and income equality. In this way, particular distributions and measures are found to have a natural pairing. While no final conclusions can be drawn about the relative usefulness of various distributions as descriptors of income pattern, we believe that this approach does provide insights into the modelling and analysis of income data that would not otherwise be obvious through traditional model building or model fitting.

2. INEQUALITY AND ENTROPY MAXIMIZATION

Let the random variable X describe personal incomes and let X have density function $f(x)$, $c_L<x<\infty$, and distribution function $F(x) = P(X\leq x)$. We take the distribution to be absolutely continuous and unbounded from above since this assumption is considered to be a reasonable approximation in studies of this kind. However, only incomes above the lower bound c_L are recorded. In general the probability of reporting is a function of past income levels, but this cannot be handled in a static analysis. Nevertheless, fairly complete reporting may be assumed above some threshold. In the present study we assume that the data are fully and consistently recorded for all $X>c_L$.

The social welfare associated with a distribution F is $W(F) = \int U(x)\, dF(x)$ where $U(x)$ is the utility derived from an income x. With $\mu=E(X)$ fixed and assuming that $U(x)$ is strictly concave, welfare would be maximized by the perfectly equal distribution of income for which $P(X=\mu) = 1$. However, this situation is not achievable, since certain income differentials must be maintained to provide incentives. Also, the presence of some degree of inequality allows mobility of incomes, and subsequent changes in the rank order of rewards for particular jobs, to ensure that the more demanding jobs are performed. Mobility cannot be modelled in the static framework and this, in turn, weakens the social welfare argument for a static optimum at complete equality. It is interesting to note that Pareto's (1897) original analysis recognized that political unrest can follow from too extreme a distribution of income, be it too equal or too unequal. Thus, we might argue that society, by its collective actions, determines both the average level of income to be enjoyed and the degree of inequality which it regards as necessary.

Within those constraints, the distribution of income may be assumed to settle to a steady form. This does *not* mean individual incomes remain fixed, but rather that there is an overall stochastic equilibrium describing the pattern of income at any point in time. Thus, the entropy-maximizing distributions derived below may be regarded as limiting distributions which are approached once the long-term income level and degree of inequality are agreed.

2.1 The Notion of Entropy. In his review of entropy, Kendall (1973) quotes Boltzmann's original statement concerning the second law of thermodynamics:

> The theorem that the negative entropy decreases through molecular collisons amounts to this: that the distribution of molecular velocities in a gas occasioned by collisions approaches the most probable.

In its many variations the second law has been interpreted as saying that the energy for change within a system tends to be dissipated. Therefore, we shall examine the distribution of income in the framework of maximizing entropy subject to the conditions already imposed, recognizing that the evolutionary nature of income patterns over time means that the system is unlikely to settle completely.

The most commonly used class of entropy measures may be defined (Rao, 1965, p. 142) as

$$e_\gamma = \int f(x)\,|1-f^\gamma(x)|\;dx/\gamma, \qquad -1<\gamma<\infty, \tag{1}$$

although the Shannon measure

$$e_0 = -\int f(x)\,\log f(x)dx \tag{2}$$

has attracted most attention and it is the negative of e_0 to which Boltzmann's quotation refers. Thus, with H as the inequality measure, our general problem may be expressed as

$$\begin{aligned} &\text{maximize:} && e_\gamma \\ &\text{subject to:} && \int f(x)dx = 1, \quad f(x)\geq 0, \\ & && E(X) = \mu, \quad \text{and } H = \text{constant.} \end{aligned}$$

The problem may be further generalized by using other measures of income level in place of $E(X)$; for example the minimum possible income might be specified. This formulation of the problem enables us to characterize particular distributions. The methods follow those of Rao (1965, pp. 140-143) and the details are omitted; the results are given in Table 1. The income inequality measure may be defined as any scale-free function of the statistics specified in the constraints - preferably one preserving the Lorenz ordering. The second column of the table indicates possible choices.

It follows quite generally (Rao, 1965, pp. 141-142) that when e_0 is maximized, subject to the linearly independent constraints

$$E\{h_j(X)\} = \beta_j', \quad j=1,\cdots,k,$$

the resulting distribution has density function proportional to $\exp\{\Sigma\, \beta_j\, h_j(x)\}$ where the parameters $\{\beta_j\}$ depend upon the $\{\beta_j'\}$. This is a k-parameter version of the exponential family so it follows directly that, given a sample of n independent observations, $\underset{\sim}{x}'=(x_1,\cdots,x_n)$, the sufficient statistics are

$$\sum_{i=1}^{n} h_j(x_i), \quad j=1,\cdots,k.$$

Thus, in a sense, we have come full circle. Once the constraints are specified, we have a unique entropy maximizing distribution, while the sample functions corresponding to the expressions in the constraints are sufficient statistics for the unknown parameters.

Of the distributions mentioned in the Introduction, the results in Table 1 offer some justification for the gamma and lognormal laws, but we have found no plausible constraints to generate the Champernowne form. Numerous authors have pointed out that the inequality measure associated with the lognormal does not in general preserve the Lorenz ordering. However, it has been shown (Creedy, 1977) that actual violations of the ordering are likely to be rare in the context of income distributions. Also, it will be shown on another occasion (see Taillie, 1981) that the variance of the logarithm does respect the Lorenz ordering within each of several familiar classes of distributions including the gamma, reciprocal gamma, Weibull, half-normal, lognormal, and Pareto; this is demonstrated for the lognormal in Hart (1975).

At first sight, Table 1 would also appear to lend support to the Pareto law. The necessary constraint of a positive lower

TABLE 1. Distributions resulting from maximizing e_0 *subject to certain constraints* (c_L = *lower bound*, AM = *arithmetic mean*, GM = *geometric mean*, HM = *harmonic mean*).

Constraints	Inequality Measure	Distribution
GM and $c_L > 0$	GM/c_L	Pareto
AM and GM	AM/GM	gamma
AM and variance	$\text{variance}/(AM)^2$	normal-truncated to (c_L,∞)
GM and variance of log X	variance of log X(1)	lognormal
HM and GM	GM/HM(1)	reciprocal gamma
AM and HM	AM/HM	$f(x) \propto \exp\{-\beta_1 x-\beta_2 x^{-1}\}$(2)
AM and E(X log X) or GM_1(3)	$GM_1-\log(AM)$	$f(x) \propto \exp\{\beta_1 x-\beta_2 x \log x\}$

(1) These inequality measures preserve the Lorenz ordering within the corresponding entropy maximizing family but not in general.

(2) This is suggestive of, but not equal to, the inverse Gaussian distribution, see Folks and Chhikara (1978). The inverse Gaussian also has AM and HM as sufficient statistics and has been used to describe labor turnover (Whitmore, 1979), but not so far as we are aware, for incomes.

(3) GM_1 denotes the geometric mean of the first moment distribution with density function $f_1(x) = xf(x)/E(X)$, $x>0$, so that $GM_1 = E(X \log X)/E(X)$.

bound is in direct opposition* to entropy maximization, though, and could not be maintained by natural forces, requiring instead some sort of government or societal intervention. The Pareto law thus represents a highly unstable income pattern, at least within the poorer segment of the population. It has been suggested that actual distributions do approximate the Paretian forms at high income levels. To the extent that this is true, the reciprocal gamma, with its Pareto-like upper tail and more reasonable lower tail, may represent a more attractive choice.

Further, it has been shown for U. S. data (Gastwirth, 1972) that, even in the upper tail, the performance of the Pareto distribution has been deteriorating in recent years, suggesting that empirical support may also be waning.

Thus far we have considered only the Shannon entropy e_0. With the same constraints as above, Rao (1965, p. 143) has shown that the distribution maximizing e_γ, $\gamma \neq 0$, has a density of the form

$$f(x) = \{\beta_0 + \beta_1 h_1(x) + \cdots + \beta_k h_k(x)\}^{1/\gamma}.$$

A translated form of the Pareto distribution results from an AM constraint when $-1/2 < \gamma < 0$. The only way in which the Champernowne distribution can be generated is to consider the random variable $Y = X^\alpha$ and then take $h_1(y) = y$ when $\gamma = -1/2$. This seems to be stretching the credibility of the method. No other familiar or plausible distributions seem to be available for $\gamma \neq 0$.

3. ENTROPY AS A MEASURE OF INEQUALITY

The continuous entropy measures defined in equations (1) and (2) are not suitable indices of inequality since they lack the basic requirement of scale invariance. Moreover, it follows directly from the definition (1) that e_γ is translation invariant. It is apparent that any inequality measure should decrease under a forward translation of the income distribution since the Lorenz curve for $X+a$, $a>0$, is a convex linear combination of the Lorenz curve of X and the 45-degree line, so that the level of inequality is always reduced. See Kakwani (1977) for a fuller discussion of results on ordering Lorenz curves.

*Formally, the differential effect upon e_0 of an infinitesmal transfer into the forbidden region is a plus infinity.

Although continuous entropy is unsatisfactory, Renyi's discrete entropy (see Hart, 1975) has been used to generate inequality measures, according to the following argument.

Suppose that n people receive incomes $x_1, \cdots, x_n$ and let the *ith* person have share $y_i = x_i/\Sigma x_i$, where the sum is over $i=1,\cdots,n$. Then the Renyi entropy measure is e^*_γ where

$$\gamma e^*_\gamma = -\log(\Sigma y_i^{\gamma+1}) \tag{3}$$
$$= -\log(\Sigma x_i^{\gamma+1}) + (\gamma+1)\log(\Sigma x_i).$$

It is readily shown that e^*_γ is a simple transformation of e_γ but defined for the first moment distribution of X. If we write $\alpha_\gamma = \Sigma x_i^\gamma/n$, it follows that

$$\gamma e^*_\gamma = -\log(\alpha_{\gamma+1}/\alpha_1^{\gamma+1}) + \gamma \log n$$

and we see that e^*_γ is dependent upon n, an unsatisfactory state of affairs. If we use

$$\gamma e^{**}_\gamma = -\log[\Sigma(x_i/\bar{x})^{\gamma+1}]$$
$$= -\log(\alpha_{\gamma+1}/\alpha_1^{\gamma+1}) \tag{4}$$

the difficulty disappears. This is precisely the suggestion made by Hart (1975) and (4) can be written in terms of expected values as

$$e^{**}_\gamma = -\frac{1}{\gamma}\log\,[E(X^{\gamma+1})/\{E(X)\}^{\gamma+1}]. \tag{5}$$

Transformations and constants apart, it can be seen that many of the entries in Table 1 are of this form.

The exact function of the sufficient statistics used to measure inequality is a matter of taste, but it is clear from the earlier discussion that the function must be scale-invariant. The original entropy measures derived from (3) do not meet that criterion nor do those based upon e^*_γ using the first moment distribution. The authors concur with the conclusion reached in Hart (1975) that entropy does not provide suitable measures of income inequality.

4. CONCLUSIONS

Since empirical tests of fit of alternative distributions to income data may be rather weak, we have examined the possibility of linking particular distributions and measures of inequality by use of entropy. The distributions in common use are found to link naturally with particular measures of inequality and this may serve to cast light upon the choice of both inequality measure and descriptive distribution.

The authors are grateful to Professor P. E. Hart and the referees for helpful comments on an earlier version of this paper.

REFERENCES

Aitchison, J. and Brown, J. A. C. (1957). *The Lognormal Distribution*. Cambridge University Press.

Champernowne, D. G. (1973). *The Distribution of Income*. Cambridge University Press.

Creedy, J. (1977). The principle of transfers and the variance of logarithms. *Oxford Bulletin of Economics and Statistics*, 39, 152-158.

Folks, J. L. and Chhikara, R. S. (1978). The inverse Gaussian distribution and its statistical applications. *Journal of the Royal Statistical Society, Series B*, 40, 263-289.

Gastwirth, J. L. (1972). The estimation of the Lorenz curve and the Gini index. *Review of Economics and Statistics*, 54, 306-316.

Hart, P. E. (1975). Moment distributions in economics: an exposition. *Journal of the Royal Statistical Society, Series A*, 138, 423-434.

Kakwani, N. C. (1977). Applications of Lorenz curves in economic analysis. *Econometrica*, 45, 719-727.

Kendall, M. G. (1973). Entropy, probability and information. *International Statistical Review*, 41, 59-68.

Ord, J. K. (1975). Statistical models for personal income distributions. In *Statistical Distributions in Scientific Work, Vol. 2*. G. P. Patil, S. Kotz and J. K. Ord, eds. Reidel, Dordrecht-Holland. Pages 151-158.

Pareto, V. (1897). *Cours d'Economie Politique*. Lausanne.

Rao, C. R. (1965). *Linear Statistical Inference and its Applications*. Wiley, New York.

Salem, A. B. Z. and Mount, T. D. (1974). A convenient descriptive model of income distribution: the gamma density. *Econometrica*, 42, 1115-1127.

Steindl, J. (1965). *Random Processes and the Growth of Firms: A Study of the Pareto Law*. Griffin, London.

Stiglitz, J. E. (1969). Distribution of income and wealth among individuals. *Econometrica*, 37, 383-397.

Taillie, C. (1981). Lorenz ordering within the generalized gamma family of income distributions. In *Statistical Distributions in Scientific Work,* C. Taillie, G. P. Patil and B. Balderssari, eds. Reidel, Dordrecht-Holland. Vol. 6, 181-192.

Theil, H. (1967). *Economics and Information Theory.* North Holland, Amsterdam.

Whitmore, G. A. (1979). An inverse Gaussian model for labour turnover. *Journal of the Royal Statistical Society, Series A,* 142, 468-478.

[*Received July* 1980. *Revised November* 1980]

RELATIONSHIPS BETWEEN INCOME DISTRIBUTIONS FOR INDIVIDUALS AND FOR HOUSEHOLDS

J. K. ORD*

University of Warwick, England

G. P. PATIL and C. TAILLIE

Department of Statistics
The Pennsylvania State University
University Park, Pennsylvania 16802 USA

SUMMARY. Comparisons of income may be based upon distributions either for individuals or for households. The conditions under which these two distributions are the same (apart from a scale factor) are examined. In particular, we find that the condition is met when the household size follows a negative binomial distribution only if income is gamma distributed.

KEY WORDS. aggregation, binomial, gamma, income distribution, martingale, negative binomial, Poisson, stable distribution.

1. INTRODUCTION

Surveys of personal income have long been used to assess the extent of inequality (cf. Ord, 1975; Hart, 1975). However, a variety of problems arise, practical and theoretical, when we try to implement such an analysis and interpret the results. Three such problems are as follows: (i) Should we use a particular distribution to represent the pattern of incomes and, if so, does this carry any implications concerning the choice of inequality measure? (ii) If data are available only for incomes above some threshhold level (e.g. minimum income for tax liability)

---
* Present Address: Department of Management Science, The Pennsylvania State University, University Park, Pennsylvania 16802 USA.

C. Taillie et al. (eds.), Statistical Distributions in Scientific Work, Vol. 6, 203-210.

can the data still be used to estimate inequality in the population at large? (iii) If data are collected for individuals, can we say anything about inequality of incomes among households, or vice-versa?

In Ord, Patil and Taillie (1981), the first problem was examined using entropy arguments when natural pairings of various distributions and inequality measures arise, e.g. the gamma and the ratio of the geometric mean to the arithmetic mean. In Ord, Patil and Taillie (1980) the second problem was considered and it was shown that truncation critically affects the inequality measure unless the underlying distribution is the Pareto. The purpose of the present paper is to examine the third question listed above.

Surveys of individual incomes may record data either for complete households or for individuals (heads of households?). The data from each kind of survey may then be used to assess the extent of inequality between incomes. The question which such assessments prompt is whether inequality measures evaluated from one type of data give a representation of inequality for the other. We shall consider this problem in the following way: Let X denote the head of household income and U the combined income for the other family members so that total income is

$$Y = X + U \tag{1}$$

where we assume that X and U are independent. Under what conditions may we say that the distributions of Y and X are of the same form, apart from a scale change, so that inequality measures based on data for one of these variables are indicative of the inequality pattern for the other? The general question remains open, but the results given below provide a partial answer and demonstrate the severe restrictions which must be imposed before X and Y can be treated more or less interchangeably. We assume, of course, that U is not trivial; otherwise, the problem is trivial.

In Section 2 we demonstrate that once the distribution of U is specified then there exists one and only one distribution for X for which Y and a multiple of X are equal in distribution. When the scaling constant and the distribution of X are specified, at most one distribution for U exists.

In Section 3, we assume that the rest of the household consists of N members where N is a random variable defined on the non-negative integers. By restricting the distribution for the incomes of household members to be a "similar" form to that of X we find that a negative binomial distribution for N and a gamma distribution for Y satisfy (1).

2. EQUIVALENCE OF INDIVIDUAL AND HOUSEHOLD INCOME DISTRIBUTIONS

We assume that $Y = X + U$ as before and look at the distributions of Y and βX where β is a scale shift parameter ($\beta > 1$ from the context of the problem). We now show that a fixed β and a specified distribution for U uniquely determine the distribution of X. The theorem could be proved by more elementary methods, but we prefer a proof based on martingales. The reader is referred to Heyde (1972) for a survey of the theory and application of martingales.

Theorem 1. Suppose U has a specified distribution with a finite non-zero mean and let $\beta > 1$ be a given real number. There exists one and only one distribution X for which Y and βX are equal in distribution. In this case, $E(X) = E(U)/(\beta-1)$ is finite so that X is uniquely determined by U and $E(X)$.

Proof. (Uniqueness). Let $U_1, U_2, U_3, \cdots$ be independent realizations of U and write $\overset{d}{=}$ for equality in distribution. From (1),

$$X \overset{d}{=} \beta^{-1}X + \beta^{-1}U_1 \overset{d}{=} \beta^{-1}(\beta^{-1}X + \beta^{-1}U_2) + \beta^{-1}U_1$$

$$\overset{d}{=} \beta^{-2}X + \beta^{-1}U_1 + \beta^{-2}U_2.$$

Iterating this argument, we find that for $n = 1,2,3,\cdots$

$$X \overset{d}{=} X/\beta^n + \sum_{i=1}^{n} U_i/\beta^i.$$

Letting $n \to \infty$, the first term X/β^n converges to zero with probability one since $\beta > 1$. Thus

$$X \overset{d}{=} \sum_{i=1}^{\infty} U_i/\beta^i, \tag{2}$$

which implies that the distribution of X is determined by β and the distribution of U. The right side of (2) has finite expectation, so $E(X) < \infty$ and (1) gives the formula $(\beta-1)E(X) = E(U)$.

(Existence). Let $U_0, U_1, U_2, \cdots$ be independent realizations of U. The sequence S_n, $n = 1,2,3,\cdots$, is a submartingale where $S_n = \sum_{i=1}^{n} U_i/\beta^i$. By the sub-martingale convergence theorem, the series $X = \sum_{i=1}^{\infty} U_i/\beta^i$ converges with probability one. Consider $Y = U_0 + X$. We have $\beta^{-1}Y = \beta^{-1}U_0 + \beta^{-1}\sum_{i=1}^{\infty} U_i/\beta^i \overset{d}{=} \sum_{i=1}^{\infty} U_i/\beta^i = X$ so that $Y \overset{d}{=} \beta X$.

Remark 1. The problem is more difficult when U has infinite expectation. Dr. A. G. Pakes (personal communication) has pointed out that, for $\beta > 1$, convergence of the sequence S_n holds provided $E[\log +|U_i|] < \infty$. Thus, for example, existence and uniqueness is established over the whole domain of attraction of the stable law.

So far we have shown that X is uniquely determined by β and U. The converse problem is more interesting. Given X, when does there exist a random variable U such that the independent sum $X + U$ is, in distribution, a scalar multiple, say β of X? Letting $L_X(s)$ and $L_U(s)$ denote the Laplace transforms of X and U respectively, the Laplace transforms derived from (1) yield

$$L_U(x) = L_X(\beta s)/L_X(s). \tag{3}$$

Since $X \geqslant 0$ and $U \geqslant 0$, these transforms exist for all $s \geqslant 0$. It follows that U, *when it exists*, is uniquely determined by X and β. The question of existence is difficult: given X, when is the right hand side of (3) a Laplace transform for *some* $\beta > 1$? It is doubtful that this question has a useful answer, although P. Levy has characterized the distributions for which the right hand side of (3) is a Laplace transform for *all* $\beta > 1$; see Feller (1971, p. 589). Given the intractability of the general problem, in the next section we restrict attention to particular cases which are of interest in the context of aggregating household incomes.

3. AGGREGATION FOR HOUSEHOLDS OF RANDOM SIZE

In addition to the head of the household, we now suppose that the rest of the household contains N members, where N is a random variable defined on the non-negative integers. Set $P(N = n) = p_n$ and let $G(s) = \sum_{n=1}^{\infty} p_n s^n$ denote the probability generating function. We now suppose that the *ith* member of the household earns an amount U_i which is a random variable and that the $\{U_i\}$ are independent and identically distributed with a density function whose Laplace transform is $L(s)$. Then it follows that $U = \sum_{i=1}^{N} U_i$ has a Laplace transform

$$L_U(s) = G\{L(s)\}. \tag{4}$$

The head of the household will usually have a larger income than other members but we might still expect that the distributions of X is in the same family as that for the $\{U_i\}$. In particular, we shall suppose that

$$L_X(s) = \{L(s)\}^m. \tag{5}$$

This does not restrict the form of $L(s)$ when m is an integer, but if we require (5) to hold for all $m > 0$ then we must assume that the distribution of X is infinitely divisible. Equation (3) now assumes the form

$$G\{L(s)\} = \{L(\beta s)/L(s)\}^m. \tag{6}$$

This expression is still very general, but we may examine particular cases of G to obtain explicit results.

Theorem 2. Let the $\{U_i\}$ be non-degenerate, independent and identically distributed with a density function whose Laplace transform is $L(s)$. The distribution of X is independent of the $\{U_i\}$ and has Laplace transform $L_X(s) = \{L(s)\}^m$, $m > 0$. The distribution of N, the number in the household other than the head, has probability generating function $G(s)$. Further $L(s)$ is assumed not to depend upon the parameters of $G(s)$.

(i) If N is binomial(n,θ), $n \geqslant 1$, $0 < \theta < 1$, no general solution to equation (6) exists.

(ii) If N is Poisson(λ), $\lambda > 0$, the unique solution to (6) requires $L(s)$ as a function of β.

(iii) If N is negative binomial(k,θ), $k > 0$, $0 < \theta < 1$, the unique solution of (6) is the exponential distribution which is independent of β but requires $k = m$.

Proof. Let $M(s) = L(-s)$ be the moment generating function for U_i and let $\psi(s) = \log M(s) = \Sigma\kappa_j s^j/j!$ be the cumulant generating function. Then equation (6) yields

$$\log G\{M(s)\} = m\psi(\beta s) - m\psi(s). \tag{7}$$

Let μ and σ^2 be the mean and variance of N. Differentiating (7) once and then twice with respect to s and setting $s = 0$ yields

$$\mu\kappa_1 = m(\beta-1)\kappa_1 \quad \text{or} \quad \mu = m(\beta-1) \tag{8}$$

and $\mu\kappa_2 + \sigma^2\kappa_1^2 = m(\beta^2-1)\kappa_2.$ (9)

From (8), the right hand side of (9) reduces to $\mu(\beta+1)\kappa_2$ and the revised version of (9) is

$$\beta\kappa_2 = (\sigma^2/\mu)\kappa_1^2. \tag{10}$$

(i) Substituting $\mu = n\theta$ and $\sigma^2 = n\theta(1-\theta)$ into (10) yields $\beta\kappa_2 = (1-\theta)\kappa_1^2$. Letting θ approach 1 we find that β goes to zero which contradicts the requirement that $\beta \geq 1$. Hence no general nondegenerate form for $M(s)$ can exist.

(ii) Substituting $\mu = \sigma^2 = \lambda$ into (8) and (10) yields

$$\lambda/m = \beta-1 \tag{11}$$

and $\beta\kappa_2 = \kappa_1^2$.

Successively differentiating (7) and using (11) we obtain $(\beta-1)\mu_j' = (\beta^j-1)\kappa_j$, $j = 1,2,\cdots$; where $\mu_j' = E(U^j)$. Either way it is evident that, once β is fixed by (11), the moments sequence for U is uniquely determined in terms of β. (Since the moments must satisfy these conditions for (6) to hold, existence of κ_1 implies existence of all κ_j, $j > 1$.)

(iii) $G(s) = \{(1-\theta)/(1-\theta s)\}^k$. The first two cumulant relations yield

$$k\theta/(1-\theta) = m(\beta-1) \tag{12}$$

and $\beta\kappa_2 = \kappa_1^2/(1-\theta).$ (13)

Varying θ with the $\{\kappa_j\}$ fixed implies from (12) and (13) that $\beta = 1/(1-\theta)$ and $k = m$. Substituting the results back into (7) we obtain the equation

$$(1-\theta)/\{1-\theta M(s)\} = M\{s/(1-\theta)\}/M(s).$$

Differentiating w.r.t. θ and then setting $\theta = 0$ we obtain the differential equation

$$sM'(s) + M(s) - \{M(s)\}^2 = 0 \qquad (14)$$

with the boundary condition $M(0) = 1$. The unique moment generating function from (14) is $M(s) = 1/(1-\gamma s)$.

Remark 2. Since we know that $X \geqslant 0$, $M(s)$ exists for all $s \leqslant 0$ so the proofs do not *require* existence of moments beyond the first.

Remark 3. The converse to (iii) is well known. That is, if $M(s) = 1/(1-\gamma s)$ then equation (6) is satisfied if and only if N is negative binomial(m,θ).

Remark 4. It is not known whether the moments sequence derived for the Poisson generates a valid probability distribution for all, or even any, values of $\beta > 1$. In any event, the dependence of the $\{\kappa_j\}$ upon β makes the result of limited interest.

Remark 5. If we restrict attention to single fixed values of θ in (i) and (iii) it may be possible to find other distributions which staisfy (6). However, their implicit dependence upon the value of θ chosen makes them of limited interest.

Remark 6. In (iii), $L_X(s) = L(s)$ when $m = 1$ and N then follows the geometric distribution.

It is evident from the statement of Theorem 2 that rather restrictive conditions are necessary if equation (6) is to hold. This is reinforced by the observation that, when there are *exactly* r other members of the household, equation (7) reduces to

$$(m+r)\psi(s) = m\psi(\beta s) - m\psi(s).$$

For $\beta > 1$ and $X \geqslant 0$ the only solution to this equation is the degenerate distribution with $\beta = (2m+r)/m$ and $\kappa_j = 0$, $j \geqslant 2$.

In conclusion, it would appear that very strong assumptions are necessary before inequality measures based upon individual and upon household incomes can be compared directly.

REFERENCES

Feller, W. (1971). *An Introduction to Probability Theory and Its Applications, Vol. II*. Wiley, New York.

Hart, P. E. (1975). Moment distributions in economics: an exposition. *Journal of the Royal Statistical Society, Series A, 138*, 423-434.

Heyde, C. C. (1972). Martingales: A case for a place in the statistician's repertoire. *Australian Journal of Statistics*, 14, 1-9.

Ord, J. K. (1975). Statistical models for personal income distributions. In *Statistical Distributions in Scientific Work, Vol. 2*, G. P. Patil, S. Kotz, and J. K. Ord, eds. Reidel, Dordrecht-Holland. Pages 151-158.

Ord, J. K., Patil, G. P., and Taillie, C. (1980). Truncated distributions and measures of income inequality. (submitted for publication).

Ord, J. K., Patil, G. P. and Taillie, C. (1981). The choice of a distribution to describe personal incomes. In *Statistical Distributions in Scientific Work*, C. Taillie, G. P. Patil, and B. Baldessari, eds. Reidel, Dordrecht-Holland. Vol. 6, 193-201.

[*Received July* 1980. *Revised November* 1980]

SPIKE INTERVAL DISTRIBUTIONS FOR NEURONS AND RANDOM WALKS WITH DRIFT TO A FLUCTUATING THRESHOLD

M. E. WISE*

Physiology Laboratory
Wassenaarseweg 62
Leiden, Netherlands

SUMMARY. Data from two large series of spike interval histograms have been studied, from cerebral neurons in rabbits and respiratory neurons in cats. These are skew distributions with highly variable sample sizes often in the thousands. The unimodal ones and many components of bimodal ones look like gamma or random walk type distributions. The log log plots suggest negative powers of time in long tails of the probability density functions (pdf's). The interpretation is in terms of random walks undergone by an action potential until it reaches a firing threshold, which can fluctuate. Further evidence for this comes from serial correlations, the form of the whole pdf and of its tail, the pattern shown by the cumulants, and especially from the values of the various powers found. A general form for the pdf is derived, and the effects on the cumulants are considered for χ^2 fluctuations. Of the various numerical analyses tried, those based on the log log plots showed the most promise. These also involve some statistical technical problems. A general plea is made for more explorations of data at a fairly elementary level in a two and fro process between theory, modelling and numerical analyses.

KEY WORDS. Spike intervals, neurons, action potentials, fluctuating thresholds, mixed random walks with drift, negative powers of time, Pareto laws, generalized inverse gaussians, log log plots.

---

* Permanently seconded from the J. A. Cohen Institute of Radiopathology and Radiation Protection.

C. Taillie et al. (eds.), Statistical Distributions in Scientific Work, Vol. 6, 211–231.

1. INTRODUCTION

Single neurons can be excited and their action potentials recorded. These can be regarded as point events, which are then called spikes. The time intervals between spikes during a single series of observations on one neuron almost always yield positively skew distributions, of all degrees, from near symmetry to ones with extremely long tails.

The actual frequencies and their variability in one experiment or in a series provide valuable biological information, but most neurologists are convinced that the forms of these distributions, if they could be well interpreted, would provide much more. Probabilistic models were being developed 15 to 20 years ago. Broadly speaking, these lead, as so often happens with skew distributions, to gamma functions, in this case from summation of exponential distributions of time intervals from Poisson processes, and to random walk type distributions.

Experimental techniques have been improving remarkably all this time. Also, facilities for storing and sorting the data in computers have become very much better. It is mostly nerves controlling sensory processes that are studied. The main series discussed in this paper are those of Blom (1969) in the somatosensory cerebral cortex of the rabbit. We also give data obtained from respiratory neurons in cats (Smolders and Folgering, 1977). Another recent series, for which the results may well be applicable, is on taste cells on the feet of flies (v.d. Molen *et al.*, 1978).

What has lagged behind is work on the interface between theory and observations. Fienberg (1974), in a good, clear review, says: "The bridge between applied probabilists who are responsible for the proliferation of stochastic models in this area and practicing physiologists who study neural activity is barely passable and without substantial structure. Thus, it should come as no surprise that few of the models proposed have received substantial empirical validation." He therefore wishes "to encourage statisticians to develop methods suitable for fitting of these models to actual data, and for discriminating between competing models."

What I think is needed still more than this from *biometricians* is exploration of the data in a to and fro process between observations and theories. This can be done with fairly elementary methods. I believe these point conclusively to random walk processes, and not to Poisson point processes, for explaining the variability of spike intervals. There are also various indications that are weak in themselves but combine to suggest a particular kind of random walk process to a slowly fluctuating threshold.

2. SPIKE INTERVALS AND RANDOM WALKS

Briefly, the random walk is undergone by an action potential. This is an event occurring in neuronal membranes in which potential changes are evoked by excitatory stimuli, which can be regarded as positive, and inhibitory ones, which are then negative, in random order. These are added until a certain potential level is reached, called the "firing threshold," at which the event is started. Although the event has a certain time duration, for theoretical purposes it is supposed to behave as a Dirac function. In the simplest possible case, then, the spike interval distribution is an inverse gaussian one for first arrival at this threshold, namely:

$$y(t) = At^{-3/2}\exp\{-\varphi(t/\mu + \mu/t)\}. \qquad (1)$$

Here μ is the mean of the distribution, φ is a dimensionless shape parameter, and A is a normalizing constant.

This suggests looking at a plot of $\log y$ *v* $\log t$. This was done by Herz *et al.* (1964). Many of their distributions then became linear from times just greater than the mode to far down in the tail. But the slopes were very variable. Only a few of them were close to 1.5 as equation (1) would suggest

In Blom's extensive series, the input data consisted of 400 time intervals, with variable units of time determined by the experimental conditions, namely $2^n \times 0.0003125$ seconds, $n = 0$ to 7. The numbers of spikes recorded varied from about 50 to 80000. Many distributions were bimodal but sometimes the first mode was very early, and was only visible as a perturbation of the main distribution. This had one of two main types. Often it was moderately skew (as Figure 1), as in equation (1) with $\varphi \simeq 3$ or more. When curves like this fit equation (1) they are barely distinguishable from lognormal or gamma pdf's (Wise, 1975). Many other curves in the series were very skew (as Figure 2) and often then the log log plot consisted mostly of a straight line. Most of the decreasing part of the pdf $y(t)$ thus fitted a negative power function:

$$y(t) \simeq At^{-W} \qquad (2)$$

with varying values of W. For the two largest samples, W was close to $2\frac{1}{2}$.

The biological significance of negative power laws is still not generally realized. The statistical analysis was carried out at the time when there was a strong preference for gamma distributions (Oosterhoff and Blom, 1969). Yet although many distributions looked very regular, it was found impossible to fit the

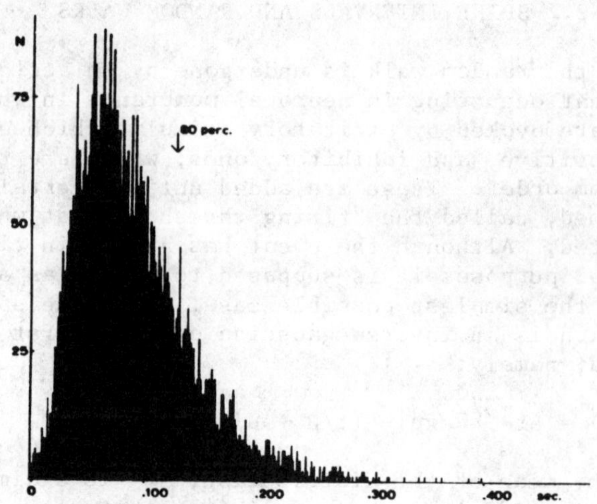

FIG. 1: A typical broad interval histogram, neuron 56 *in Blom's series. In the corresponding log log plot no straight part was seen. Number of intervals* 5939, 80*th percentile at* 103 *time units, unit of time* .00125 *sec. Mean* 91.42 *ms*;. *s.d.* 51.42 *ms. (By Courtesy of J. L. Blom, from Figure* 23 *in his thesis.)*

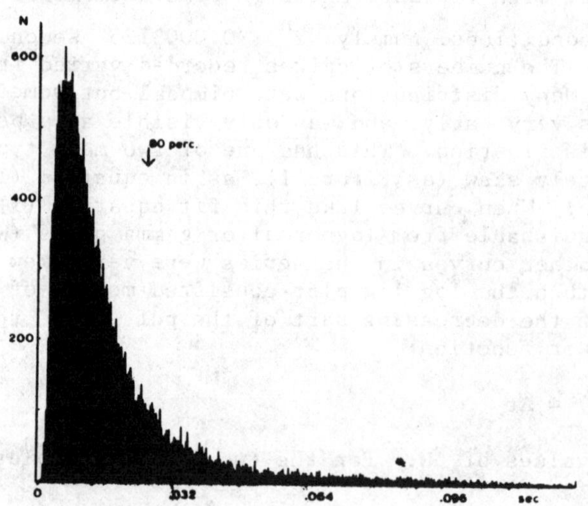

FIG. 2: A skew interval histogram, neuron 69 *in Blom's series.* 30490 *intervals, frequencies obtained from* 0 *to* 400 *units of time, unit* 0.3125 *ms. Mean* 18.55 *ms, s.d.* 19.32 *ms. See also Table* 3 *and Figure* 3. *(By courtesy of J. L. Blom, from Figure* 7 *in his thesis.)*

whole of any curve. By truncating the tail at the 80*th* percentile fits that *looked* almost acceptable were found; even so, there were small but systematic deviations.

The statistical moments revealed more positive patterns. In the 300 interval histograms for "normal" neurons, means and standard deviations were almost in proportion. Information yielded by the higher moments is discussed in the next section.

Finally, Blom and Oosterhoff investigated serial correlations r_1 to r_{10} between successive spike intervals. The method of sorting the data into interval histograms sacrifices such information. Table 1 shows the first 5 coefficients for their random sample from the normal neurons. The results suggest small, irregular fluctuations extending over a few spikes, possibly in the time scale parameter. However, since most of the coefficients are small, during the (short) periods of steady state the spikes probably occur at random. Possible observable effects of this on the pdf for the interval distribution, and on its cumulants, will be considered in the next section.

3. FLUCTUATING MEANS OR THRESHOLDS

3.1 A Possible Pdf. We can start with the pdf (almost the same as in equation (1)):

$$y(t) = at^{-W} \exp\{-\varphi(t/\mu + \mu/t)\} \qquad (3)$$

and allow W to take values other than $3/2$. When $W > 3/2$, $y(t)$ can to good approximation be interpreted as a distribution of first passage times for particles undergoing a random walk with drift through a *mixed* medium (Wise, 1971, 1975). More recently (3) has been called the *generalized inverse gaussian distribution* (Barndorff-Nielsen *et al.*, 1978). These authors, too, interpret it, in a more abstract approach, as a distribution of "first hitting times."

If either the threshold or the mean passage time fluctuates, the main effect of this will be of course to change μ; however for the Wiener process, with $W = 3/2$, φ/μ remains constant. This is a good approximation in the more general case. So the simplest generalization that is possible is to make $\varphi\mu$ a random variable, which comes only into the factor $\exp(-\varphi\mu/t)$. Hence we can replace this factor by $\exp(-\theta\varphi\mu/t)$ and let θ be a random variable with mean 1. We then obtain a new distribution by keeping everything else in equation (3) unchanged, including the normalizing factor a. In fact a is a function of φ and μ,

TABLE 1: Serial correlation coefficients obtained by Blom (1969) *for* 23 *neurons in his group* 1, *with* 400 *intervals per neuron.*

Celnr.	r_1	r_2	r_3	r_4	r_5
009	0.099	0.086	0.057	0.112	0.056
020	0.189	0.202	0.115	0.121	0.061
031	-0.017	0.248	0.029	0.052	0.146
033	-0.013	0.025	-0.096	-0.116	-0.014
038	0.096	0.015	-0.058	-0.042	0.028
069	0.002	0.000	0.100	0.066	-0.007
077	0.146	0.045	0.005	0.034	0.032
078	0.142	0.116	0.043	-0.017	-0.013
079	0.225	0.197	0.087	0.037	0.077
080	0.041	0.082	-0.008	0.006	-0.028
082	-0.101	0.075	0.059	0.177	-0.050
098	0.043	0.007	0.006	0.044	-0.020
099	-0.069	0.038	0.133	0.071	0.087
104	0.277	0.119	0.087	0.125	0.016
105	0.093	0.125	0.067	0.120	0.118
106	0.133	0.077	0.023	-0.062	0.015
107	0.170	0.094	0.056	0.060	0.067
117	0.037	0.056	-0.016	-0.054	-0.050
129	0.005	0.014	-0.022	-0.013	-0.056
131	-0.014	0.000	0.077	0.061	-0.014
133	0.150	0.074	0.080	0.058	0.024
135	0.100	0.152	0.052	-0.007	0.025
136	0.054	0.037	-0.086	0.074	0.047
r>0	18	23	17	16	14
r<0	5	0	6	7	9

For r_6 to r_{10} the numbers of negative values were 8, 7, 3, 6, and 11 out of 23 respectively.

which complicates the interpretations, but clearly any generalization obtained in this way will still correspond to one particular distribution of θ.

Since φ and μ cannot be negative in this context we can try a χ^2 distribution for $\varphi\mu$. The corresponding pdf for θ is then

$$f(\theta) = F\theta^{\gamma-1}\exp(-\gamma\theta) \tag{4}$$

where the normalizing factor is:

$$F = \gamma^{\gamma}/(\gamma-1)! \simeq \{\gamma/(2\pi)\}^{\frac{1}{2}}\exp\{\gamma - 1/(12\gamma)\} \tag{5}$$

when Stirling's approximation is accurate enough. Then the spike interval pdf becomes:

$$Y(t) = aJ(t,\gamma)t^{-W}\exp(-\varphi t/\mu)$$

where $J(t,\gamma) = F\int_0^{\infty}\theta^{\gamma-1}\exp\{-\theta(\gamma + \varphi\mu/t)\}d\theta$. Putting

$$\psi = 1/(\gamma + \varphi\mu/t);\ \theta = u\psi$$

we soon obtain:

$$J = F\int_0^{\infty}\psi^{\gamma}u^{\gamma-1}e^{-u}\,du = F\psi^{\gamma}(\gamma - 1)! = (\psi\gamma)^{\gamma}$$

and a closed form for the time interval pdf:

$$Y(t) = at^{-W}\{1 + c/(\gamma t)\}^{-\gamma}\exp(-\beta t) \tag{6}$$

where $c = \varphi\mu$ and $\beta = \varphi/\mu$. Equation (3) is then the limit of (6) as $\gamma \to \infty$.

In general the tail is very similar to that of (3) but the curve has a broader top. The early part is also different from (3); when t is small enough:

$$\{1 + c/(\gamma t)\}^{-\gamma} \simeq (\gamma t/c)^{\gamma},\ Y(t) \simeq at^{\gamma-W}(\gamma/c)^{\gamma}\exp(-\beta t) \tag{7}$$

so that $Y(t)$ begins as a gamma function, which so many researchers have taken as the basic form for the spike interval histogram. However the approximation (7) is valid only over a very short period.

For curve fitting it was found best to start with the limiting form (3). Several histograms with very many spikes and with good fits to the upper tails indeed had broader tops than could be fitted by (3). The generalized form (6) was equally good in the (long) tail, and fitted back to a slightly earlier time near to the mode. The pdf for *fluctuations*, however, could take many other forms that that of equation (4). Even so, this pdf could be a useful addition to the repertoire.

3.2 Cumulants. In Blom's collection of interval histograms, moments and cumulants up to the fourth were calculated. Considering how skew most of these distributions are, it is very surprising that these yielded any intelligible patterns at all, but these were found. The effect of fluctuating thresholds on their cumulants was therefore worth studying. This has been done for a distribution with cumulants k_n, which we write as εf_n such that the first four are:

$$\varepsilon/\beta,\ \varepsilon\alpha/\beta^2,\ \varepsilon\alpha(1+\alpha)/\beta^3 \ \text{ and } \ \varepsilon\alpha(1+\alpha)(2+\alpha)/\beta^4. \quad (8)$$

Here $1/\beta$ is the time parameter and α is a dimensionless form parameter. This is the generalized form of the inverse gaussian distribution, for which $\alpha = \frac{1}{2}$ and for mixed random walks $\alpha < \frac{1}{2}$. Then equation (3) is a fairly good approximation to the pdf (Wise, 1971, 1975). In what follows ε is replaced by $\theta\varepsilon$ where θ is a random variable with cumulants c_n. Writing the cumulants of the distribution of arrival times as ℓ_1, ℓ_2, ℓ_3 with fluctuations (that are independent of this arrival time) we calculate the expected values of t^n. They are obtained from binomial-type series relating moments m to cumulants k, namely:

$$m_1 = \bar{t} = k_1 \qquad m_3 = \overline{t^3} = k_3 + 2k_2m_1 + m_2 \quad (9)$$
$$m_2 = \overline{t^2} = k_2 + k_1m_1 \qquad m_4 = \overline{t^4} = k_4 + 3k_3m_1 + 3k_2m_2 + k_1m_3.$$

Then

$$\ell_1 = \bar{t} = f_1\varepsilon\bar{\theta} = f_1\varepsilon c_1 \qquad \overline{\theta^2} = c_2 + c_1\bar{\theta} = c_2 + c_1^2$$
$$\overline{t^2} = \varepsilon f_2\bar{\theta} + \varepsilon^2 f_1^2\overline{\theta^2} \qquad \ell_2 = \overline{t^2} - \bar{t}^2 = \varepsilon f_2 c_1 + \varepsilon^2 f_1^2 c_2.$$

Similarly from the expressions for $\overline{t^3}$ and $\overline{t^4}$, by using (9) in both directions we obtain:

$$\ell_3 = \varepsilon f_3 x_1 + 3\varepsilon^2 f_1 f_2 c_2 + \varepsilon^3 f_1^3 c_3$$

$$\ell_4 = \varepsilon f_4 c_1 + \varepsilon^2(3f_2^2 + 4f_1f_3)c_2 + 6\varepsilon^3 f_2 f_1^2 c_3 + \varepsilon^4 f_1^4 c_4.$$

Two particular cases are worth giving in full. If θ is distributed normally with unit mean c_1 and variance c_2 equal to σ^2 (i.e. $c_3 = c_4 = 0$), we have:

$$\ell_1 = \varepsilon/\beta; \qquad \ell_2 = (\varepsilon\alpha + \varepsilon^2\sigma^2)/\beta^2;$$

$$\ell_3 = \varepsilon\alpha(1 + \alpha + 3\varepsilon\sigma^2)/\beta^3 \quad \text{and}$$

$$\ell_4 = \{\varepsilon\alpha(1 + \alpha)(2 + \alpha) + \varepsilon^2\sigma^2(4\alpha + 7\alpha^2)\}/\beta^4. \tag{10}$$

The χ^2 distribution, equation (4) has cumulants $1, 1/\gamma, 2/\gamma^2, 6/\gamma^3, \cdots$ which yield:

$$\ell_1 = \varepsilon/\beta; \qquad \ell_2 = \varepsilon(\alpha + \varepsilon/\gamma)/\beta^2$$

$$\ell_3 = \varepsilon\{\alpha(1 + \alpha) + 3\varepsilon\alpha/\gamma + 2\varepsilon^2/\gamma^2\}/\beta^3$$

$$\ell_4 = \varepsilon[\alpha(1 + \alpha)(2+\alpha) + \{3\alpha^2 + 4\alpha(1 + \alpha)\}\varepsilon/\gamma + 12\alpha\varepsilon^2/\gamma^2 + 6\varepsilon^3/\gamma^3]/\beta^4.$$

As described previously (Wise, 1975) some dimensionless functions of the cumulants provide quick tests on whether this type of random walk function is at all possible. These functions are:

$$g_1 = k_2/k_1^2; \; g_2 = k_3k_1/k_2^2 \;\text{ and }\; g_3 = k_4k_2/k_3^2 \tag{11}$$

Hence if there are no fluctuations:

$$g_1 = \alpha/\varepsilon; \; g_2 = 1 + 1/\alpha \;\text{ and }\; g_3 = (2 + \alpha)/(1 + \alpha)$$

Using capital letters for the corresponding g's with fluctuations, if these are normally distributed we obtain:

$$G_1 = \alpha/\varepsilon + \sigma^2; \; G_2 = \alpha(1 + \alpha + 3\varepsilon\sigma^2)/(\alpha + \varepsilon\sigma^2)^2$$

so that G_2 *decreases* continuously from $1 + 1/\alpha$ to zero as σ^2 increases. For the χ^2 distribution of fluctuations we have:

$$G_1 = \alpha/\varepsilon + 1/\gamma, \; G_2 = 1 + \varepsilon/(\varepsilon + \alpha\gamma) + \alpha/(\alpha + \varepsilon/\gamma)^2 \tag{12}$$

$$G_3 = \frac{(\alpha + \varepsilon/\gamma)\{\alpha(1+\alpha)(2+\alpha) + \varepsilon\alpha(4+7\alpha)/\gamma + 12\varepsilon^2\alpha/\gamma^2 + 6\varepsilon^3/\gamma^3\}}{\{\alpha(1+\alpha) + 3\varepsilon\alpha/\gamma + 2\varepsilon^2/\gamma^2\}^2}$$

Typical values of α should be between 0.3 and 0.5; if $\alpha > 0.5$ the underlying mechanism of mixed random walks in drift is appreciably different. The factor ε multiplying the cumulants is dimensionless; possible numerical values seem to lie between 0.5 and 0.15 for interval histograms. Under these conditions G_3 does not vary much with γ, but the effect of γ upon G_2 is interesting. As $1/\gamma$ increases from zero (zero means no fluctuations) G_2 passes through a minimum where $\gamma = \varepsilon/(2 - \alpha)$, then $G_2(\text{min}) = 2 - \alpha/4$. So for mixed random walks in series, there should be an absolute lower limit of G_2 equal to 1.875.

4. NUMERICAL ANALYSES

Direct curve fitting to (3) and (6) was not successful. One reason for this became evident only when log log plots were made of the tails. It was very seldom that a single distribution was present when the sample was large enough for more than one of them to be distinguished. But there were many successes in fitting considerable parts of distributions, which will be discussed later.

The cumulants and moments of every distribution in the series had already been calculated (by Blom). Using these, it was sometimes possible to fit (3) using existing programs (Wise 1971), but again there were systematic deviations and fitting by moments is certainly not optimal in such skew distributions. But the key problem here is what to fit rather than how to fit.

The numerical values of dimensionless functions of observed cumulants defined in equation (11) however yielded interesting patterns, especially for g_2.

They were calculated for every normal neuron with more than 800 intervals, and in three other groups with more than 2800 intervals. Table 2 shows the results. Those for the normal neurons, group 1 in the table, are particularly interesting. Although we would expect the sampling errors of the g_2's to be large, there is clearly a sudden increase in their frequency-density close to the predicted limit of 1.875. In groups 2, 3 and 4 a convulsive drug was applied in different ways as shown, and the 4 groups are clearly all different in relation to the distribution of the g_2's. The quantitative analyses by Blom and Oosterhoff (1969) in terms of moments and truncated gamma functions showed no clear difference. But we are still some

TABLE 2: Distributions of the function g_2 *of observed cumulants defined in equation (11) in Blom's 4 experimental groups. The* g_2 *column gives the center of each frequency interval, boundaries at* 0.875, 1.125, 1.375,···,4.875. *The actual* g_2 *values outside this range are shown, as is the predicted threshold for* χ^2 *fluctuations at* 1.875 *(horizontal dashed line).*

Group 1: normal neurons, number of intervals N ⩾ 800 *and* ⩾ 2800.
Group 2: Methionine sulfoxamine (M.S.O.) *for inducing convulsions administered intravenously,* N ⩾ 2800.
Group 3: M.S.O. *applied to the cortical surface,* N ⩾ 2800.
Group 4: M.S.O. *injected into the ventrical system,* N ⩾ 2800.

Group:	1	1	2	3	4
N:	⩾800	⩾2800	⩾2800	⩾2800	⩾2800
g_2					
				0.73	-0.51, 0.24
<0.875				0.78	-0.49, 0.71
1	0	0	0	1	2
1.25	4	2	0	1	2
1.5	9	2	6	5	0
1.75	8	2	5	3	2
2	19	5	1	1	3
2.25	16	5	6	3	3
2.5	7	5	7	1	2
2.75	17	10	5	4	5
3	6	4	3	2	3
3.25	7	5	2	1	5
3.5	2	0	2	3	2
3.75	5	5	2	1	3
4	3	2	0	3	3
4.25	4	1	0	2	0
4.5	2	2	2	0	1
4.75	0	0	0	2	1
>4.875	6.12	5.77, 6.02	5.33, 6.82	5.20, 5.68, 5.81 7.49, 15.15	5.79, 6.13, 6.42 6.46, 7.14, 7.59 7.71, 10.27, 12.65
Total	110	52	43	40	50

way from a physiological interpretation. Very low g_2 values could correspond to markedly bimodal histograms. Very high ones correspond to very low values of α and to random walks with drift through a very variable medium (Wise, 1975). Values of $g_2 \geqslant 3$ can be interpreted in this way without fluctuations, $g_2 = 3$ would corresponding to "drift" through a uniform medium (the Wiener process) and for $1.875 < g_2 < 3$ the fluctuations have to be assumed.

5. PLOTTING THE FREQUENCY DENSITY CURVES

This exploration has gone from advanced to simple methods, partly because computer plotters have become so much better. At present more information than from all the more sophisticated analyses seems to be obtainable from log log plots of observed frequency densities from the mode onwards.

The input data had been stored on paper tapes in sets of 400 consecutive numbers, namely the frequencies of occurrence $f(n)$ for $n = 1$ to 400 units of time interval. An obvious problem was what to do about the zero frequencies, because we need to know about the form of the whole of the observable tail. It was decided to count back from $n = 400$ and accumulate values of $f(n)$ until $\Sigma f(n) \geqslant 4$ and then stop. The mean of the n's weighted by $f(n)$ was regarded as the location of this frequency class, with density 4 or more divided by the class width. As soon as $f(n) \geqslant 4$ for all n less than a certain n_1, we obtain the usual frequency density function fdf with x values of the midpoint of each class. It was hoped that any bias in the curve fitting would be small. When straight lines were fitted they looked reasonable, but no allowance was made for random errors in both x and y.

A more serious loss was that the distribution of

$$X^2 = \Sigma(Y_{obs} - Y_{fit})^2/Y_{fit}$$

is clearly no longer that of χ^2 for frequencies obtained in this way. This showed up very clearly with the most successful fits to straight lines, with X^2 values often well below the number of degrees of freedom.

In many of the normal neurons the mode occurred very early and the whole of the log log plot beyond the mode nearly fitted a striaght line. Figures 3 and 4 show these plots for the two experiments with the most spikes, more than 30,000 and 80,000 respectively. Although the histograms look so regular, the

TABLE 3: Data on the interval histogram for neuron 69 *(Figures* 2 *and* 3*). There were* 30497 *spikes observed during* 10 *minutes. The frequencies* y *of time intervals* n *from* 1 *to* 400 *units were obtained; unit of time was* 0.0003125 *sec. From intervals* 19 *to* 179*,* y *was fitted by:*

$$\log y = b_0 + b_1 \log T + b_2 T - \gamma \log(1 + \lambda/T)$$

where T = n - 1.5*, and*

$$b_0 = 16.472,\ b_1 = -2.4776,\ b_2 = -0.00358,\ \lambda = 0.59295,\ \gamma = 90.$$

Interval n *corresponds to times between* n-1 *and* n *units. A "dead time" of* 1 *unit was allowed for, hence* T *is* 1 *less than the mid point of the corresponding interval. Beyond the mode the function is very nearly of the form of equation (3) with* $\phi\mu = \gamma\lambda = 53.37$.

The table shows components of $X^2 = \Sigma(y - yfit)^2/yfit$ *in each decade (with two intervals extra in the first decade);* $x^2(10)$ *sum of all* 10 *terms,* $x^2(1)$ = *one term,* Σy *and* Σy_{fit} *from summing the* 10 *frequencies observed and expected.*

n-1		$X^2(10)$	Σy_{obs}	Σy_{fit}	$X^2(1)$
19	30	11.7	6345	6516.9	4.53
31	40	9.5	4105	4151.0	0.51
41	50	13.0	3077	3007.3	1.62
51	60	18.6	2248	2184.1	1.87
61	70	9.3	1643	1614.5	0.53
71	80	15.3	1190	1217.3	0.65
81	90	12.6	960	936.6	0.58
91	100	16.5	710	731.5	0.63
101	110	14.3	581	580.1	0.00
111	120	13.3	481	467.1	0.41
121	130	19.3	407	380.2	1.89
131	140	15.0	309	312.9	0.05
141	150	12.7	255	259.8	0.09
151	160	9.9	223	217.6	0.13
161	170	14.2	181	183.6	0.04
171	180	6.0	157	155.9	0.01
		211.2			13.54

With 156 degrees of freedom $P(X^2 > 211.2) < 0.005$.

The complete data for part of the above table were:

n-1	81	82	83	84	85	86	87	88	89	90
y_{obs}	103	102	89	109	114	104	79	98	74	88
y_{fit}	104.7	102.0	99.4	96.9	94.4	92.1	89.8	87.6	85.4	83.3

n-1	91	92	93	94	95	96	97	98	99	100
y_{obs}	105	77	66	63	69	74	73	75	53	55
y_{fit}	81.3	79.3	77.4	75.6	73.8	72.1	70.4	68.8	67.2	65.6

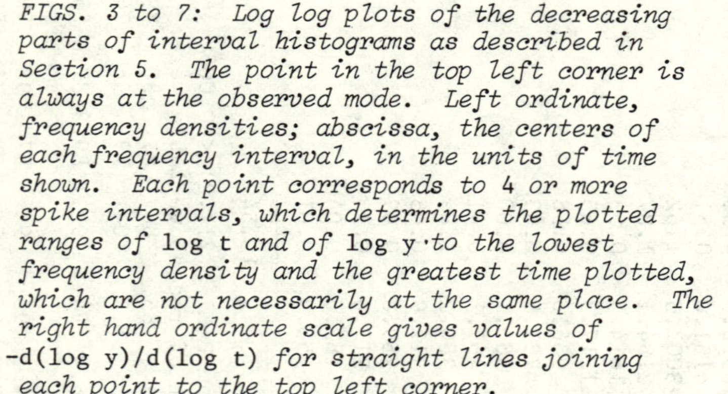

FIGS. 3 to 7: Log log plots of the decreasing parts of interval histograms as described in Section 5. The point in the top left corner is always at the observed mode. Left ordinate, frequency densities; abscissa, the centers of each frequency interval, in the units of time shown. Each point corresponds to 4 or more spike intervals, which determines the plotted ranges of log t *and of* log y *to the lowest frequency density and the greatest time plotted, which are not necessarily at the same place. The right hand ordinate scale gives values of* -d(log y)/d(log t) *for straight lines joining each point to the top left corner.*

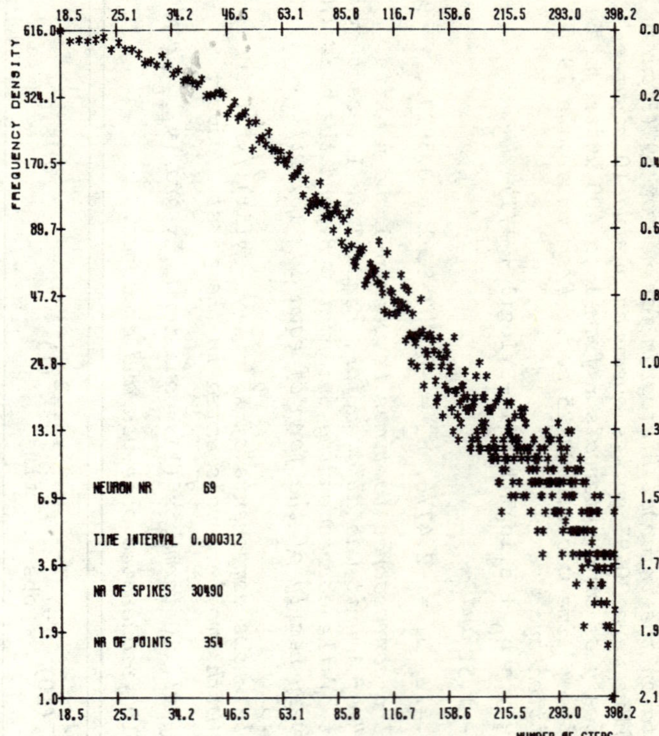

FIG. 3: Neuron 69 as in Figure 2 and Table 3.

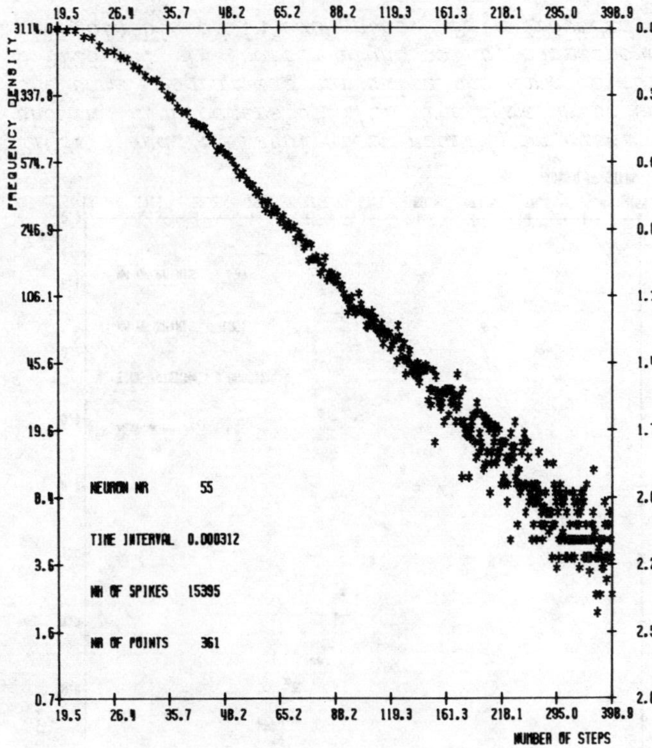

FIG. 4: The largest sample in Blom's series. The number of spike intervals N *is counted in the plotting program, but only* 2^{15} *positive and* 2^{15} *negative integers were available, hence* N = 65536 + 15395 = 80931.

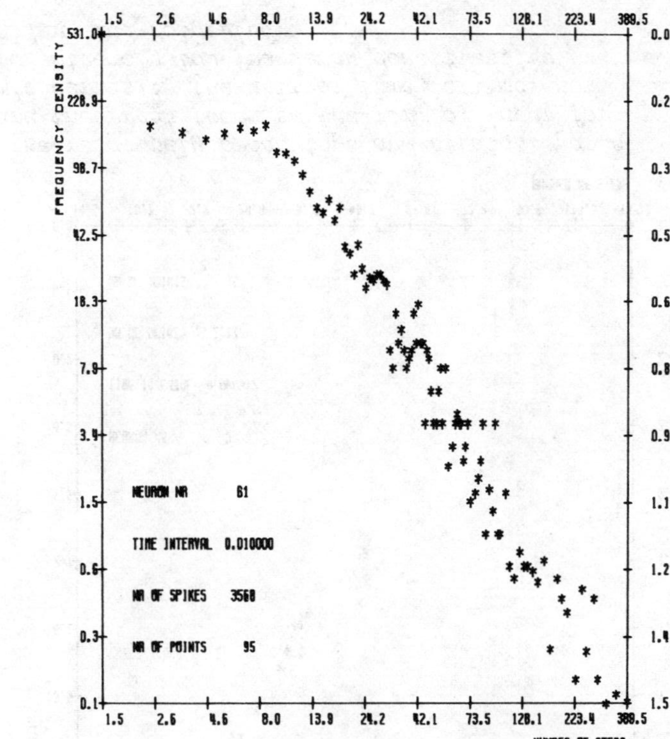

FIG. 5: A bimodal histogram with a very early mode and a long straight part corresponding to a power of t *near to -2. Data from Blom's series.*

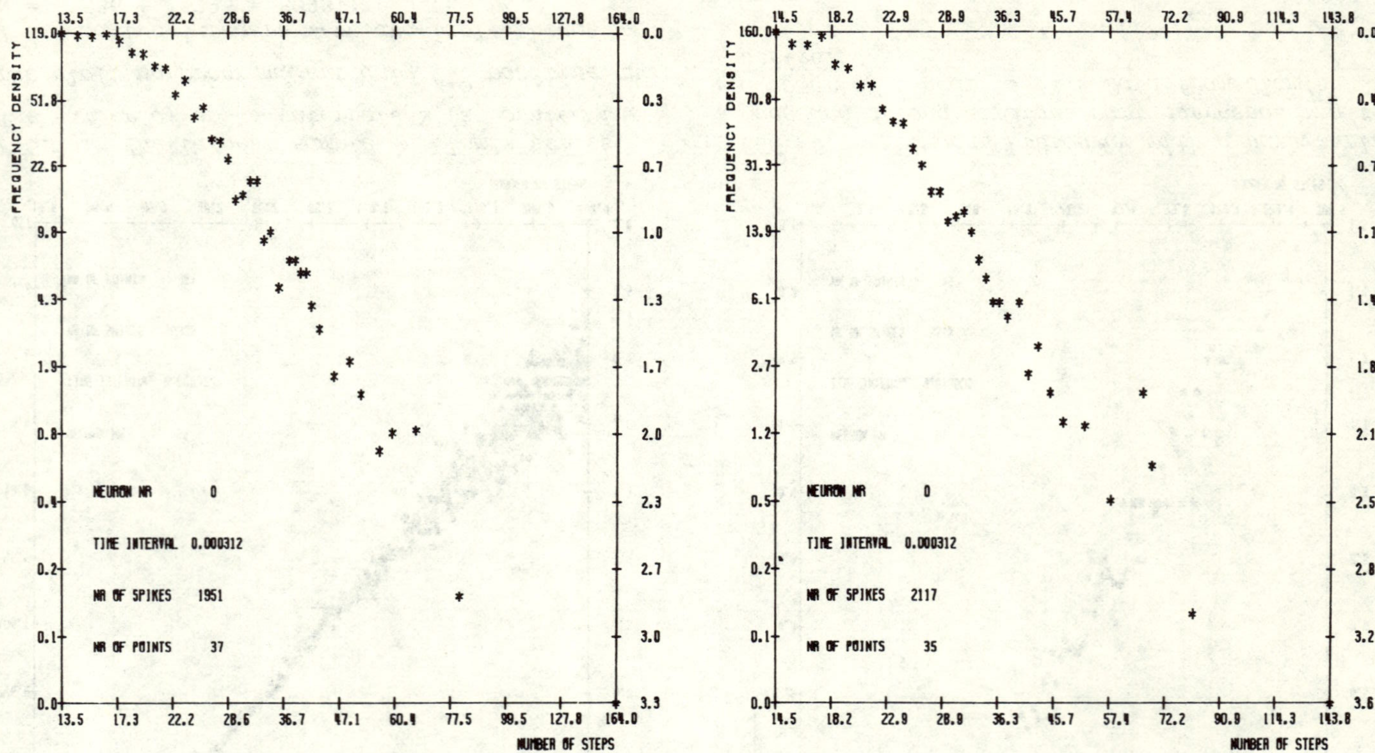

FIGS. 6 and 7: Respiratory neurons in cats. Note the relatively broad tops and straight parts showing high powers of 1/t. The time units shown are incorrect (because the form of input for the same program was not quite the same as for Blom's series): the unit of time is 1 ms, not 0.3125 ms. The bottom right point, corresponding to extremely low frequency densities, is probably biased in each plot. Data provided by Smolders and Folgering.

detailed plot of the tail shows that the underlying distribution is a mixture. For the 400 frequency classes, with a mode at about 20, the second distribution begins to show at about 200 units of time. From just before the mode to about 180, Table 3 shows that there was a good fit to the random walk type function derived in Section 3. The simplest form, equation (3), would also fit over most of this range.

The detailed calculations showed that the X^2 values were inconsistent. When frequency classes were grouped with widths of 5 or 10 units the X^2 values pointed to fully acceptable fits, but the fluctuations within these groups were too large, so that when X^2 was calculated for every unit, it was much (very significantly) too large. There appeared to be no periodic component in the spike intervals. Errors made at random in measuring the time intervals would only spread the whole distribution out. Possibly the recording process was such that the actual boundaries between the class intervals differed slightly from the exact ones throughout one experiment. This effect was in fact only observed in experiments with the smallest possible unit of time, namely 0.0003125 sec. In that case it is of no interest biologically, but it provides in any case an obvious warning over rejecting a fit because X^2 is too large, namely not to do this without first looking at a table of observed and predicted frequencies.

It was never possible to fit a large sample to the whole of one curve; the increasing part of the curve, even when the mode was early, was usually broader than the fit predicted. Sometimes another mode appeared very early, forming part of a secondary distribution that had little effect on the tail (Figure 5).

From almost any theory or model there should be an end exponential, just as for clearance curves. Yet more often than not this could not be found by curve fitting. Possibly in many of these cases it could only be detected beyond the observable range. In other cases the shift of origin effect may apply, as shown previously (Wise 1971, 1975). If

$$y(t) = Ax^{-W}\exp\{-\varphi(x + 1/x)\} \quad \text{where} \quad x = (t-t_0)/\mu \quad \text{and} \quad t_0 > 0,$$

$\log y(t)$ as a function of $\log t$ can have two points of inflexion beyond the mode, and a nearly straight portion with slope steeper than $-W$, over a longer period than when $t_0 = 0$. For the spike intervals t_0 can arise both from the random walk theory and from a possible dead time after one spike, before the potentials begin to increase.

All attempts to estimate t_0 as an extra parameter failed; in most cases a dead time was estimated to be slightly less than the time of the shortest spike interval observed.

This shift of origin effect could in any case explain the most striking feature of the log log plots. Whenever there was a long tail in a distribution that *looked* fairly regular this was a straight line.

These plots have now been made for more than 100 neurons in Blom's series. As mentioned, these fall into two classes, the broad ones and the long tailed ones. The broad ones do not look very different from any moderately skew curve such as for Figure 1. The complete series form a remarkable procession, for which of course I have no space, in any case they need more analyses. Readers who have seen carnivals in towns in Europe where many Roman Catholics live might be reminded of such processions. The trend is for ever more monstrous forms to follow one another. If plots for the whole series could be obtained (there are technical difficulties in retrieving the rest of these data) this would provide a very unusual challenge to statisticians' skills!

The same plot program has been applied to 60 plots for spike interval histograms for respiratory neurons in cats (Smolders and Folgering, 1977). These histograms are inevitably mixtures, since, for technical convenience, those observed during both breathing in and breathing out were recorded. Then the absolute frequencies must vary periodically but probably in proportion. It is easily seen that any observed power function of time in a steady state fdf will still be observed, but over different periods. The surprise was that not only did so many log log plots clearly show such power functions, but that they occurred in tails of "broad" distributions, in contrast to most of those in Blom's series. Also the slopes were much higher, 3½ to 7. It seems we have to accept that power laws can occur with much higher values of W than any previously observed or expected. Figures 6 and 7 show typical examples.

A disturbing but inevitable conclusion seems to be that very many theoretical papers on modelling spike interval distributions would and should not have been written if the authors had known about these power laws.

6. CONCLUSION

At Trieste, negative power laws observed in tails of pdf's were discussed in several contexts, particularly in econometrics (see e.g. Arnold, 1981; McDonald, 1981). One point found to be common with the neuron problem was that the power laws are robust (e.g. to many factors disturbing models and theories). Interpreting them in terms of random walks with drift began at least 80 years after Pareto's discovery gave his name to them, yet as was pointed out his original power was $-3/2$, as for the simplest case of random walks with drift! It seems a good example where there should be cross fertilization in these very different fields of application.

The powers $-W$ actually observed — when this was so — varied considerably, as we have seen. The lowest values of W found so far are about 1. For $1 < W < 1.5$ a random walk interpretation is possible, but it is more complex than for $W > 1.5$ (Wise, 1975). In the distribution of W's the frequency density does seem to increase markedly round about 1.5. For the respiratory neurons studied by Smolders and Folgering there were not many with $W < 3.0$, whilst it was seldom that any in Blom's series had $W > 3.0$.

Probably progress will be made through semi-empirical laws, for example if and when *distributions* of W values are found to correspond to different groups of experiments as was found with the g_2's as in Table 2. This includes the many cases where the distributions are clearly mixtures. So far only main components in the tail have been well fitted, and over a range of time that seems best judged by the eye looking at a log log plot.

On the theoretical side, random walk with drift mechanisms were introduced some time ago (see e.g. Gerstein and Mandelbrot, 1964; Johannesma, 1969). Non-steady state conditions are more common than not, but it seems they do not invalidate numerical analyses of the long tails of these distributions. However some workers prefer to study spectral analyses and serial correlations rather than interval histograms; De Kwaadsteniet (1980) considers a model in which two different steady states alternate.

To sum up the factors consistent with mixed random walks with drift to a fluctuating threshold or with a fluctuating time scale, these are:

1. The power laws, especially when $W > 1.5$.
2. The variability of W, with most values $\geqslant 1.5$.
3. The patterns shown in Table 2, with the proviso that nothing is known about the pdf for the supposed fluctuations, the choice of χ^2 was arbitrary.
4. Where distributions are unimodal, the tops are broader than those for inverse gaussians or the generalized inverse gaussians.

5. The majority of small positive serial correlation coefficients as opposed to negative ones.

All these are the results of back and forth studies between data and models.

ACKNOWLEDGEMENTS

Dr. J. L. Blom and Dr. F. D. J. Smolders have provided much of their data on computer tapes and together with B. L. van Zomeren in this laboratory have well coped with the problems of converting the data to the PDP 11/70 in use here. B. L. van Zomeren organized the storage of these sets of data and made log log plots of over 200 of the distributions. Their help was indispensible.

REFERENCES

Arnold, B. C. (1971). *Pareto Distributions and Applications*. Monograph in preparation.

Barndorff-Nielsen, O., Blaesild, P. and Halgreen, C. (1978). First hitting time models for the generalized inverse gaussian distribution. *Stochastic Processes and Their Applications*, 7, 49-54.

Blom, J. L. (1969). *Quantitative aspects of neuronal activity in the cerebral cortex of the rabbit*. Thesis, Amsterdam University, Netherlands.

Fienberg, S. E. (1974). Stochastic models for single neuron firing trains: a survey. *Biometrics*, 30, 399-427.

Gerstein, G. L. and Mandelbrot, B. (1964). Random walk models for the spike activity of a single neuron. *Biophysics Journal*, 4, 41-68.

Herz, A., Greutzfeldt, O. and Fuster, J. (1964). Statistische Eigenschaften der Neuronaktivität im ascendierenden visuellen System. *Kybernetik*, 2, 61-71.

Johannesma, P. I. M. (1969). *Stochastic neural activity*. Thesis, Nijmegen University, Netherlands.

Kwaadsteniet, J. W. de (1980). *Statistical analysis and stochastic modelling of neuron spike-train activity*. Thesis, Free University of Amsterdam, Netherlands.

McDonald, J. B. (1981). Some issues associated with the measurement of income inequality. In *Statistical Distributions in Scientific Work*, C. Taillie, G. P. Patil, and B. Baldessari, eds. Reidel, Dordrecht-Holland. Vol. 6, 161-179.

Molen, J. N. v.d., Meulen, J. W. v.d., Kramer, J. J. de and Pasveer, F. J. (1978). Computerized classification of taste cell responses. *Journal of Comparative Physiology A*, 128, 1-11.

Oosterhoff, J. and Blom, J. L. (1969). Estimation of the parameters of a truncated gamma distribution in the case of grouped observations. Appendix to Blom (1969), see above.

Smolders, F. D. J. and Folgering, H. Th. M. (1977). *Actions and interactions of* CO_2 *and* O_2 *on the controlling system of the lung ventilation.* Thesis, Nijmegen University, Netherlands.

Wise, M. E. (1971). Skew probability curves with negative powers of time and related to random walks in series. *Statistica Neerlandica,* 25, 159-180.

Wise, M. E. (1975). Skew distributions in biomedicine including some with negative powers of time. In *Statistical Distributions in Scientific Work, Vol. 2,* G. P. Patil, S. Kotz, and J. K. Ord, eds. Reidel, Dordrecht-Holland. Pages 241-262.

[*Received May* 1980. *Revised October* 1980]

PROBABILITY DISTRIBUTIONS ARISING FROM THE ASCERTAINMENT AND THE ANALYSIS OF DATA ON HUMAN FAMILIES AND OTHER GROUPS

JON STENE

Institute of Statistics
University of Copenhagen
Studiestraede 6
DK-1455 Copenhagen K, Denmark

SUMMARY. The paper presents an exposition of probability distributions arising from the selection or ascertainment of data on human families or groups and from the statistical analysis of such data consisting of a number of families. It is demonstrated how different methods of ascertainment, characterized by certain parameter structures, generate different distributions for the selected data. Certain properties of these distributions are studied, especially the type of statistical inference one can make about their parameters. It is demonstrated that the reliability of the conclusions depends heavily on how detailed the available information is. The following types of available information for each family is considered: (i) number of children and number of affected children; (ii) number of children, number of affected children and number of probands; and (iii) number of children, number of affected children, and birth orders of probands among affected children.

KEY WORDS. ascertainment models, size-biased sampling, convolutions of weighted binomial distributions, human genetics, segregation analysis, family data, selection procedures, inferential separation, ancillarity, M-ancillarity.

C. Taillie et al. (eds.), Statistical Distributions in Scientific Work, Vol. 6, 233–264.

1. INTRODUCTION

In this paper we will study the construction of probability models for data consisting of groups of individuals selected through the information that some of its members possess a certain, rare character. We will consider different types of information which may be available in connection with such data and demonstrate that the inference which can be made about the original distribution and about the data collection procedures from the available data, depends heavily on how detailed the provided information is. The models and methods which have been applied to such data, especially in human genetics, will be discussed. It will be demonstrated that the information which has usually been provided, may be insufficient for making reliable inference about the models and its parameters. How such inference can be made, when adequate information is available, will be demonstrated.

2. GENERAL BACKGROUND ABOUT THE DATA

In real life our data consist often of groups of individuals or units. Our random variables are either the number of individuals or units in a group or the number of the individuals within such a group possessing a certain character. Examples are the number of children in a family, the number of children with a certain disorder in a family of a given size, the size of youth groups of a certain type, etc. In the real life situations we want to consider, the data have been obtained through a two-stage procedure: (i) a random mechanism generating the original probability distribution and (ii) a procedure for selecting the data, where the probability for the group to be included in the data, may depend on the number of individuals in the group, the number possessing a certain character and on group specific properties such as socio-economic status, geographical localization of the group, etc.

The groups are selected or ascertained through information about one or a few of the individuals in the group. Examples are families with children which may have been selected through one of their children attending a certain school, families where at least one of the children may have a certain inherited disease and the ascertainment takes place through a hospital registry for children having this disease. When an individual belonging to such a group has been selected, information about the other members of the group is ascertained.

These types of data selection are usually not planned in the same sense as survey sampling investigations, where each unit in the population has an *a priori* known probability for being included in the sample. By the data selection procedures to be considered here, the probabilities for the groups to be included in the data, are usually unknown, but the probabilities will be assumed to satisfy certain structural relations. Contrary to what is the case in classical survey sampling, we will assume an original distribution for the groups in the population. Referring to the terminology in modern sampling theory, we will consider *superpopulation models* and *superpopulation parameters* (Cassel, *et al.*, 1977).

The probability distributions of the random variables in the selected data will usually be different from those of the original distributions and they will depend on the actual data selection procedure through certain parameters characterizing the procedure. This fact has been demonstrated in a large number of papers by Weinberg (1912a,b, 1928), Haldane (1932, 1938), Fisher (1934), Bailey (1951, 1961), Smith (1956, 1959), Morton (1959, 1962, 1969), Li (1970), Elandt-Johnson (1970, 1971a,b, 1975), Stene (1970a,b,c, 1975, 1977, 1978a,b, 1979) and Thompson and Cannings (1979) to name some of the authors who have discussed these problems in human genetics and Bailey (1956, pp. 101-104, 165; 1975, pp. 231, 263) who has discussed them in the theory of epidemics. As mentioned in papers by Rao (1965, 1977), Patil and Ord (1976) and Patil and Rao (1977, 1978) similar problems have also cropped up in other areas. These five last-mentioned papers are mainly devoted to the study of structural properties of the probability distribution for one single group selected through such a procedure, which these authors have called *size-biased sampling*. The effects of several different selection procedures have been studied.

In the real life situations referred to above, the ultimate aim has been to make inference about the original distribution and its parameters, e.g., the probability that a child have a certain inherited disorder in a family where the trait is inherited. The parameters characterizing the data selection procedure are incidental or nuisance parameters.

We will investigate the conditions which the data and the models have to satisfy in order that adequate inference can be drawn. We will also demonstrate how such inference can be carried out for samples of different structures. Problems in human genetics will form our empirical background. Therefore, we use the binomial distribution as the original distribution for our data. However, the problems considered here have a wide generality and the results can easily be transferred to other subject matter areas and to other original distributions for the data.

3. DEFINITIONS AND BASIC ASSUMPTIONS

Since human genetics will be our empirical reference area, we use common terms in that area. By a *family* is meant a group of individuals who are either biologically related or are mutually connected in other ways. The term family as defined here can mean a youth group, the children belonging to the same school class, a group of workers doing the same type of work in a factory, etc. The individuals in our groups have one and only one of two possible characters, one of which is of particular interest and usually rare in the population, e.g., an inherited disorder. An individual possessing the character of interest is said to be an *affected* individual. Individuals lacking this character are called *normal*.

Individuals being affected are assumed to be Bernoulli events which occur with probability θ. More details are given in Stene (1977). From this assumption it follows that the probability that a family with s children has r affected ones, is

$$\binom{s}{r}\theta^{r}(1-\theta)^{s-r}; \qquad r=0,1,\ldots,s. \tag{1}$$

This is the *original distribution* of our data. We will assume that θ is the same for all families that are able to have affected children. The general case where θ may vary, has been considered by Morton (1962, 1969, 1971), Elandt-Johnson (1970, 1971a,b) and Thompson and Cannings (1979).

In human genetics, the parents will usually be normal, but one or both may possess such properties that they may have children with a certain disorder. This means that the families will usually be detected because one or more of the *children are affected*. The *observable conditions of the offspring* are used as instruments for the ascertainment of information regarding the *hidden conditions of the parents* and properties of these parental conditions, e.g., the parameter θ.

By a *method of ascertainment* we mean a sampling procedure by which data about traits in families are collected.

A *proband* is an affected child who *either* has been detected independently of the other members of the family and through whom information about the conditions of the other members of the family is ascertained *or* is a member of a *proband group*. A *proband group* is a group of two or more affected children who have been detected simultaneously and independently of the other members of the family and through which group information about the conditions of the other members of the family is ascertained. The *ascertainment probability of an affected child* is the conditional probability that this affected child will be a proband. Finally, the *ascertainment probability of a family* is the conditional probability that this family will be ascertained.

4. DATA AND NOTATION

4.1 Types of Information. The following types of information may be available about human family data: (1) the number of families in the sample, (2) the number of children in each family, (3) the number of affected children in each family, (4) the number of affected and normal children of each sex in each family, (5) the number of probands in each family, (6) the number of probands of each sex in each family, (7) birth orders of affected and normal children, and (8) birth orders of probands among affected children.

Only a few of the items given above are normally reported, usually only the first three items. We will consider and classify the different models after the amount of information they utilize and demonstrate which type of inference can be done by a certain type of information about the data.

4.2 Notation When Only the Total Number of Children and the Number of Affected Children are Reported.

s: Number of children in the family.

S: Maximum number of children in any family in the sample.

N_s: Number of families in the population with s children and being able to have children with the disorder.

r: Number of affected children in the family.

a_{sr}: $(r=0,\dots,s;\ s=1,\dots,S)$ Number of families in the sample with s children, r of whom are affected.

$a_{s.}=\Sigma_{r=1}^{s} a_{sr}$: Number of families with s children with at least one affected.

$s_{.}=\Sigma_{s=1}^{S} s a_{s.}$: Number of children in the sample.

$a_{.r}=\Sigma_{s=r}^{S} a_{sr}$: Number of families with r affected children.

$r_s=\Sigma_{r=1}^{s} r a_{sr}$: Number of affected children in families with s children.

$r_{.}=\Sigma_{s=1}^{S} r_s=\sum_{r=1}^{S} r a_{.r}$: Number of affected children in the sample.

$\ln_{.}=\Sigma_{r=1}^{S} a_{.r} \ln(r)$

$\underset{\sim}{A}$: The data set as given in the triangular data matrix in Table 1.

$\underset{\sim}{a}_{s.}$: The vector $(a_{1.},\dots,a_{S.})$.

$\underset{\sim}{a}_{.r}$: The vector $(a_{.0},\dots,a_{.S})$ or the vector $(a_{.1},\dots,a_{.S})$.

$$c_0(\underset{\sim}{A}) = \prod_{s=1}^{S} \left\{ a_{s.}! \prod_{r=0}^{s} \binom{s}{r}^{a_{sr}} / a_{sr}! \right\}; \quad c_1(\underset{\sim}{A}) = \prod_{s=1}^{S} \left\{ a_{s.}! \prod_{r=1}^{s} \binom{s}{r}^{a_{sr}} / a_{sr}! \right\}.$$

(In (4) and (7) $a_{s.}$ means number of families with s children).

TABLE 1: Data matrix for a sample where only s *and* r *are recorded for each family.*

	s	r: 1	2	3	...	S	Number of families with s children
Number of children	1	a_{11}					$a_{1.}$
	2	a_{21}	a_{22}				$a_{2.}$
	3	a_{31}	a_{32}	a_{33}			$a_{3.}$
	...	...	...	...	...	...	...
	S	a_{S1}	a_{S2}	a_{S3}	...	a_{SS}	$a_{S.}$
Number of families with r affected children		$a_{.1}$	$a_{.2}$	$a_{.3}$	...	$a_{.S}$	

4.3 Notation When Also the Number of Probands is Reported for Each Family.

m : Number of probands in the family.

a_{srm} : $(m=1,\dots,r;\ r=1,\dots,s;\ s=1,\dots,S)$. Number of families in the sample with s children, r of whom are affected and m of these again are probands.

$a_{sr.} = \Sigma_{m=1}^{r} a_{srm}$: Number of families with s children, r of whom are affected.

$a_{s..} = \Sigma_{r=1}^{s} a_{sr.}$: Number of families with s children.

$a_{.rm} = \Sigma_{s=r}^{S} a_{srm}$: Number of families with r affected children, m of whom are probands.

$a_{.r.} = \Sigma_{s=r}^{S} a_{sr.}$

$= \Sigma_{m=1}^{r} a_{.rm}$: Number of families with r affected children.

$a_{..m} = \Sigma_{r=m}^{S} a_{.rm}$: Number of families ascertained through m probands.

$r_{.} = \sum_{r=1}^{S} r a_{.r.}$: Number of affected children in the sample.

$m_{.} = \sum_{s=1}^{S} \sum_{r=1}^{s} \sum_{m=1}^{r} m a_{srm}$: Number of probands in the sample.

$\underset{\sim}{B} = \{a_{srm};\ m=1,\ldots,r;\ r=1,\ldots,s;\ s=1,\ldots,S\}$. The sample given as a data matrix.

$\underset{\sim}{a}_{s..}$: The vector $(a_{1..},\ldots,a_{S..})$.

$\underset{\sim}{a}_{.r.}$: The vector $(a_{.1.},\ldots,a_{.S.})$.

$c(\underset{\sim}{B},m) = \prod_{s=1}^{S} \{a_{s..}!\ \prod_{r=1}^{s} \prod_{m=1}^{r} \binom{s}{r}^{a_{srm}} / a_{srm}!\}$.

4.4 Notation When Birth Order Data are Given Regarding the Affected Children.

i : Birth order of the proband among the affected children when there is only a single proband in the family.

a_{sri}: $(i=1,\dots,r;\ r=1,\dots,s;\ s=1,\dots,S)$. Number of families in the sample with s children, r of whom are affected and where the family's only proband has birth order i among the affected children.

$a_{sr.} = \sum_{i=1}^{r} a_{sri}$: Number of families with s children, r of whom are affected.

$a_{s..} = \sum_{r=1}^{s} \sum_{i=1}^{r} a_{sri}$: Number of families with s children.

$a_{.r.} = \sum_{s=r}^{S} \sum_{r=1}^{r} a_{sri}$: Number of families with r affected children.

$a_{..i} = \sum_{s=i}^{S} \sum_{r=i}^{s} a_{sri}$: Number of families where the proband has birth order i among the affected children.

$r_{.} = \sum_{r=1}^{S} ra_{.r.}$: Number of affected children in the data.

$\ln. = \sum_{r=1}^{S} a_{.r.} \ln(r)$

$\underset{\sim}{D} = \{a_{sri};\ i=1,\dots,r;\ r=1,\dots,s;\ s=1,\dots,S\}$. The sample given as a data matrix.

$\underset{\sim}{a}_{s..}$: The vector $(a_{1..},\dots,a_{s..})$.

$\underset{\sim}{a}_{.r.}$: The vector $(a_{.1.},\dots,a_{.s.})$.

$$c(\underset{\sim}{D},i) = \prod_{s=1}^{S} \{a_{s..}!\prod_{r=1}^{S}\prod_{i=1}^{r} \binom{s}{r}^{a_{sri}}/a_{sri}!\}.$$

5. MODELS BASED ON INFORMATION ON s AND r

5.1 Introductory Remarks. The models considered in this section are based on data where only s, the number of children, and r, the number of affected children, are recorded for each family. Most *ascertainment models*, as they are called in connection with human genetics (see, e.g., Elandt-Johnson, 1971b), or *models for size-biased sampling* with the binomial distribution as original distribution, as these models have been called recently (see, e.g., Rao and Patil, 1978), belong to this group.

The assumptions regarding the ascertainment of families made in the present paper are so far as I know more general than any made previously, especially by allowing the ascertainment probability for a family to vary from family to family, even when the number of children and the number of affected children are kept constant. This type of assumption takes into account variations from family to family caused by family specific factors, such as socio-economic conditions, ethnic origin, geographical localization of family home, etc. In most cases occurring in practice, families are not ascertained unless there is at least one affected child in it, i.e., $r \geqslant 1$. However, in certain cases families with $r=0$ may also be included in the data after an investigation of the parents.

5.2 A General Model. Assumptions About the Ascertainment.

1. The families are ascertained independently of each other.
2. The conditional probability for the ascertainment of family ν having s children, r of whom are affected, is given by $\lambda_\nu \pi_{sr}$, where the parameter π_{sr} depends on s and r only and is constant for all families with s children, r of whom are affected, and where the parameter λ_ν is independent of θ, s and r, but depends on family specific factors, such as geographical localization of family home and socio-economic, ethnic and religious factors, may vary from family to family.

An assumption of this type was first presented in Stene (1977), but only special versions of π_{sr} were considered there. The subscript ν in λ_ν is a label for family ν. From this assumption and the assumption in Section 3 about the original distribution we get by multiplying (1) with $\lambda_\nu \pi_{sr}$ that the conditional probability that family ν has r affected children and has been ascertained, given that it has s children in all, is

$$\lambda_\nu \pi_{sr} \binom{s}{r} \theta^r (1-\theta)^{s-r}; \qquad r=0,1,\ldots,s. \tag{2}$$

Disregarding the number of affected children in the family we get the following probability for family ν with s children to be ascertained

$$\lambda_\nu \sum_{r=0}^{s} \pi_{sr} \binom{s}{r} \theta^r (1-\theta)^{s-r} = \Lambda_\nu \ .$$

The probability that family ν had not been ascertained is $\Lambda'_\nu = 1-\Lambda_\nu$. Let us confine ourselves to families being able to have children with the disorder and having s children each. Let there be N_s such families in the population. For these families the original distribution (1), the binomial distribution with parameters θ and s, is assumed to be the same. The probability that we have ascertained exactly the families $\nu_1,\ldots,\nu_n$, but no others, is

$$\prod_{\substack{\nu=\nu_i \\ i=1,\ldots,n}} \Lambda_\nu \cdot \prod_{\substack{\nu=1 \\ \nu\neq\nu_i;\ i=1,\ldots,n}}^{N_s} \Lambda'_\nu \ .$$

The conditional probability that family ν_1 has $r(\nu_1)$ affected children,..., ν_n has $r(\nu_n)$ affected children, given that just these families have been ascertained, is a product of n factors, the one for family ν is

$$P_{sr} = \frac{\pi_{sr}\binom{s}{r}\theta^r(1-\theta)^{s-r}}{f_1(s,(\pi_{sr}))} \ ; \qquad r=0,1,\ldots,s. \qquad (3)$$

where

$$f_1(s,(\pi_{sr})) = \sum_{j=0}^{s} \pi_{sj}\binom{s}{j}\theta^j(1-\theta)^{s-j} \ .$$

Notice that the parameter λ_ν has been eliminated. This means the family label of family ν is an irrelevant information for the parameters θ and the π_{sr}'s. This is the case for all families $\nu_1,\ldots,\nu_n$ which have been selected. This means that we can disregard the information that just these families have been ascertained and only consider the number of the affected children in these families, each of which has s children. We can do this because we have defined the ascertainment probability as a product $\lambda_\nu \pi_{sr}$, where π_{sr} and λ_ν satisfy the stated conditions. The expression P_{sr} in (3) can be interpreted as the conditional probability that an arbitrary family in the sample has r affected children given that it has s children in all and that it has been ascertained according to a method which is in accordance with $\lambda_\nu \pi_{sr}$. Obviously, $P_{s0}+\cdots+P_{ss} = 1$.

When we confine ourselves to families with s children, we consider the joint conditional probabilities for the number of affected children in the different families, given that there are n families with s children in the sample. Hereafter, let $n=a_{s.}$ in accordance with the notation in Section 4.2. The above-mentioned conditional probability equals

$$\prod_{r=0}^{s} P_{sr}^{a_{sr}}$$

for the actual set of families. Since the a_{sr}'s are the minimal sufficient statistics for the P_{sr}'s the families with s children, we consider the conditional probability for $a_{s0},\ldots,a_{ss}$ given $a_{s.}$ which is the multinomial distribution with parameters $P_{s0},\ldots,P_{ss}$ and $a_{s.}$. By considering the conditional probability for all the a_{sr}'s in Table 1 given the $a_{s.}$'s, we get the probability

$$P\{\underset{\sim}{A}|\underset{\sim}{a}_{s.}\} = \prod_{s=1}^{S} a_{s.}! \prod_{r=0}^{s} \frac{P_{sr}^{a_{sr}}}{a_{sr}!}$$

$$= c_0(\underset{\sim}{A}) \frac{\theta^{r.}(1-\theta)^{s.}}{\prod_{s=1}^{S} f_1(s,\theta,(\pi_{sr}))^{a_{s.}}} \prod_{s=1}^{S} \prod_{r=0}^{s} \pi_{sr}^{a_{sr}}$$

$$= c_0(\underset{\sim}{A}) \frac{\left(\frac{\theta}{1-\theta}\right)^{r.} \prod_{s=1}^{S} \prod_{r=0}^{s} \pi_{sr}^{a_{sr}}}{\prod_{s=1}^{S} f_2(s,\theta,(\pi_{sr}))^{a_{s.}}} \tag{4}$$

where $f_2(s,\theta,(\pi_{sr})) = \sum_{r=0}^{s} \pi_{sr}\binom{s}{r}\left(\frac{\theta}{1-\theta}\right)^r$.

From the conditional probability (4) we notice that the minimal sufficient statistics for the parameters in the conditional distribution given $a_{s.}$ is $r.$ and the a_{sr}'s.

5.3 Known Functions of Unknown Parameters Characterizing the Method of Ascertainment. In this section we will demonstrate how different specifications of π_{sr} by known functions of unknown parameters are leading to different probability models for the data.

Consider the following multiplicative specification of the parameter π_{sr}

$$\pi_{sr} = \sigma_s \cdot \mu_r \tag{5}$$

where the parameter σ_s depends on s, the number of children in the family, only, and μ_r depends on r, the number of affected children in the family, only. If we in (3) introduce (5) for π_{sr}, we get

$$P_{sr} = \frac{\mu_r\binom{s}{r}\left(\frac{\theta}{1-\theta}\right)^r}{f_3(s,\theta,(\mu_r))} \qquad r=0,\ldots,s \tag{6}$$

where $f_3(s,\theta,(\mu_r)) = \sum_{j=0}^{s} \mu_j\binom{s}{j}\left(\frac{\theta}{1-\theta}\right)^j$.

Notice that σ_s has been eliminated. This means that any information about more extensive sampling of families with certain numbers of children than others is irrelevant information. Because σ_s is eliminated, we consider only specifications of μ_r in the sequel. The probability (4) takes here the form

$$P\{\underset{\sim}{A}|\underset{\sim}{a}_{s.}\} = c_0(\underset{\sim}{A}) \cdot \frac{\left(\frac{\theta}{1-\theta}\right)^{r.} \prod_{r=0}^{S} \mu_r^{a_{.r}}}{\prod_{s=1}^{S} f_3(s,\theta,(\mu_r))^{a_{s.}}} . \tag{7}$$

In (6) let $\mu_0=0$, $\mu_r=1-(1-\pi)^r$ where $(r=1,\ldots,S)$ and where $(0<\pi\leqslant 1)$. By this specification (6) and (7) take the forms

$$P_{sr} = \frac{\binom{s}{r}\theta^r(1-\theta)^{s-r}[1-(1-\pi)^r]}{1-(1-\theta\pi)^s}$$

$$= \frac{\binom{s}{r}\left(\frac{\theta}{1-\theta}\right)^r (1-\theta)^s[1-(1-\pi)^r]}{1-(1-\theta\pi)^s} \quad (r=1,\ldots,s) \tag{8}$$

$$P\{\underset{\sim}{A}|\underset{\sim}{a}_{s.}\} = c_1(\underset{\sim}{A}) \frac{\left(\frac{\theta}{1-\theta}\right)^{r.} (1-\theta)^{s.} \prod_{r=1}^{S} [1-(1-\pi)^r]^{a_{.r}}}{\prod_{s=1}^{S} [1-(1-\theta\pi)^s]^{a_{s.}}} . \tag{9}$$

The model (8) with all $\sigma_s=1$ in (5) and all $\lambda_\nu=1$ has been considered and applied by many authors. The interpretation of the model will be discussed in Section 6.2. The model (8) can be obtained from (19) by disregarding the number of independent probands. It was originally proposed by Weinberg (1928) and later studied by Fisher (1934), Bailey (1951, 1961), Smith (1959), Morton and his colleagues (see Morton, 1969, 1962, 1969, and Morton, *et al.*, 1971), Elandt-Johnson (1970, 1971a,b), Selvin (1975), Stene (1977) and Thompson and Cannings (1979) in human genetics and by Rao (1965, 1977) and Patil and Rao (1977, 1978) where references are given to applications in other areas.

Let μ_r in (5) be given by $\mu_0=0$, $\mu_r=r^\alpha$, where $\alpha\geqslant 0$, and $(r=1,\ldots,S)$ whence (6) takes the form

$$P_{sr} = \frac{\binom{s}{r} r^{\alpha} \left(\frac{\theta}{1-\theta}\right)^{r}}{f_4(s,\theta,\alpha)} \; ; \qquad r=1,\ldots,s \tag{10}$$

where $f_4(s,\theta,\alpha) = \sum_{j=1}^{s} \binom{s}{j} j^{\alpha} \left(\frac{\theta}{1-\theta}\right)^{j}$.

An interpretation of this model will be given in Section 7.2. The model (10) can be obtained from (24) by disregarding the birth order of the proband among the affected children. For $\lambda_\nu=\lambda$ (constant) and $\sigma_s=1$ this model has been considered by Rao (1965), Patil and Ord (1976) and Patil and Rao (1978). The more general form was considered by Stene (1978b). Here (7) takes the form

$$P\{\underset{\sim}{A}|\underset{\sim}{a}_{s.}\} = c_1(\underset{\sim}{A}) \frac{\left(\frac{\theta}{1-\theta}\right)^{r.} e^{\alpha.\ln.}}{\prod_{s=1}^{S} f_f(s,\theta,\alpha)^{a_{s.}}} . \tag{11}$$

Let m be a known integer and let in (5) $\mu_r=0$, for $r<m$; $\mu_r = \binom{r}{m}^{\alpha_m}$ for $r \geq m$, where the parameter $\alpha_m \geq 0$. Hence (6) takes the form

$$\frac{\binom{s}{r}\binom{r}{m}^{\alpha_m} \left(\frac{\theta}{1-\theta}\right)^{r}}{f_m(s,\theta,\alpha_m,m)} \; ; \qquad r=m,\ldots,s. \tag{12}$$

where $f_m(s,\theta,\alpha_m,m) = \sum_{j=m}^{s} \binom{s}{j}\binom{j}{m}^{\alpha_m} \left(\frac{\theta}{1-\theta}\right)^{j}$. The interpretation of the model is given in Section 7.3. This model has been considered by Stene (1978a, 1979). For $m=1$ we are back to the model (10).

If in (8) we choose $\pi=1$ or in (10) $\alpha=0$ we get the important special case, the zero-truncated binomial

$$\binom{s}{r} \frac{\theta^r(1-\theta)^{s-r}}{1-(1-\theta)^s} \; ; \qquad r=1,\ldots,s. \tag{13}$$

If instead in (8) $\pi \to 0$ or in (10) we choose $\alpha=1$, then we get

$$\binom{s-1}{r-1}\theta^{r-1}(1-\theta)^{s-r}; \qquad r=1,\ldots,s. \tag{14}$$

Important special cases of (12) are $\alpha_m=0$, which gives a left truncated binomial truncated at $(m-1)$, and $\alpha_m=1$, which can be rewritten as

$$\frac{\binom{s}{r} r\cdot(r-1)\ldots(r-m+1)\left(\frac{\theta}{1-\theta}\right)^r}{\sum_{j=m}^{s}\binom{s}{j} j(j-1)\ldots(j-m+1)\left(\frac{\theta}{1-\theta}\right)^j}; \tag{15}$$

which is an example of a *factorial moment distribution* considered by Patil and Ord (1976). The interpretations of (13), (14), and (15) will be given in Sections 7.2 and 7.3.

In Section 6 it will be shown that inference about the parameters π and θ can be separated and in Section 7 similarly for the parameters α and θ. However, for none of the models considered so far are there any possibilities to separate the inference about the parameters characterizing the method of ascertainment from the inference about θ. The reason is that we, in no case, can find a statistic ancillary for either set of parameters.

A property of (9) is that $P\{\underset{\sim}{A}|\underset{\sim}{a}_{s.}\}>0$ for $\pi>1$ and of (11) that $P\{\underset{\sim}{A}|\underset{\sim}{a}_{s.}\}>0$ for $\alpha<0$. A consequence of these properties is that the maximum likelihood estimates will not exist for many sets of data. The traditional method has been to determine those parameter values (θ,π) in (9) where $(0\leqslant(\theta,\pi)\leqslant 1)$ or those parameter values (θ,α) in (11) where $(0\leqslant\theta\leqslant 1,\alpha\geqslant 0)$ which maximize $P\{\underset{\sim}{A}|\underset{\sim}{a}_{s.}\}$. By this procedure our maximum likelihood estimates will often be border values. In many cases the results will be unsatisfactory, which the following three examples may indicate.

Example 1. Selvin (1975) analyzed a large set of data consisting of 411 families with albino children published by Haldane (1938). The number s of children in the families varied from 2 to 15 and Selvin considered a model of the type (9) with π depending on s, i.e., π_s. He estimated θ and $\pi_2,\ldots,\pi_{15}$ jointly by the maximum likelihood method and got the results $\hat{\theta}=0.325$ and $\hat{\pi}_2=0.79$, $\hat{\pi}_3=1.0$, $\hat{\pi}_4=0.84$, $\hat{\pi}_5=1.0$, $\hat{\pi}_6=0.34$, $\hat{\pi}_7=1.0$, $\hat{\pi}_8=0.80$, $\hat{\pi}_9=1.0$,

$\hat{\pi}_{10}=1.0$, $\hat{\pi}_{11}=0.51$, $\hat{\pi}_{12}=1.0$, $\hat{\pi}_{13}=0.33$, $\hat{\pi}_{14}=0.94$, $\hat{\pi}_{15}=0.89$. The large variation between the estimates $\hat{\pi}_s$ and the fact that many of them possess the boundary value $\hat{\pi}_s=1.0$ is quite disturbing. These observations indicate that the estimation is unstable owing to the properties of the model.

Example 2. If we apply model (11) to the same data, we get the border values $\hat{\theta}=0.31$ and $\hat{\alpha}=0$. (If we removed the restriction $\alpha \geqslant 0$ we would get a negative estimate of α.)

Example 3. The data in Table 2 were originally published in Stene (1978b). They were simulated for fixed $a_{2.}=18$, $a_{3.}=12$, and $a_{4.}=6$ and $\theta=0.50$ and $\alpha=0.75$. By the use of Fisher's scoring method we get the border values $\hat{\theta}=0.63$ and $\hat{\alpha}=0$. (Unrestricted use of Fisher's scoring method would have produced the values $\theta^*=0.90$ and $\alpha^*=-1.870$.) These results must be considered as unsatisfactory.

A more thorough discussion of the parameter estimation for these models will be given elsewhere.

TABLE 2: The data from Stene (1978b) *classified after number of children* s, *and number of affected children* r.

	s	r = 1	2	3	4	$a_{s.}$
Number	2	11	7			18
of	3	4	5	3		12
Children	4	0	1	4	1	6
	$a_{.r}$	15	13	7	1	36

5.4 Models Where the μ_r*'s are Known.* In models where π_{sr}'s or the μ_r's are completely known, there are only one unknown parameter, namely θ. The form of the model depends on the actual values of the π_{sr}'s or μ_r's. In the conditional distribution given the $a_{s.}$'s the minimal sufficient statistic for θ is $r.$. The estimation of θ can be carried out from the relevant model by means of the maximum likelihood method. For (9) with π known, the maximum likelihood equation is $G(\hat{\theta})=0$, where

$$G(\theta) = \frac{r.}{\theta(1-\theta)} - \frac{s.}{(1-\theta)} - \sum_{s=1}^{S} a_{s.} \frac{s\pi(1-\theta\pi)^{s-1}}{1-(1-\theta\pi)^s}$$

and the Fisher information is

$$I(\theta) = \frac{s.}{(1-\theta)^2} + \sum_{s=1}^{S} a_{s.} \frac{\theta(1-2\theta)[1-(1-\pi)(1-\theta\pi)^{s-1}]}{\theta^2(1-\theta)^2[1-(1-\theta\pi)^s]}$$

$$- \sum_{s=1}^{S} a_{s.} \frac{s\pi^2(1-\theta\pi)^{s-2}}{[1-(1-\theta\pi)^s]^2} \{s-[1-(1-\theta\pi)^s]\}$$

(see Elandt-Johnson, 1971b, pp. 471-472). The estimation is easily done by Fisher's scoring method $\hat{\theta}=\theta^*+G(\theta^*)/I(\theta^*)$ where θ^* is a trial value.

For model (11) with known α the situation is similar. Fisher's scoring method is here

$$\hat{\theta} = \theta^*+G_1(\theta^*,\alpha)/I_{11}(\theta^*,\alpha) \tag{16}$$

where $G_1(\theta,\alpha)$ and $I_{11}(\theta,\alpha)$ are given below.

$$G_1(\theta,\alpha) = \frac{1}{\theta(1-\theta)} \left[r.- \sum_{s=1}^{S} a_{s.} \frac{f_{10}(s,\theta,\alpha)}{f_4(s,\theta,\alpha)} \right]$$

$$I_{11}(\theta,\alpha) = \frac{1}{\theta^2(1-\theta)^2} \sum_{s=1}^{S} a_{s.}\left[\frac{f_{20}(s,\theta,\alpha)}{f_4(s,\theta,\alpha)} - \left(\frac{f_{10}(s,\theta,\alpha)}{f_4(s,\theta,\alpha)}\right)^2\right]$$

where $f_{10}(s,\theta,\alpha) = \sum_{r=1}^{s} \binom{s}{r} r^{\alpha+1} \left(\frac{\theta}{1-\theta}\right)^r$,

$$f_{20}(s,\theta,\alpha) = \sum_{r=1}^{s} \binom{s}{r} r^{\alpha+2} \left(\frac{\theta}{1-\theta}\right)^r .$$

Example 4. Consider again the data in Table 2. In Stene (1978b) we estimated by means of birth order data $\hat{\alpha}=0.754$. Although we could not reject the hypothesis $\alpha=1$ against $\alpha<1$, and hence used model (14) as the basis for our estimation of θ, we will here consider the estimated value $\alpha=0.754$ as known and estimate θ by (16). As a trial value for θ use $\theta^*=0.5$. Hence we get $G_1(0.5,0.754)=8.39026$ and $I_{11}(0.5,0.754)=242.32211$. By (16) we get $\hat{\theta}=0.53462$, i.e., the correction $\hat{\theta}-\theta^*=0.03462$. Using $\hat{\theta}$ as our new trial value, our new correction is -0.00015 and we choose as our maximum likelihood estimate $\hat{\theta}=0.534$ for θ. Further is $I_{11}(0.534,0.754)=245.8$ and hence $\mathrm{var}(\hat{\theta})=1/245.8= 0.0046$. Compare these results with those in Example 3.

The estimation is particularly simple when the model is (14). The maximum likelihood estimator is

$$\hat{\theta} = (r. - \sum_{s=1}^{S} a_{s.})/(s. - \sum_{s=1}^{S} a_{s.}); \ \mathrm{var}(\hat{\theta}) = \theta(1-\theta)/(s. - \sum_{s=1}^{S} a_{s.}).$$

For model (13) the estimation of θ is more complicated. In addition to the maximum likelihood method given by, e.g., Elandt-Johnson (1971b, pp. 472-473) a number of more or less efficient methods have been developed, the first one by Weinberg (1912a,b). A review has been given by Li (1970). An investigation of small-sample properties of several of these estimators has been carried out by Thomas and Gart (1971). Tables for facilitating the calculation of the variances for such estimators have been provided by Finney (1949) and by Li and Mantel (1968). Tables for use in testing the genetically important hypotheses $\theta=0.25$ and $\theta=0.50$ have been provided by Smith (1956), Maynard-Smith, *et al.* (1962) and by Elandt-Johnson (1971b).

Model (12) with $\alpha_m=0$ and $m=2$ has been applied by Chung, *et al.* (1959) and $\alpha_m=0$ and $m=2,3$ by Stene (1970a,b,c). In Stene (1970a) estimation and test methods for data consisting of families ascertained through proband groups for various, but known, sizes have been provided.

6. MODELS BASED ON INFORMATION ON s, r AND m

6.1 Introductory Remarks. The models considered in this section are based on data where s, the number of children, r, the number of affected children, and m, the number of probands, have been recorded for each family. Few models of this type are available yet. The only well-known one is that considered in Section 6.2.

Let the ascertainment probability corresponding to that in Section 5.2 be $\lambda_\nu\pi_{srm}$ where π_{srm} depends on s, r and m only and λ_ν is family dependent as in Section 5.2. Here we will consider the multiplicative case where π_{srm} is specified analogously to (5) as $\pi_{srm}=\sigma_s\xi_{rm}$ with ξ_{rm} depending on r and m only. Then we get the conditional probability that family ν has r affected children and m of these probands given that it has s children in all and has been ascertained by this method, is

$$P_{srm} = \frac{\xi_{rm}\binom{s}{r}\left(\frac{\theta}{1-\theta}\right)^r}{g_1(s,\theta,(\xi_{rm}))} \quad ; \quad 1\leqslant m\leqslant r\leqslant s \qquad (17)$$

where $$g_1(s,\theta,(\xi_{rm})) = \sum_{i=1}^{s}\sum_{j=1}^{i}\xi_{ij}\binom{s}{i}\left(\frac{\theta}{1-\theta}\right)^i .$$

Notice that both λ_ν and σ_s are eliminated. In certain cases we may envisage that $m=0$, but that situation will be left out. The conditional probability for the data matrix $\underset{\sim}{B}$ in Section 4.3, given $a_{s..}$, is

$$P\{\underset{\sim}{B}|\underset{\sim}{a}_{s..}\} = c(\underset{\sim}{B},m) \frac{\left(\frac{\theta}{1-\theta}\right)^{r.} \prod_{r=1}^{S} \prod_{m=1}^{r} \xi_{rm}^{a_{.rm}}}{\prod_{s=1}^{S} g_1(s,\theta,(\xi_{rm}))^{a_{s..}}} . \tag{18}$$

It is not possible to carry out separate inference for the parameters θ and the ξ_{rm}'s. A joint estimation of θ and the ξ_{rm}'s will not be possible without imposing further restrictions on the ξ_{rm}'s.

6.2 *Weinberg's Incomplete Multiple Ascertainment Model.* Let ξ_{rm} be specified in the following way

$$\xi_{rm} = \binom{r}{m}\pi^m(1-\pi)^{r-m} \qquad 1 \leq m \leq r.$$

By this specification (17) takes the form

$$P_{srm} = \frac{\binom{s}{r}\left(\frac{\theta}{1-\theta}\right)^r (1-\theta)^s \binom{r}{m}\left(\frac{\theta}{1-\theta}\right)^m (1-\pi)^r}{1-(1-\theta\pi)^s} ; \quad 1 \leq m \leq r \leq s \leq S \tag{19}$$

and (18) takes the form

$$P\{\underset{\sim}{B}|\underset{\sim}{a}_{s..}\} = c(\underset{\sim}{B},m) \cdot \frac{\left(\frac{\theta}{1-\theta}\right)^{r.} \left(\frac{\pi}{1-\pi}\right)^{m.} (1-\theta)^{s.}(1-\pi)^{r.}}{\prod_{s=1}^{S} [1-(1-\theta\pi)^s]^{a_{s..}}} . \tag{20}$$

The model (19) was suggested by Weinberg (1928) for the case that $\lambda_\nu=1$ for all ν and $\sigma_s=1$ for all s. *The interpretation of* (19) *is the conditional probability that a family in the sample has* r *affected children with* m *of these being probands who have been selected independently, each with the same probability* π *of being a proband, given that the family has* s *children and has been ascertained by this method.* The parameter π is usually assumed to be the same for all affected children in the family and is often assumed to be the same for all families in the population as well. The model (20) is based on these assumptions. The assumption that π is constant for all families, has been discussed extensively in the literature (see references in Section 5.3). *If* m *is not recorded, the corresponding probabilities are* (8) *and* (9).

Our ultimate aim is to make inference about the parameter θ. From (20) we notice that $r.$ is a canonical statistic both for $\theta/(1-\theta)$ and for $(1-\pi)$. Therefore, we cannot find directly a conditional distribution which is parametrized by θ only. The only two ways to make inference about θ will be either a two-stage procedure, where we make inference first about π and then about θ, or a joint estimation of both parameters. Separate inference about the two parameters may be based on the following factorization of (20):

$$P\{\underset{\sim}{B}|\underset{\sim}{a}_{s..}\} = P\{\underset{\sim}{B}|\underset{\sim}{a}_{s..},\underset{\sim}{a}_{.r.}\}\cdot P\{\underset{\sim}{a}_{.r.}|\underset{\sim}{a}_{s..}\} . \tag{21}$$

The first factor is

$$P\{\underset{\sim}{B}|\underset{\sim}{a}_{s..},\underset{\sim}{a}_{.r.}\} = \frac{c(\underset{\sim}{B},m)}{c(\underset{\sim}{a}_{s..},\underset{\sim}{a}_{.r.},m)} \frac{\left(\frac{\pi}{1-\pi}\right)^{m.}(1-\pi)^{r.}}{\prod_{r=1}^{S}[1-(1-\pi)^{r}]^{a_{.r.}}} \tag{22}$$

where $$c(\underset{\sim}{a}_{s..},\underset{\sim}{a}_{.r.},m) = \sum \prod_{s=1}^{S} a_{s..}! \prod_{r=1}^{s}\prod_{m=1}^{r} \frac{\binom{s}{r}^{a_{srm}}}{a_{srm}!}$$

with summations over all a_{srm}'s satisfying the conditions

$$\sum_{r=1}^{s}\sum_{m=1}^{r} a_{srm} = a_{s..},\quad \sum_{s=r}^{S}\sum_{m=1}^{r} a_{srm} = a_{.r.} \qquad (r,s=1,\ldots,S).$$

The second factor is (compare with (9))

$$P\{\underset{\sim}{a}_{.r.}|\underset{\sim}{a}_{s..}\} =$$

$$c(\underset{\sim}{a}_{s..},\underset{\sim}{a}_{.r.},m)\left(\frac{\theta}{1-\theta}\right)^{r.}(1-\theta)^{s.}\frac{\prod_{r=1}^{S}[1-(1-\pi)^{r}]^{a_{.r.}}}{\prod_{s=1}^{S}[1-(1-\theta\pi)^{s}]^{a_{s..}}} \quad . \qquad (23)$$

Notice that (22) is parametrized by π only, while (23) depends on both θ and π. In the model (22) the statistic $\underset{\sim}{a}_{.r.}$ is M-*ancillary* for π. This is because (23), the conditional marginal distribution of $\underset{\sim}{a}_{.r.}$ given $\underset{\sim}{a}_{s..}$, is *non-informative* with regard to π in the sense that for any observation $\underset{\sim}{a}_{.r.}$ and the given model (23) one is unable to say that some value π_0 of π is more credible than any other. The reason is that for $\pi=\pi_0$, we can find a value of θ such that $\underset{\sim}{a}_{.r.}$ is a mode point of (23) and for any other $\pi\neq\pi_0$ we will also be able to find a value of θ such that $\underset{\sim}{a}_{.r.}$ is a mode point. This means that the model (23) gives a perfect fit to data for any π, (Barndorff-Nielsen, 1978, p. 48). This conclusion can be drawn because (23) is strongly unimodal, which may be proved by the same type of arguments as used by Pedersen (1975a,b) and by Barndorff-Nielsen (1978, p. 211 ff.). The result of these arguments is that all available information regarding π is contained in (22), which does not contain any available information regarding θ.

The information about θ is present in (23). The statistic $\underset{\sim}{a}_{.r.}$ is M-*sufficient* for θ. The model (23) does not contain, as mentioned above, any available information regarding π. However, π in the form of $\mu_r=1-(1-\pi)^r;(r\geqslant 1)$ is a structural parameter in (23). Inference about θ cannot be obtained before π (or μ_r) is known. When π is known, it turns out that $r_.$, which is a linear function of $\underset{\sim}{a}_{.r.}$, is the minimal M-sufficient statistic regarding θ.

By the principle of ancillarity (Barndorff-Nielsen, 1978, p. 34) it follows that inference about π should be carried out in the conditional distribution given $\underset{\sim}{a}_{s..}$ and $\underset{\sim}{a}_{.r.}$. From (22) it follows that in this conditional distribution $m.$ is the minimal B-sufficient statistic for π. The maximum likelihood equation is $G_B(\hat{\pi})=0$ where

$$G_B(\pi) = \frac{m.}{\pi(1-\pi)} - \frac{r.}{1-\pi} - \sum_{r=1}^{S} a_{.r.} \frac{r(1-\pi)^{r-1}}{1-(1-\pi)^r}$$

and the Fisher-information

$$I_B(\pi) = \frac{r.}{(1-\pi)^2} + \sum_{r=1}^{S} a_{.r.} \frac{r\pi}{1-(1-\pi)^r}$$

$$- \sum_{r=1}^{S} a_{.r.} \left[\frac{r(r-1)(1-\pi)^{r-2}}{1-(1-\pi)^r} - \left(\frac{r(1-\pi)^{r-1}}{1-(1-\pi)^r} \right)^2 \right] .$$

The maximum likelihood estimator is found by iteration in Fisher's scoring method $\hat{\pi}=\pi^*+G_B(\pi^*)/I_B(\pi^*)$ as usual. When we either have estimated π or have tested a hypothesis about π in this distribution, the chosen value of π is then considered as known. Hence the μ_r's, where $\mu_r=1-(1-\pi)^r$ for $r\geq 1$ will be known. By this procedure the *structure* of our ascertainment model, specified by the μ_r's, has been determined without the presence of irrelevant information regarding the unknown parameter θ.

By the principle of sufficiency (Barndorff-Nielsen, 1978, p. 35) we can now make inference about θ by means of the model (23) and the statistic r. This can be done by the method indicated in the beginning of Section 5.4.

It should be noticed that we have carried out our separate inference on the basis of the factorization (21). We could also have used the factorization $P\{\underset{\sim}{B}|r.\}\cdot P\{r.|a_{s..}\}$, but then the estimation of π would have been more complicated.

By separating the inference about the two parameters θ and π the inference about each parameter is based on that part of the model where the relevant information regarding this parameter is concentrated, and irrelevant information stemming from the other parameter is excluded (see Barndorff-Nielsen, 1978, ch. 4). On the contrary, joint inference about both parameters, which has been recommended by many authors, e.g., Morton (1959, 1962, 1969) and Elandt-Johnson (1971b), will have the effect that the inference regarding each of the parameters will be affected by irrelevant information stemming from the other one. The conclusions will, therefore, be less clear-cut than in case of separate inference. Therefore, joint inference should be avoided.

7. MODELS BASED ON INFORMATION ON s, r, AND THE BIRTH ORDERS OF PROBANDS AMONG AFFECTED CHILDREN

7.1 Introductory Remarks. The models considered in this section are based on data where s, the number of children, r, the number of affected children, m, the number of probands, and the *birth orders of the probands among the affected children* are reported for each family. As far as I know, the only models of this type which have been considered yet, are those in Stene (1977, 1978a,b, 1979). For data of this type it is also possible to construct general models analogously to what we did in Sections 5.2 and 6.1. Instead we will consider the models proposed in Stene (1978b, 1979).

7.2 Families With a Single Proband. Let the conditional probability that family ν with s children, r of whom are affected, be ascertained through the child who has birth order i among the affected children, be $\lambda_\nu \sigma_s$ for $(i=1)$ and $\lambda_\nu \sigma_s [i^\alpha-(i-1)^\alpha]$ for $(i=2,\ldots,r)$, where λ_ν and σ_s have the same meaning as before. The parameter $\alpha \geq 0$. By going through similar arguments as we did in Section 5.2, we get that *the conditional probability that the family has* r *affected children and the* ith *affected child is the proband given that the family has* s *children and is ascertained in this way, is*

$$P_{sr1} = \frac{1}{r^\alpha} \cdot \frac{\binom{s}{r} r^\alpha \left(\frac{\theta}{1-\theta}\right)^r}{f_4(s,\theta,\alpha)} ; \qquad i=1; \qquad 1 \leq r \leq s$$

$$P_{sri} = \frac{i^\alpha-(i-1)^\alpha}{r^\alpha} \cdot \frac{\binom{s}{r} r^\alpha \left(\frac{\theta}{1-\theta}\right)^r}{f_4(s,\theta,\alpha)} ; \qquad i=2,\ldots,r; \qquad 2 \leq r \leq s \qquad (24)$$

where $f_4(s,\theta,\alpha)$ is given in (10). *Disregarding the birth order we get the conditional probability* (10) *that the family has* r *affected children given that it has* s *children in all and is ascertained in this way.*

From (24) we notice that the conditional probability that the proband has birth order i among the affected children, given that these are r affected children and the family has been ascertained through a single proband, is

$$1/r^{\alpha} \text{ for } (i=1) \text{ and } [i^{\alpha}-(i-1)^{\alpha}]/r^{\alpha} \text{ for } (i=2,\ldots,r). \tag{25}$$

The parameter α *formalizes a connection between the birth order of child number* i *among the affected children and the probability for this child to be the proband of the family.*

We notice from (25) that if $\alpha=0$, then the eldest affected child is the proband with probability one. This means that if the family has been ascertained, then it has been ascertained through the first affected child. If $\alpha=1$, then each affected child has the same, not necessarily small, probability of being the proband. We mentioned in Section 5.3 that $\alpha=0$ corresponded to the zero-truncated binomial distribution (13), and the $\alpha=1$ corresponded to (14). If we instead consider (19) and (8) then $\pi=1$, which leads to (13), means that each affected child is a proband with probability one, while $\pi \to 0$, which leads to (14), means that an affected child has a very small probability of being a proband, and usually there will be only a single proband in each family in the ascertained data. These differences in interpretation of (13) and (14) may cause difficulties. If there is only a single proband in each family and its birth order is disregarded, the use of (19) will lead to (14). If, however, the proband is always the eldest child, then (13) should have been used. This wrong choice of model will lead to heavily biased estimates of θ.

The conditional probability for the data matrix $\underset{\sim}{D}$ in Section 4.4 given $\underset{\sim}{a}_{s..}$ is

$$P\{\underset{\sim}{D}|\underset{\sim}{a}_{s..}\} = c(\underset{\sim}{D},i) \cdot \frac{\prod_{i=2}^{S} [i^{\alpha}-(i-1)^{\alpha}]^{a_{..i}}}{\prod_{s=1}^{S} f_4(s,\theta,\alpha)^{a_{s..}}} \left(\frac{\theta}{1-\theta}\right)^{r.}$$

Analogously to what we did in Section 6.2 we can separate the inference of α from that of θ by factorizing $P\{\underset{\sim}{D}|\underset{\sim}{a}_{s..}\}$ in the following way

$$P\{\underset{\sim}{D}|\underset{\sim}{a}_{s..}\} = P\{\underset{\sim}{D}|\underset{\sim}{a}_{s..},\underset{\sim}{a}_{.r.}\}\cdot P\{\underset{\sim}{a}_{.r.}|\underset{\sim}{a}_{s..}\} .$$

The first factor is

$$P\{\underset{\sim}{D} | \underset{\sim}{a}_{s..}, \underset{\sim}{a}_{.r.}\} = \frac{c(\underset{\sim}{D}, i)}{c(\underset{\sim}{a}_{s..}, \underset{\sim}{a}_{.r.}, i)} \; \frac{\prod_{i=2}^{S} [i^{\alpha} - (i-1)^{\alpha}]^{a_{..i}}}{\exp(\alpha \cdot \ln.)}$$

where $c(\underset{\sim}{a}_{s..}, \underset{\sim}{a}_{.r.}, i) = \sum \prod_{s=1}^{S} a_{s..}! \prod_{r=1}^{s} \prod_{i=1}^{r} \frac{\binom{s}{r}^{a_{sri}}}{a_{sri}!}$

where the summation takes place over all a_{sri}'s satisfying the conditions

$$\sum_{r=1}^{s} \sum_{i=1}^{r} a_{sri} = a_{s..}, \quad \sum_{s=r}^{S} \sum_{i=1}^{r} a_{sri} = a_{.r.} \qquad (r, s = 1, \ldots, S).$$

The second factor is (compare with (11))

$$P\{\underset{\sim}{a}_{.r.} | \underset{\sim}{a}_{s..}\} = \frac{c(\underset{\sim}{a}_{s..}, \underset{\sim}{a}_{.r.}, i)}{\prod_{s=1}^{S} f_4(s, \theta, \alpha)^{a_{s..}}} \left(\frac{\theta}{1-\theta}\right)^{r.} \cdot e^{\alpha . \ln.} .$$

In the same way as in Section 6.2 we find that $\underset{\sim}{a}_{.r.}$ is M-ancillary for α in the first factor above and that $r.$ is M-sufficient for θ in the second factor. By the use of the principle of ancillarity inference about α should be carried out in $P\{\underset{\sim}{D} | \underset{\sim}{a}_{s..}, \underset{\sim}{a}_{.r.}\}$. This is done in Stene (1978b). Introducing the determined value of α in $P\{\underset{\sim}{a}_{.r.} | \underset{\sim}{a}_{s..}\}$ we make inference about θ by means of (16) by the use of the principle of sufficiency.

7.3 Families With m *Probands*. Let the family have m probands who have been detected simultaneously. By arguing analogously to what we did in the beginning of Section 7.2, we find that if a family with r affected children has been ascertained through a group of m probands, the conditional probability that these have the birth orders $j_1, \ldots, j_m$ among the affected children may be specified in the following way (Stene, 1979)

$$\prod_{k=1}^{m} \frac{\left[\binom{j_k}{k}^{\alpha_k} - \binom{j_k-1}{k}^{\alpha_k}\right]}{\binom{j_{k+1}-1}{k}^{\alpha_k}} \tag{26}$$

where all $\alpha_k \geq 0$ and where $k \leq j_k \leq j_{k+1}-1$, $(k=1,\ldots,m-1)$ and $m \leq j_m \leq r$ and $j_{m+1}-1=r$. *The parameter* α_k *formalizes a connection between the birth order of child number* j_k *among the affected children and the probability that this child is the* kth *eldest one in the group of* m *probands, through which the family has been ascertained.* If $\alpha_1=\ldots=\alpha_m=0$ and we make the convention that $0^0=0$, then the ascertainment takes place through the m eldest affected children with probability one, and if $\alpha_1=\ldots=\alpha_m=1$ all possible groups of m affected children have the same probability of being the proband group. In this case (26) is reduced to $1/\binom{r}{m}$. In Stene (1979) it is demonstrated that *the conditional probability that a family has* r *affected children given that it has* s *children and has been ascertained through the simultaneous information about* m *probands, is*

$$\frac{\binom{s}{r}\binom{r}{m}^{\alpha_m}\left(\frac{\theta}{1-\theta}\right)^{r}}{f_m(s,\theta,\alpha_m)} \quad (m \leq r \leq S) \tag{27}$$

where $f_m(s,\theta,\alpha_m) = \sum_{j=m}^{s} \binom{s}{j}\binom{j}{m}^{\alpha_m}\left(\frac{\theta}{1-\theta}\right)^{j}$

which depends, in addition to θ, only on *the parameter* α_m, *which formalizes a connection between the birth order of one affected child in the family and the probability that this child is the youngest member of the proband group of size* m. This means that *the ascertainment probability of the family depends on the birth order among the affected children of the youngest member of the proband group of size* m, but the ascertainment probability does not depend directly on the birth orders of any other members of this group. Model (27) is identical to (12).

The interpretation of the special case $\alpha_m=0$ of (12), is that the youngest member of the proband group of size m has birth order m among the affected children, and the interpretation of (15) is that all affected children with birth orders among the affected children at least equal to m have the same probability of being the youngest member of the proband group of size m.

In order to make inference about α_m we consider the conditional probability that the family has r affected children and the youngest member of the proband group has birth order j_m, given that the family has s children and has been ascertained through a proband group of m members is

$$P_{srm} = \binom{r}{m}^{-\alpha_m} \frac{\binom{s}{r}\binom{r}{m}^{\alpha_m}\left(\frac{\theta}{1-\theta}\right)^r}{f_m(s,\theta,\alpha_m)} \; ; \quad j_m=m$$

$$P_{srj_m} = \frac{\left[\binom{j_m}{m}^{\alpha_m} - \binom{j_m-1}{m}^{\alpha_m}\right]}{\binom{r}{m}^{\alpha_m}} \; \frac{\binom{s}{r}\binom{r}{m}^{\alpha_m}\left(\frac{\theta}{1-\theta}\right)^r}{f_m(s,\theta,\alpha_m)} \; ; \quad j_m=m+1,\ldots,r \tag{28}$$

These expressions are analogous to those we considered in (24) and by an analogous procedure we may demonstrate that the inference on α_m should take place in the conditional distribution given $\underset{\sim}{a}_{s..}$ and $\underset{\sim}{a}_{.r.}$.

8. CONCLUDING REMARKS

In the preceding sections it has been demonstrated that the possibilities for making inference about the unknown parameters depend on how detailed the provided information is. When only s and r are known, there are no possibilities for separating the inference about the different parameters and by joint maximum likelihood estimation, the estimates will not exist for many of the possible observations, such that one has to relay on border values of the estimates. When the number of probands was known, then reliable inference about both parameters in the model (19) (and (8)) could be made, and when the birth order of the (youngest) proband among the affected individuals was known then reliable inference about both parameters in (24) (and (10)) or (27) (and (12)) could be made.

This means that we need an additional type of information for each new parameter we want to estimate. When the relevant information was available, we could make separate inference about the ascertainment model specified by the actual parameter and θ, the parameter of interest. This type of two-stage procedure is preferable to joint inference about the parameters.

The birth orders of the probands among the affected individuals is more detailed information than the number of probands only. By this more detailed information it is possible to investigate if (19) is a realistic model for the data. This can be done by means of the methods mentioned in Sections 7.2 and 7.3 as indicated for special cases in the discussion after (25).

REFERENCES

Bailey, N. T. J. (1951). The estimation of frequencies of recessives with incomplete multiple selection. *Annals of Eugenics*, 16, 215-222.

Bailey, N. T. J. (1957). *The Mathematical Theory of Epidemics*. Griffin, London.

Bailey, N. T. J. (1961). *Introduction to the Mathematical Theory of Genetic Linkage*. Clarendon Press, Oxford.

Bailey, N. T. J. (1975). *The Mathematical Theory of Infectious Diseases*. Griffin, London.

Barndorff-Nielsen, O. (1978). *Information and Exponential Families in Statistical Theory*. Wiley, Chichester.

Cassel, C. M., Särndal, C. E., and Wretman, J. H. (1977). *Foundations of Inference in Survey Sampling*. Wiley, New York.

Chung, C. S., Robison, O. W., and Morton, N. E. (1959). A note on deaf mutism. *Annals of Human Genetics*, 23, 357-366.

Elandt-Johnson, R. C. (1970). Segregation analysis for complex models of inheritance. *American Journal of Human Genetics*, 22, 129-144.

Elandt-Johnson, R. C. (1971a). Complex segregation analysis, II. Multiple classification. *American Journal of Human Genetics*, 23, 17-32.

Elandt-Johnson, R. C. (1971b). *Probability Models and Statistical Methods in Genetics*. Wiley, New York.

Elandt-Johnson, R. C. (1975). Segregation analysis: an overview. In *Proceedings of the 8th International Biometric Conference*, L. C. A. Corsten and T. Postelnicu, ed. The Publishing House of the Academy of the Socialist Republic of Romania, Bucharest. Pages 313-323.

Finney, D. J. (1949). The truncated binomial distribution. *Annals of Eugenics*, 14, 319-328.

Fisher, R. A. (1934). The effect of methods of ascertainment upon the estimation of frequencies. *Annals of Eugenics*, 6, 13-25.

Haldane, J. B. S. (1932). A method for investigating recessive characters in man. *Journal of Genetics*, 25, 251-256.

Haldane, J. B. S. (1938). The estimation of the frequencies of recessive conditions in man. *Annals of Eugenics*, 8, 256-262.

Li, C. C. (1970). The incomplete binomial distribution. In *Mathematical Topics in Population Genetics*, K. Kojima, ed. Springer Verlag, Berlin. Pages 337-366.

Li, C. C. and Mantel, N. (1968). A simple method of estimating the segregation ratio under incomplete ascertainment. *American Journal of Human Genetics*, 20, 61-81.

Maynard-Smith, S., Penrose, L. S., and Smith, C. A. B. (1961). *Mathematical Tables for Research Workers in Human Genetics*. Churchill, London.

Morton, N. E. (1959). Genetic tests under incomplete ascertainment. *American Journal of Human Genetics*, 11, 1-16.

Morton, N. E. (1962). Segregation and linkage. In *Methodology in Human Genetics*, J. Burdette, ed. Holden-Day, San Francisco. Pages 17-52.

Morton, N. E. (1969). Segregation analysis. In *Computer Applications in Human Genetics*, N. E. Morton, ed. University of Hawaii Press, Honolulu. Pages 129-139.

Morton, N. E., Yee, S., and Lew, R. (1971). Complex segregation analysis. *American Journal of Human Genetics*, 23, 602-611.

Patil, G. P. and Ord, J. K. (1976). On size-biased sampling and related form-invariant weighted distributions. *Sankhya*, 38B, 48-61.

Patil, G. P. and Rao, C. R. (1977). The weighted distributions: A survey of their applications. In *Applications of Statistics*, P. R. Krishnaiah, ed. North-Holland, Amsterdam. Pages 383-405.

Patil, G. P. and Rao, C. R. (1978). Weighted distributions and size-biased sampling with applications to wildlife populations and human families. *Biometrics*, 34, 179-189.

Pedersen, J. G. (1975a). On strong unimodality of two-dimensional discrete distributions with applications to M-ancillarity. *Scandinavian Journal of Statistics*, 2, 99-102.

Pedersen, J. G. (1975b). On strong unimodality and M-ancillarity with applications to contingency tables. *Scandinavian Journal of Statistics*, 2, 127-137.

Rao, C. R. (1965). On discrete distribution arising out of methods of ascertainment. In *Classical and Contagious Discrete Distributions*, G. P. Patil, ed. Statistical Publishing Society, Calcutta. Pages 320-332. Reprinted in *Sankhya*, 27B, 311-323.

Rao, C. R. (1977). A natural example of a weighted binomial distribution. *American Statistician*, 31, 24-26.

Selvin, S. (1975). Testing the Mendelian segregation ratio under incomplete ascertainment. *Human Heredity*, 25, 194-203.

Smith, C. A. B. (1956). A test for segregation ratios in family data. *Annals of Human Genetics*, 20, 257-265.

Smith, C. A. B. (1959). A note on the effects of method of ascertainment on segregation ratios. *Annals of Human Genetics*, 23, 311-323.

Stene, J. (1970a). Analysis of segregation patterns between sibships within families ascertained in different ways. *Annals of Human Genetics*, 33, 261-283.

Stene, J. (1970b). Comparison of segregation ratios of families ascertained in different ways. *Annals of Human Genetics*, 33, 395-412.

Stene, J. (1970c). Statistical inference on segregation ratios for D/G-translocations, when the families are ascertained in different ways. *Annals of Human Genetics*, 34, 93-115.

Stene, J. (1975). Sibships ascertained through a twin pair with at least one affected twin. *Annals of Human Genetics*, 38, 361-364.

Stene, J. (1977). Assumptions for different ascertainment models in human genetics. *Biometrics*, 33, 523-527.

Stene, J. (1978a). Ascertainment models human families selected through several affected individuals. *Biometrics*, 34, 455-457.

Stene, J. (1978b). Choice of ascertainment model I. Discrimination between single-proband models by means of birth order data. *Annals of Human Genetics*, 42, 219-229.

Stene, J. (1979). Choice of ascertainment model II. Discrimination between multi-proband models by means of birth order data. *Annals of Human Genetics*, 42, 493-505.

Thomas, D. G. and Gart, J. J. (1971). Small sample performance of some estimators of the truncated binomial distribution. *Journal of the American Statistical Association*, 66, 169-177.

Thompson, E. A. and Cannings, C. (1979). Sampling schemes and ascertainment. In *Proceedings of the Snowbird Conference on Coronary Disease* (August, 1978), C. F. Sing and M. H. Skolnik, eds. Alan R. Liss, Inc., New York. Pages 363-382.

Weinberg, W. (1912a). Über Methode und Fehlerquellen der Untersuchung auf Mendelsche Zahlen beim Menschen. *Archiv für Rassen- und Gesellschafts-Biologie*, 9, 165-174.

Weinberg, W. (1912b). Zur Vererbung der Anlage der Bluterkrankheit mit methodologischen Ergänzungen meiner Geschwistermethode. *Archiv für Rassen- und Gesellschafts-Biologie*, 9, 694-709.

Weinberg, W. (1928). Mathematische Grundlagen der Probandenmethode. *Zeitschrift für induktive Abstammungs- und Vererbungslehre*, 48, 179-228.

[*Received May* 1980. *Revised October* 1980]

A STOCHASTIC MODEL FOR THE STUDY OF THE DISTRIBUTION OF CHROMOSOME ABERRATIONS IN HUMAN AND ANIMAL CELLS EXPOSED TO RADIATION OR CHEMICALS

KONANUR G. JANARDAN

Sangamon State University
Springfield, Illinois 62708 USA

DAVID J. SCHAEFFER

Illinois Environmental Protection Agency
2200 Churchill Road
Springfield, Illinois 62706 USA

RUSSEL J. DuFRAIN

Oak Ridge Associated Universities
Oak Ridge, Tennessee 37830 USA

SUMMARY. A biologically motivated stochastic model is presented for the study of variation in the number of chromosome aberrations which are induced by chemical or physical agents in human and animal cells. The parameters L and M of the model are both dispersion sensitive when M is not equal to L, the effect of M appears to increase the dispersion of the aberrations relative to the Poisson distribution. Over- (under-) dispersion is characterized by the ratio M over L being greater (less) than unity. The theoretical distribution of the number of aberrations is shown to be an infinite geometric mixture of Poisson distributions; and approaches the ordinary Poisson distribution as M approaches L.

KEY WORDS. stochastic model, damage and restitution rates, chromosome aberrations, radiation and chemical data, human and mammalian dosimetry, over-dispersion, under-dispersion, modified poisson process.

C. Taillie et al. (eds.), Statistical Distributions in Scientific Work, Vol. 6, 265–277.

1. INTRODUCTION

Damage to genetic material can be caused by various chemical and physical insults. One way of visualizing this damage is through analysis of chromosome aberrations such as chromosome and chromatid deletions, dicentrics, chromatid exchanges, and others. The study of chromosome aberration induction by these agents has evolved over the past several decades, and discrete distribution models have been developed as one method of analysis for the data generated in experimental studies. Most of these theoretical distribution models have relied on the Poisson, binomial, or negative binomial distribution of observed events (Savage, 1970; Kohler *et al.*, 1976; Yankovenko *et al.*, 1976) to describe the expected frequency distribution of aberrations in a collection of analyzed cells. These intracellular distributions of lesions can give additional information about the underlying mechanism of the interaction of the damaging agent with cellular components. Recently, Janardan and Schaeffer (1977), Janardan *et al.* (1979), and DuFrain *et al.* (1980) have proposed the use of the Lagrangian Poisson Distribution (LPD) as a model of over-, under-, or no dispersion relative to the Poisson. Schaeffer *et al.* (1980) have shown that this model can be formally linked through statistical mechanics to thermodynamics, and that one parameter of the LPD is the equilibrium constant when the genetic system analyzed is at equilibrium with the insult. Thus, the LPD is capable of providing biologic and chemical information unavailable from other distribution models. However, certain assumptions which are used in the development of the LPD may be restrictive of the more complex biologic phenomena occuring in living cells which are stressed by environmental insult. Further, because both spontaneous and induced lesions represent stochastic processes, a need has arisen for the development of a new model that takes the stochastic nature of the biologic system into account.

2. DEVELOPMENT OF STOCHASTIC MODEL

2.1 Biologic Motivation. Many studies have shown that the production of chromosome aberrations have complex time and does components. The interaction of these appears to be such that the production or repair of subsequent damage in an already damaged cell may proceed with a different rate than did the initial damage induction. The underlying mechanisms responsible for these differences are not readily apparent from the available data, but probably relate more to the activation or inactivation of restitution mechanisms than to fundamental changes in the manner in which damages are produced, although the latter cannot at present be ruled out.

Cytogeneticists have continually observed that chromosome aberrations are either over-dispersed or under-dispersed relative to the Poisson distribution (Savage, 1970). However, no single model, with the exception of the LPD, covers cases with under-dispersion, over-dispersion, or the same dispersion, as the Poisson expectation. As we continue to use the LPD in our work, however, the assumption of a constant rate of aberration production or repair may sometimes be restrictive of the actual biologic system's response. Hence, a theoretical need has arisen for modifying some of the underlying assumptions of the LPD, in certain cases.

2.2 *Postulates*. The Poisson process assumes that: (i) the probability that one aberration is observed in a cell during an infinitesimal interval of time dt is $\lambda dt + O(dt)$; (ii) the probability that no aberration is found during the infinitesimal interval dt is $1-\lambda dt + O(dt)$; and (iii) the probability that more than one aberration occurs during dt is $O(dt)$. We add to these Poisson postulates the following two for the development of our stochastic model: (iv) the probability that a new aberration is sustained during dt by the cell already having $k(\geq 1)$ aberrations is $\mu dt + O(dt)$; and (v) the probability that no aberration is found during dt in the cell already having $k(\geq 1)$ aberration is $1-\mu dt + O(dt)$.

2.3 *Model*. Let $X(t)$ denote the number of aberrations sustained by a cell during a period $(0,t)$ and let $P_n(t)$ denote the probability that there are n aberrations in the cell at time t. Then $P_n(t) = \Pr\{X(t)=n\}$, and the above assumptions give Chapman-Kolmogorov's forward equations:

$$P_0'(t) = \lambda P_0(t), \qquad P_1'(t) = -\mu P_1(t) + \lambda P_0(t),$$

$$P_n'(t) = -\mu P_n(t) + \mu P_{n-1}(t) \quad \text{for } n \geq 2.$$

This system of differential equations is solved using the probability generating function (p.g.f.) technique. This gives $P_0(t) = \exp(-L)$ and, for $n \geq 1$,

$$P_n(t) = LM^{n-1}(M-L)^{-n} e^{-L}(1-e^{(L-M)}\{\sum_{j=0}^{n-1} (M-L)^j/j!\}), \tag{1}$$

where $L = \lambda t$ and $M = \mu t$. This probability function can be seen to be an infinite geometric mixture of Poisson distributions. Setting $(L/M) = 1-p$ where $M > L > 0$ such that $0 < p < 1$, we can write (1) as

$$P_n(t) = \sum_{k=n}^{\infty} (1-p)\, p^{k-n} e^{-M} M^k/k!, \quad \text{for } n \geq 1.$$

The mean and variance of $X(t)$ are readily shown to be

$$\mu_1' = M + (1-M/L)(1-e^{-L})$$

$$\mu_2 = M^2 - \mu_1' + (1-e^{-L})(1-M/L)(1-2M/L) + M(3-2M/L).$$

3. ESTIMATION

We can obtain the estimates of L and M either by maximum likelihood or by the method of moments. Both of these methods require iterative determination of the parameter estimates. However, an easy way of estimating these parameters is to employ the method of mean and zero frequency (Anscombe, 1950) in which the observed mean $\bar{x}$ and the relative frequency of cells with no aberrations are equated to their expected values. In addition to being easy to compute, these estimators have high asymptotic efficiency relative to maximum likelihood estimators.

First we estimate $L = \lambda t$ for each value of t, using the observed proportions of cells with no abberations, from the equation

$$\tilde{L} = -\ln f_0 .$$

Then, we estimate $M = \mu t$ by setting the mean of the distribution to the sample mean, $\bar{x}$, of aberration per cell:

$$\tilde{M} = \tilde{L}\,(\bar{x}-1 + f_0)/(\tilde{L}-1 + f_0) .$$

Using standard techniques, the variances of these estimators for a sample of size n are derived as:

$$\mathrm{Var}(\tilde{L}) = (1-P_0)/n\,P_0 .$$

$$\mathrm{Var}(\tilde{M}) = n^{-1}[L/(L-1+P_0)]^2(\mu_2 - p_0\mu_1'c + P_0(1-P_0c^2),$$

where $P_0 = P_0(t)$ and $c = 1-(M/L) + M(1-P_0)/P_0L^2$. These theoretical variances may be estimated by substituting $\tilde{L}$, $\tilde{M}$ and f_0 in place of L, M and P_0, respectively.

These mathematical results are applied to data obtained from human and mammalian cytogenetic experiments which are described in the next section.

4. EXPERIMENTAL

4.1 Human Cytogenetic Dosimetry: Radiation from ^{241}Am. Human lymphocyte cultures were exposed to 43.8, 87.7, 175.3, or 350.6 nCi/ml ^{241}Am for 1.7 h. This gave doses of 0.85, 1.71, 3.42 or 6.84 rad, as described by DuFrain *et al.* (1979). The study clearly shows that dicentrics per cell, in cells damaged by alpha particle radiation, are over-dispersed when compared to the expectation from a Poisson distribution. However, this over-dispersion did not adversely affect the fitting of the dose response data to a Poisson regression, as discussed in detail by Frome and DuFrain (1979).

4.2 Human Cytogenetic Dosimetry: Chromotid Exchanges in Leukocytes by an Alkylating Agent. Kohler *et al.* (1976) analyzed the intercellular distributions of chromatid exchanges (CE) induced in human leukocytes of one male blood donor in vitro after 24-h treatment with the bifunctional alkylating agent 2,5-bis (methoxyethoxy)-3,6-bis-ethyleneiminio-p-benzoquinone (A-139) in concentrations of 10^{-7} up to 10^{-5} M. The experiments also showed that the distribution of lesions was over-dispersed with respect to the expected Poisson values. This prompted the authors to suggest the negative binomial as an alternative model for the description of the observed dispersion.

4.3 Mammalian Cytogenetic Dosimetry: Lesions in Cells of Rabbits Induced by Streptonigrin (NSC-45383). Female rabbits were dosed intravenously with Streptonigrin (supplied by Drug Development Branch NCI) to determine if various cell types are differentially sensitive to aberration induction by a direct acting chemical agent. Bone marrow cells from a femur and lymphoblasts from a mesenteric lymph node were directly prepared for cytogenetic evaluation and scored microscopically for abnormalities. Detailed evaluations of 7000 metaphases, included enumeration of the minimum number of chromatid breaks necessary to cause the observed abnormalities. This data then served as the data base for statistical evaluations (DuFrain *et al.*, 1979). Again, the dispersion was greater than expected from a Poisson distribution and the authors tested the geometric distribution, negative binomial distribution, and the LPD.

5. DISCUSSION

Tables 1-4 provide the experimental results and the fits to the data from, and the parameters of, the Poisson, LPD and stochastic models. The tabulated parameters are: λp (Poisson);

TABLE 1: Human cytogenetic dosimetry: Radiation from ^{241}AM. Exposure in rads.

Exposure	D&R Cells	Frequency of Cells with D & R							Model and Parameters
0.85	20/400	385	11	3	1	0	0	0	Observed
		380.5	19.0	0.5	.0	–	–	–	Poisson: 0.05
		384.3	12.6	2.3	0.6	–	–	–	LPD: 0.04011, 0.1978*
		385.0	10.9	3.3	0.7	–	–	–	Stochastic: 0.03822, 0.6627, 17.3
1.71	17/200	187	10	2	1	–	–	–	Observed
		183.7	15.6	0.7	0				Poisson: 0.085
		186.6	10.7	2.0	0.5	–	–	–	LPD: 0.615, 0.1864*
		187.0	9.7	2.7	0.5	–	–	–	Stochastic: 0.06721, 0.6086, 9.06
3.42	36/200	176	15	6	3	0	0	0	Observed
		167.1	30.1	2.7	0.2	–	–	–	Poisson: 0.18
		173.9	19.4	4.6	1.4	–	–	–	LPD: 0.1396, 0.2242
		176.0	15.1	6.4	1.9	–	–	–	Stochastic: 0.1278, 0.9722, 7.61
6.84	69/200	159	23	13	2	2	0	1	Observed
		141.6	48.9	8.4	1.0	0.1	0	0	Poisson: 0.345
		156.7	28.5	8.8	3.3	1.4	0.6	0.3	LPD: 0.2441, 0.2926
		159.0	22.2	12.0	4.8	1.5	0.4	0.1	Stochastic: 0.2294, 1.3156, 5.73

$*\lambda_2 = 0$ ($p > .05$)

TABLE 2: Human cytogenetic dosimetry: Chromatid exchanges in 373 leukocytes induced by 0.04g 2, 5-bis(Methoxyethoxy)-3,6-bis-ethyleneimino-p-benzoquinone.

Frequency of Cells with CE							
0	1	2	3	4	5	6	Model and Parameters
302	38	21	7	3	1	1	Observed
267.5	88.9	14.8	1.6	0.1	0	0	Poisson: 0.3324
295.6	50.9	15.7	6.0	2.5	1.2	0.6	LPD: 0.2325, 0.3005
302.0	36.7	21.2	9.1	3.0	0.8	0.2	Stochastic: 0.2112, 1.4423, 6.83

TABLE 3: Mammalian cytogenetic dosimetry: Lesions in rabbits lymphoblasts induced by streptonigrin (NSC-45383). Exposure in ug/kg.

Exposure	0	1	2	3	4	5	≥6	Model and Parameters
0	585	15	-	-	-	-	-	Observed
	585.2	14.6	-	-	-	-	-	Poisson: 0.025
	585.0	15.0	-	-	-	-	-	LPD: 0.0253, 0.1190
	585.0	15.0	-	-	-	-	-	Stochastic: 0.0253, 0.0, --
30	404.0	80	13	3	-	-	-	Observed
	397.3	91.4	10.5	0.8	-	-	-	Poisson: 0.23
	403.7	80.6	13.3	2.1	-	-	-	LPD: 0.2140, 0.0695*
	404.0	79.2	14.7	1.8	-	-	-	Stochastic: 0.2132, 0.3832, --
60	413	124	42	14	5	0	2	Observed
	372.5	177.6	42.3	6.7	0.8	0.1	0	Poisson: 0.4767
	417.3	119.5	39.5	14.4	5.5	2.2	1.6	LPD: 0.3632, 0.2380
	413.0	114.8	51.3	16.1	3.9	0.8	0.1	Stochastic: 0.3735, 0.9969, 2.67
75	200	57	30	7	4	0	2	Observed
	170.8	96.2	27.1	5.1	0.7	0.1	0	Poisson: 0.5633
	199.6	61.7	22.5	91	3.9	1.7	1.5	LPD: 0.4076, 0.2765
	200.0	53.8	29.7	11.8	3.6	0.9	0.2	Stochastic: 0.4055, 1.2929, 3.19
90	155	83	32	14	13	2	1	Observed
	126.1	109.3	47.4	13.7	3.0	0.5	0.1	Poisson: 0.8667
	155.5	80.2	35.9	15.8	6.9	3.1	1.1	LPD: 0.6572, 0.2417
	155.0	71.4	44.6	19.8	6.8	1.9	0.5	Stochastic: 0.6604, 1.4300, 2.17

$*\lambda_2 = 0$ $(p > .05)$

TABLE 4: Mammalian cytogenetic dosimetry: Lesions in rabbits bone marrow cells induced by streptonigrin (NSC-45383). Exposure in ug/kg.

Exposure	0	1	2	3	≥ 4	Model and Parameters
0	842	8	-	-	-	Observed
	842.0	7.9	-	-	-	Poisson: 0.0094
	842.0	8.0	-	-	-	LPD: 0.0095, -.0041*
	842.0	8.0	-	-	-	Stochastic: 0.0095, 0.0, --
30	464	32	4	-	-	Observed
	461.6	36.9	15	-	-	Poisson: 0.08
	463.6	33.1	2.9	-	-	LPD: 0.0755, 0.0560*
	464.0	32.3	3.7	-	-	Stochastic: 0.0747, 0.2195, --
60	525	66	9	-	-	Observed
	521.6	73.0	5.4	-	-	Poisson: 0.14
	524.3	68.3	6.8	-	-	LPD: 0.1350, 0.036
	525	66.7	8.3	-	-	Stochastic: 0.1335, 0.2348, 1.76
75	257	36	7	-	-	Observed
	253.9	42.3	3.5	-	-	Poisson: 0.1667
	256.2	38.3	4.8	-	-	LDP: 0.1577, 0.0538
	257.0	36.7	6.3	-	-	Stochastic: 0.1547, 0.3174, 2.05
90	409	61	26	3	1	Observed
	388.6	97.9	12.3	1.0	0.1	Poisson: 0.252
	404.1	73.6	16.5	4.1	1.1	LPD: 0.2129, 0.1553
	409.1	63.4	21.5	5.1	1.0	Stochastic: 0.2009, 0.7443, 3.70

*$\lambda_2 = 0$ ($p>.05$)

λ_1 and λ_2 (LPD); and L, M, M/L (stochastic model). In general, the fits follow the order: Stochastic model $\geqslant$ LPD > Poisson. Tables 1, 3 and 4 suggest that λ_p (Poisson), λ_2 (LPD), and L and M (stochastic model) increase with exposure; however, these parameters may have different meanings. Comparison of Tables 3 and 4 shows that all of the models indicate that the production of lesions in lymphoblasts is a more sensitive measure of exposure than is the production of lesions in bone marrow cells of the rabbit.

A proper interpretation of the results requires that these models be different from Poisson. The LPD reduces to Poisson when $\lambda_2=0$, and the stochastic model becomes Poisson when $L \to M$. Since no test of the significance of λ_2 is available, we develop a conditional test on λ_2 given λ_1 based on the confidence interval of λ_1 (Haight, 1967). Thus $\lambda_2=0$ if $(\tilde{\lambda}_1-\tilde{\lambda}_p)(N/\tilde{\lambda})^{1/2} \leqslant Z_{(1-\alpha)}$, and $\lambda_2 \neq 0$ otherwise. Similarly, the stochastic model is the same as the Poisson if $(\tilde{L}-\tilde{M})(N/\tilde{\lambda}_p)^{1/2} \leqslant Z_{(1-\alpha)}$. These tests show that the LPD does not differ from the Poisson for the lowest two doses in Tables 1, 3 and 4. The stochastic model is the same as the Poisson only for the controls in Tables 3 and 4.

The parameter of the Poisson, λ_p, is the mean number of lesions per cell. This parameter, λ_p (and λ_2, of the LPD), increases linearly with dose.

We have previously shown that λ_2 of the LPD is the ratio of the rates of the forward (production) and backward (restitution) processes (Janardan and Schaeffer, 1977; Janardan *et al.*, 1979). As the net rate attains equilibrium, λ_2 stabilizes and becomes the equilibrium constant for the system. This occurs in Table 3 with $\lambda_2=0.24$. The net free energy for the production of lesions in rabbit lymphoblasts is estimated (Schaeffer *et al.*, 1980) as 877 cal/mole/lesion, and 5.3 Kcal/mole, in excellent agreement with our previously reported estimates for other mutagens (Janardan *et al.*, 1979; Schaeffer *et al.*, 1980). The exposure, in terms of dose (D), dependency of λ_2 can be given as $\lambda_2 = a \ln(D+1) + b \ln(D+1)^2$ (Schaeffer *et al.*, 1980).

The parameter L of the stochastic model is a logarithmic function only of the frequency of the zero*th* class. Hence, L, like λ_p and λ_1, is a linear function of the dose.

The parameter M may be thought of as a measure of the common exposure-dependent mechanism for cellular response to damage of individual cells by a clastogenic agent. Thus, it probably relates to the rate of damage or repair after the initial insult has occured. The exposure dependency of M appears to be similar to that given for λ_2.

The parameter, M appears to be sensitive to two different factors. For chemical data, M is determined by physiological factors which affect the delivery of the chemical insult to the subsequently affected organ. Then, genetic related factors which relate to entry of the material into the cell, activation, relative rates of repair, etc. come into play. In cases where $M>L$, it appears that additional damage occurs more readily than the initial damage. This probably reflects impairment of repair systems, better delivery of the clastogen to the damaged versus undamaged DNA, or differences in the site or type of damage. The ratio of M/L increases with dose for the streptonigrin data in Tables 3 and 4. This suggests that the rate at which subsequent damages occur increases with increasing damage.

The value attained by the parameter, M in application of the theoretical model to radiation experiments may be related to the inherent energy of the radiation, and the microdosimetric (i.e., site specific delivered dose) consequences. Broadly speaking, these factors appears to parallel the physiologic (total exposure) and genetic (delivered dose) for chemical insult. The significance of these deductions is currently the subject of study. For the ^{241}Am data in Table 1, the ratio M/L decreases with increasing dose. Apparently, the efficiency with which the initial damage occurs increases faster with dose than does the propagation of subsequent damage. This may reflect a saturation of the repair process by the radiation which is only slightly affected thereafter.

6. CONCLUSION

The theoretical results and examples show that the parameters L and M of the stochastic model are both dispersion sensitive. When $M \neq L$, the effect of M is to increase the dispersion of the damages relative to the Poisson. Over-dispersion is characterized by $M/L > 1$. Although not discussed here, this model, like LPD, will handle under-dispersed damage relative to the Poisson as well. Thus for the data of Haag *et al.* (1977), λ_2 (LPD) is negative; corresponding to $M/L < 1$. Here, the rate at which the initial damage occurs increases faster with dose than does the rate of subsequent damages. Biochemical and Biophysical factors which govern whether a particular insult will result in under-, over- or appropriate dispersion relative to the expected Poisson forms an increasingly important area of current research in biology.

ACKNOWLEDGEMENTS

We thank Dr. W. Kohler for bringing the reference by Yakovenko *et al.* to our attention, and Marla Gregory for typing the manuscript.

The work of K. G. Janardan is partially supported by the Illinois Environmental Protection Agency under Contract F-50 (1979). The work of R. J. DuFrain is partially supported under Contract No. DE-AC05-760R00033 between the Department of Energy, Office of Health and Environmental Research, and Oak Ridge Associated Universities and NIH Grant HD08828. Reproduction in whole or in part is permitted for any purpose of the United States Government.

REFERENCES

Anscombe, F. J. (1950). Sampling theory of the negative binomial and logarithmic series distribution. *Biometrika*, 37, 358-382.

DuFrain, R. J., Littlefield, L. G., Joiner, E. E., and Frome, E. L. (1979). Human cytogenetic dosimetry. A dose-response relationship for alpha particle radiation from ^{241}Am. *Health Physics*, 37, 279-289.

DuFrain, R. J., Frome, E. L., and Littlefield, L. G. (1980). Dose-response induced cytogenetic abnormalities in somatic cells of the rabbit. *Environmental Mutagenesis*, 2, (in press).

Frome, E. L. and DuFrain, R. J. (1979). Estimation of the amount of exposure to an environmental clastogen using human cytogenetic dosimetry. In proceedings of *1978 DOE Statistical Symposium*, H. T. Davis, R. R. Prairie, and T. Pruit, eds. Albuquerque, N. M., US DOE CONF 78 11 08. pages 169-172.

Haag, J., Brenot, J., and Parmentier, N. (1977). Chromosomal aberrations in pig lymphocytes after neutron irradiation in vitro. *Radiation Research*, 70, 187-197.

Haight, F. (1967). *Handbook of the Poisson Distribution*. New York.

Janardan, K. G., and Schaeffer, D. J. (1977). Models for the analysis of chromosomal aberrations in human leukocytes. *Biometrical Journal*, 19, 599-612.

Janardan, K. G., Kerster, H. W., and Schaeffer, D. J. (1979). Biological applications of the Lagrangian Poisson distribution. *BioScience*, 29, 599-602.

Kohler, W., Loeschcke, V., and Obe, G. (1976). Analysis of intercellular distributions of chromatid aberrations. *Mutation Research*, 34m, 427-436.

Loeschcke, V., and Kohler, W. (1976). Deterministic and stochastic models of the negative binomial distribution and the analyses of chromosomal aberrations in human leukocytes. *Biometrishe Zeitschrift*, 18, 427-451.

Savage, J. R. K. (1980). Sites of radiation induced chromosome exchanges. In *Current Topics in Radiation Research 6*, M. Ebert and A. Howard, eds. North Holland Publ., Amsterdam. Pages 129-194.

Schaeffer, D. J., Janardan, K. G., and Kerster, H. W. (1980). Development and application of an equilibrium dose response model. *Environmental Mutagenesis*, 2, (in press).

Schaeffer, D. J., Janardan, K. G., Clark, A. C. M. and Kerster, H. W. (1980). Statistically estimated free energies of chromosome abberation production from frequency data (submitted for publication).

Yakovenko, K. N., Bochkov, N. P., and Chebotarey, A. N. (1978). The distribution of chemically induced chromosome breaks in human cells. *Biological Zentrablat*, 95, 437-450.

[*Received June* 1980. *Revised September* 1980]

A MODEL FOR THE ANALYSIS OF PLATELET SURVIVAL

DANIELA COCCHI

Institute of Statistics
University of Bologna
Bologna, Italy

SUMMARY. A new survival function with increasing failure rate and finite domain is suggested for the analysis of the residual survival of labeled platelets in hematology. The model can assume different shapes and a biological meaning may be attributed to its parameters. So, this model may be utilized in the study of labeled platelet survival as an alternative to the widely used always upward concave curves. The stationarity assumption for the platelet birth and death process, which implies a concave upward function, may therefore be checked according to the shape the new model assumes. In general for hemopathic subjects the shape of the survival platelet curve tends to be like the exponential, while the nonhemopathic subjects show a two-inflexion point curve analogous to that observed for cohort human survival.

KEY WORDS. finite domain models, platelet residual survival, reliability models, survival analysis.

1. SOME ASPECTS OF EXPERIMENTS ON PLATELET SURVIVAL

In hematological studies there is often the need to look for an adequate approximation for the "law" of platelet survival. The need arises in particular when the hematological phenomena are examined in subjects suffering from leukemia, with the aim of detecting specific pathological features. In this case some diagnostic assistance is sought through statistical analysis, in particular for making a decision on splenectomy and, as a consequence, for judging the platelet behavior of splenectomized subjects.

C. Taillie et al. (eds.), Statistical Distributions in Scientific Work, Vol. 6, 279–287.

The data on platelet survival are recorded as follows. From each individual a blood sample is drawn; the platelets of the sample are labeled by a radioactive isotope and then reintroduced into the bloodstream at a certain time t_0. A number of blood samples are taken at later times giving the n pairs

$$(t_j, L_j) \quad j = 1, \cdots, n, \tag{1}$$

where L_j indicates the radioactivity level recorded at time t_j and L_0 is the level of radioactivity at t_0.

When platelet survival is studied, a fundamental restrictive assumption is always made, which needs some discussion. It is the stationarity assumption, which mainly implies that the rates of production and destruction of platelets are always equal at each instant. We must also say that stationarity is more easily acceptable for long periods but, in the relatively short residual life span in which a group of labeled platelets is considered, the stationarity principle is hard to consider reasonable.

If the stationarity condition holds, it is possible to find a relatively simple relationship between the so-called residual survival curve $S_r(t)$, which concerns labeled platelets, and the survival curve of a cohort of platelets $S(v)$. The quantity $S_r(t)$ refers to platelets which, at the moment of labeling, have different ages, being drawn at random from those in circulation. The most important consequence of the assumption of stationarity, as it is well known (Breny, 1971), is the possibility of computing the mean life of a platelet cohort, that is

$$E(v) = \int_0^{\omega} S(v)\, dv, \tag{2}$$

where ω is the extreme age, which may be assumed infinite or not, according to the well-known Mills-Dornhorst formula

$$E(v) = 1/\mu_r(0), \tag{3}$$

where $\mu_r(0)$ is the instantaneous failure rate at time 0 of the residual life span, which, for every t, may be computed from

$$\mu_r'(t) = -S_r'(t)/S_r(t). \tag{4}$$

On the contrary, if stationarity does not hold, as is likely over short periods, there is no possibility of passing from the residual survival function to statistics regarding cohorts of platelets. In such a case it is only possible to compute, see Bergner (1962, 1964),

$$E[\mu(a)] = -S_r'(0) = f_r(0) \tag{5}$$

where $\mu(a)$ is the instantaneous elimination rate of a platelet of age a, also known as the force of mortality.

Under stationarity, moreover, it has been shown (Breny, 1971) that the survival function is necessarily concave upward. Unfortunately, the upward concavity of the residual survival curve, which is only a necessary condition for stationarity, is often considered a sufficient condition. As a consequence, upward concavity of the survival curve is always assumed in practice, but the plotting of individual survival values very often shows net changes in convexity. Therefore, the possibilities of misusing a model can be as follows: first, the choice of an upward concave curve implies stationarity even though the process is not stationary and, secondly, even when variations in convexity are evident, the inflexion points are considered as a spurious effect instead of a true characteristic. So only concave upward models are usually considered.

Sometimes models are chosen without considering whether a substantial meaning may be attributed to their parameters, and without taking into account the behavior of their instantaneous failure rate. In fact it sometimes contrasts with the more realistic assumption of an increasing failure rate, which decreases in some cases, as in some mixed linear and exponential models (Kutti and Weinfield, 1971), or tends to a constant quantity as t tends to infinity.

2. THE MULTIPLE HIT MODEL

The most frequently employed model, besides the exponential, and also the one suggested by the Panel on Diagnostic Applications of Radioisotopes in Hematology (AA.VV., 1977), is the multiple hit model. It assumes that the elimination of platelets is a consequence of a variable number of independent "hits", which platelets sustain randomly in circulation; these "shocks" are all considered to have the same intensity. Therefore the survival function depends on a parameter n, that is the number of hits necessary to eliminate platelets, occurring at the constant distance μ. The cohort survival function is the erlang distribution, see Murphy and Francis (1969, 1971),

$$S(v) = \frac{\mu^n}{(n-1)!} \int_v^\infty e^{-\mu u} u^{n-1} du = \sum_{i=0}^{n-1} \frac{e^{-\mu v}(\mu v)^i}{i!}, \quad (6)$$

and the residual survival function is

$$S_r(t) = \frac{1}{n} \sum_{i=0}^{n-1} \frac{e^{-\mu t}(\mu t)^j}{j!}. \quad (7)$$

Also this model does not escape the inconvenience of being derived under stationarity. Moreover, platelet destruction is considered here as an exogenous process rather than an endogenous one, because platelets die after suffering successive "hits" in circulation; therefore the risk of death does not depend on age.

The hypotheses needed to justify (6) and (7) are indeed very restrictive. In order to obtain a more general model, Murphy (1971) proposed a generalized erlang distribution. But this model, like other survival functions which may "adequately" fit platelet survival data, is often considered only an arbitrary mathematical construct by applied scientists.

3. AN ALTERNATIVE MODEL

As an alternative to the commonly used models mentioned above, it would be useful to have a model which takes into account the fact that the elimination of platelets is due to both endogenous and exogenous effects. In such a model the endogenous factors related to aging should have more influence as the residual life of labeled platelets increases. It is also more realistic to think that "hits" attacking platelets act with variable intensities, in some cases eliminating and in others only damaging them. We will not assume stationarity and will focus our attention on the residual survival function, which is, after all, the only phenomenon that can be recorded.

It will not be unduly restrictive to assume a finite extreme age, ω, and to adopt a failure rate which increases to infinity as t tends to ω. Starting from these assumptions the following expression (Ferreri, 1977a) may be derived

$$\mu'_r(t) = [\mu_r(t) - \lambda]^a \, \psi(t)/\alpha, \quad 0 \leq t < \omega, \quad (8)$$

where $\alpha > 0$, $a > 1$, $\lim_{t \to \omega} \mu_r(t) = \infty$, and $\mu_r(0) \geq \max(0, \lambda)$; the function $\psi(t)$ is positive and regular in the interval

$[0,\omega)$ with $\lim_{t\to\omega} \int_t^{\omega} \psi(u)du = 0$. Here $\psi(t)$ and a are to be specified while λ and α are parameters. The parameter λ is related to the exogenous factors of platelet death. It does not depend on time but, since $\mu_r(t)$ tends to infinity, its influence decreases as age increases. This agrees with what is presumed to actually happen, since the exogenous damages corresponding to "hits" that old platelets may sustain are not very relevant compared with those due to aging, and become a random component of destruction.

After having fixed $a = 2$, in Ferreri (1977a,b) the case where $\psi(t) = 1$ was studied and in Cocchi (1980) that of

$$\psi(t) = \exp[\gamma(\omega - t)]. \tag{9}$$

Here we want to draw attention to the case where $\gamma=1$, to describe the contribution of all the damages sustained by the platelets during the experiment. These damages are a decreasing function of time, equal for all the subjects considered. A constant value of γ may also be interpreted as a common biological aging, instead of a chronological one, due to the task of platelets in circulation.

The instantaneous failure rate is

$$\mu_r(t) = \lambda + \alpha/[\exp(\omega-t)-1] \tag{10}$$

which shows that the parameter α may reflect, in some sense, the biological features of the labeled platelets since a greater α value implies a faster rate of growth of $\mu_r(t)$.

The residual survival function is

$$S_r(t) = \left\{\frac{1 - \exp[-(\omega-t)]}{1 - \exp(-\omega)}\right\}^{\alpha} \exp^{(-\lambda t)}, \tag{11}$$

with $0 \leqslant t \leqslant \omega$, $\alpha > 0$, and $\lambda > -\alpha/(\exp \omega - 1)$. Calculation of the mean and variance of (11) is cumbersome, except for negative λ, we have

$$E(t) = \frac{\exp(-\lambda\omega)}{[1 - \exp(-\omega)]^{\alpha}} B_{1-\exp(-\omega)}(\alpha+1, -\lambda). \tag{12}$$

The convexity of (11) can be studied by recalling that

$$S_r''(t) = S_r(t)\ \{[\mu_r(t)]^2 - \mu_r'(t)\}. \tag{13}$$

Therefore the inflexion points are the solutions to the equation

$$\lambda x^2 - \sqrt{\alpha}\, x + \alpha - \lambda = 0 \tag{14}$$

where $x = \exp[(\omega-t)/2]$. It is easy to see that (11) may be: (a) always concave upward; (b) firstly concave downward and then concave upward, showing an inflexion point; or (c) concave upward in the beginning and the end of the residual life span but concave downward in the central region, showing two inflexion points.

Case (c) is particularly interesting being analogous to human cohort survival, and because it recalls the exponential behavior, typical of stationarity, in the beginning and the end of the curve. The difference between case (c) and exponentiality will be less evident the closer the two inflexion points.

Now the question arises of whether any inflexion points occurring in the *estimated* residual survival function are spurious. Further investigation is required to distinguish them from the proper ones, and thence to accept or reject stationarity. Thus stationarity is regarded as hypothesis to be tested and not as an assumption.

4. AN APPLICATION

The application concerns a group of 44 individuals, studied in collaboration with the Hematology Service of the Faculty of Medicine of the University of Bologna. Following the method already explained, a succession of data of type (1) was recorded on each subject, with n varying from 5 to 9. Each individual belonged to one of two groups: 17 were splenectomized and 27 were not. Each group was divided into 3 types: the non-hemopathic, those affected by chronic mieloid leukemia, or thrombocytemia.

Probably because of the small number of observations, estimation of the parameters α, λ, ω by nonlinear least squares was not successful (several algorithms failed to converge). So ω was estimated exogenously, after examining the literature, and was fixed, for each individual, at $\tilde{\omega} = 12$ days. An effort to find exogenous individual values for ω, by the method due to Vincent (1951), did not lead to appreciable differences in the estimates of the other parameters. Moreover, we do not know the value $L(0)$ corresponding to the value of stabilization, a very important parameter when we examine the biological features of our subjects. In particular it is interesting to compare it with the radioactivity value L_0 of the labeled platelets before their reintroduction into the bloodstream. On the basis of ω the estimated parameters were $\hat{\alpha}$, $\hat{\lambda}$, $\hat{L}(0)$, computed by ordinary least squares, after a logarithmic transformation.

5. RESULTS AND COMMENTS

The analysis shows that for hemopathic non-splenectomized patients the residual platelet survival curve may be concave upward (case a of Section 3) or in some cases (case c of Section 3) show two inflexion points, very often close to each other. On the contrary non-hemopathic subjects, or patients brought closer to normality by splenectomy, show a type (c) behavior of the residual curve where the concave downward trait is quite appreciable.

The occurrence of inflexion points in $S_r(t)$ was found to be frequent (Cocchi, 1980, Tables 1-4). But now the problem arises of distinguishing between proper and spurious inflexion points. Regarding the position of the inflexion points, we noticed that the passage to the final concave upward trait, ending with the death of all the labeled platelets, occurred in conjunction with a high residual age whose ordinate is between 20% and 10% of the initial radioactivity level L_1 and is slightly higher than the last observed value. We may say that the last part of $S_r(t)$ is mainly determined by the structure of the model we adopt and therefore we may consider the second inflexion point as spurious in the sense we formerly specified. The second inflexion point actually occurs in a zone in which the experimenter has already decided to censor the data recording.

According to (11), for the non-hemopathic or splenectomized subjects we have a relatively fast destruction of the platelets which do not, however, adequately work for reasons related to the suffering caused by the experiment. The length of this initial phase varies from 4 to 7 days with very few exceptions, and can hardly be considered a consequence of the stabilization process, which is supposed to conclude very close to time 0. After this phase, a slower death process begins which, in correspondence to a "normal" (in the lexian sense) value of death age, speeds up again, since the remaining platelets, strongly reduced at that value, will totally disappear within ω. The formal analogy with cohort human survival is evident.

The only statistic which may be computed without assuming the stationarity condition is (5), i.e., the average instantaneous elimination rate,

$$E[\mu(a)] = \lambda + \alpha/\exp(\omega-1).$$

Taking $\tilde{\omega} = 12$ days, we obtain $\bar{\mu}(a) \simeq \lambda$ for platelet studies.

A general comment on the analysis is that (11) may be helpful in the distinction between hemopathic or non-hemopathic (or hemopathic splenectomized) subjects by looking at the convexity of the estimated residual survival function. The Hematology Service of the University of Bologna is going to carry out an experiment (with volunteer M.D.) on non-hemopathic subjects in order to have a greater number of data for individual curves, especially in the initial portion of the curve which is the more difficult region to investigate, since it is there that stabilization occurs.

REFERENCES

AA.VV. (1977). Recommended Methods for Radioisotope Platelets Survival Studies. (The Panel on Diagnostic Application of Radiosotopes in Hematology, International Committee for Standardization in Hematology). *Blood*, 50, 1137-1144.

Bergner, P. E. E. (1962). On the stochastic interpretation of cell survival curves. *Journal of Theoretical Biology*, 2, 279-295.

Bergner, P. E. E. (1965). On stationary and non-stationary red cell survival curves. *Journal of Theoretical Biology*, 9, 366-388.

Breny, H. (1971). Fundamental formulae: I. Calculation of mean and distribution of cohort and population survival. Chap. 4, *Platelet Kinetics*, J. M. Paulus, ed. North Holland Publ. Co., Amsterdam.

Cocchi, D. (1980). Un modello di affidabilità a domino finito e relativo impiego nell'analisi della sopravvivenza piastrinica residua. *Statistica*, XL, 69-91.

Ferreri, C. (1977a). Di una classe di distribuzioni di frequenza per l'analisi della sopravvivenza residua di piastrine isotopicamente marcate in tema di affidabilità. *Statistica*, XXXVII, 120-143.

Ferreri, C. (1977b). Un'analisi biometrica della sopravvivenza residua di piastrine isotopicamente marcate. Quaderno n. 1 dell'Istituto di Statistica dell'Università di Bologna, C.L.U.E.B., Bologna.

Kutti, J., and Weinfield, A., (1971). Platelet survival in man. *Scandinavian Journal of Hematology*, 8, 336-346.

Murphy, E. A. (1971). Models of the destruction of blood platelets. Chap. 8, *Platelet Kinetics*, J. M. Paulus, ed. North Holland Publ. Co., Amsterdam.

Murphy, E. A. and Francis, M. E. (1969). The estimation of blood platelet survival: I. General principles of the study of cell survival. *Thrombosis et Diathesis Haemorragica*, XXII, 281-295.

Murphy, E. A. and Francis, M. E. (1971). The estimation of blood platelet survival: II. The multiple hit model. *Thrombobis et Diathesis Haemorragica*, XXIV, 53-80.

Vincent, P. E. (1951). La mortalité des vieillards. *Population*, 6, 181-204.

[*Received August* 1980]

EXTINCTION AND WAITING TIMES IN BIRTH-DEATH PROCESSES: APPLICATIONS TO ENDANGERED SPECIES AND INSECT PEST CONTROL

BRIAN DENNIS

Graduate Ecology Program
213 Mueller Laboratory
The Pennsylvania State University
University Park, Pennsylvania 16802 USA

SUMMARY. Stochastic birth-death processes as models of sexually reproducing populations are discussed, with emphasis on the role of familiar univariate statistical distributions. Models of mating frequency in endangered species and insect pest populations are proposed. The effects on reproduction of a paucity of matings are analyzed with respect to waiting times and extinction probabilities in stochastic population growth models and compared to deterministic cases. The probabilities of extinction in the stochastic models are the tails of familiar discrete probability distributions, with the lower critical population density corresponding to the mode of the distribution. Mating limitation of population growth produces extinction probabilities identical to those produced by constant-rate population harvesting. Distributional properties of extinction probabilities in birth-death processes are discussed with reference to weighted distributions, power series distributions, and generating functions.

KEY WORDS. extinction, birth-death processes, endangered species, insect pests, mating frequency, negative binomial, Poisson, binomial, waiting time.

1. INTRODUCTION

The evolutionary advantages of sexual reproduction are the subject of much debate among biologists (e.g., Williams, 1975). A quite obvious disadvantage of sexuality, however, has been somewhat neglected: namely, "it takes two to tango." An obligate sexual organism must encounter a fertile member of the opposite

C. Taillie et al. (eds.), Statistical Distributions in Scientific Work, Vol. 6, 289–301.

sex to reproduce (in the case of many hermaphroditic organisms such as snails, any fellow species member will suffice). A species' survival at very low densities may be tenuous due to the uncertainty of mates finding each other (Allee, 1938; Andrewartha and Birch, 1954).

Entomologists routinely exploit this "copulatory imperative" for controlling various insect pest populations at high as well as low densities. Mating in such populations can often be disrupted by releasing large numbers of sterilized males (Braumhover *et al.*, 1955; Steiner *et al.*, 1970), or inundating an infested area with the pest species' chemical sex attractant (pheremone) (Sower and Whitmer, 1977; Richerson, Brown, and Cameron, 1976).

Mathematical models of mating and growth in rare populations are themselves rare (Volterra, 1938; Philip, 1957; Mosimann, 1958). Few models of the sterile male technique have been proposed as well (Knipling, 1955; Costello and Taylor, 1975; Prout, 1978). The mathematical perspective in these models tends to be deterministic (but see Costello and Taylor, 1975). Sexual reproduction, however, would seem a natural topic for a stochastic approach.

This paper discusses stochastic birth-death processes as models of sexually reproducing populations, with emphasis on the role of familiar univariate statistical distributions. Section 2 presents birth process models of mating frequency in endangered species or pest populations yielding several traditional probability distributions. In section 3, deterministic population growth models are compared to their stochastic birth-death processes counterparts. Here, the effects of a paucity of mating encounters on reproduction are analyzed with respect to waiting times and extinction probabilities. The tails of familiar discrete probability distributions emerge in a novel context as extinction probabilities. The "critical density" (the lower point where births in the population cease compensating for deaths) acquires an interesting meaning in the stochastic models. Mating limitation of population growth is seen to have effects similar to harvesting the population. Finally, section 4 catalogues various additional properties of extinction probabilities in birth-death processes.

2. MATING FREQUENCY DISTRIBUTIONS

Stochastic fluctuations in the number of mating encounters may affect reproduction in sparse populations. The frequency of matings for a given individual female during a breeding season can be described by a discrete probability distribution as follows. Let $X(a)$ = number of matings that a female has had after searching an effective area a. The quantity a is taken as a measure of time if females remain stationary, waiting for males to arrive. Assuming that the ratio of males to females in the population remains constant, $X(a)$ can be regarded as a stochastic, homogeneous pure birth process with a rate, $\delta(x)$, roughly proportional to the population density, n. For small n it is reasonable to assume that $\delta(x)$ is a linear function of x:

$$\delta(x) = (b + cx)n. \tag{1}$$

This birth rate accommodates a variety of biological situations. Given that a female has encountered males, she is: a) more likely to mate an additional time during a small interval Δa if $c > 0$ (aggregation); b) equally likely when $c = 0$ (random); or c) less likely if $c < 0$ (regularity or satiation). From the general solution for $\Pr[X(a) = x] = p_x(a)$ in pure birth processes (Bartlett, 1978, p. 58), cases a, b, and c yield the negative binomial, Poisson, and binomial distributions, respectively (e.g. Patil and Stiteler, 1974; Boswell, Ord, and Patil, 1979):

$$p_x(a) = \binom{b/c+x-1}{x}(e^{-acn})^{b/c}(1-e^{-acn})^x, \quad x = 0,1,2,\cdots \quad (c > 0) \tag{2}$$

$$p_x(a) = e^{-abn}(abn)^x/x!, \quad x = 0,1,2,\cdots \quad (c = 0) \tag{3}$$

$$p_x(a) = \binom{-b/c}{x}(1-e^{acn})^x(e^{acn})^{(-b/c)-x}, \quad x = o,1,\cdots,-b/c) \quad (c<0). \tag{4}$$

For many species, a female must encounter only one male during a breeding season to realize full reproductive potential. Under these circumstances, the quantity of interest is the probability of finding one or more mates:

$$\Pr[X(a) \geqslant 1] = 1 - e^{-\beta n}. \tag{5}$$

Here $\beta = ab$ is a measure of the inherent mate-finding abilities of the species members. This mating probability was first proposed by Philip (1957), who derived it using Poisson process assumptions (case b). Mosimann (1958) provides some interesting estimates of β for box turtle populations. The expected number of females that mate, and presumably reproduce, is assumed proportional to $n(1-e^{-\beta n})$.

For other species, a female's reproductive rate increases with actual mating frequency. The expected mating frequencies for (2), (3), and (4) above are, respectively,

$$E[X(a)] = (b/c)(e^{acn}-1), \quad (6)$$

$$E[X(a)] = abn, \quad (7)$$

$$E[X(a)] = (-b/c)(1-e^{acn}) \quad (8)$$

The assumption that n is small is seen to be important in (6) and (7), for mating frequency could hardly increase indefinitely with increasing n due to biological constraints. In (8) an upper limit to mating frequency (satiation or saturation) is built in.

If the per individual reproductive rate is proportional to expected mating frequency, then (7) is essentially the mating model proposed by Volterra (1938). Volterra presumed that mating encounters between the sexes were analogous to bimolecular collisions of gas molecules. Note also that the functional dependence of (8) and (5) on n are similar. Thus, Philip's negative exponential mating function (5) could represent a mating frequency as well as a mating probability.

It is reasonable to assume that a population is not homogeneous with regard to the value of β. For instance, β might vary from individual to individual or from day to day due to difference in distances traversed, home ranges, weather factors affecting pheremone diffusion, or other random environmental factors. Heterogeneity in β can be represented by a continuous probability density, $f(\beta)$. The unconditional probability of mating is then, from (5),

$$\Pr[X(a) \geqslant 1 \text{ (unconditional)}] = 1 - \int_0^\infty e^{-\beta n} f(\beta) d\beta$$

$$= 1 - g(n). \quad (9)$$

The function $g(n)$ is the Laplace transform of $f(\beta)$, and occurs as a simple example of "marking" a renewal stream with a Poisson process (Rade, 1972).

The exponential density, $f(\beta) = \theta e^{-\theta\beta}$, is a likely candidate for the form of $f(\beta)$, as it describes a wide variety of stochastic phenomena without entailing a net increase in the number of model parameters. The probability of mating, from (9), is a rectangular hyperbola in n:

$$\Pr[X(a) \geqslant 1] = n/(\theta + n). \quad (10)$$

The parameter θ is the population density at which the probability of mating is $1/2$.

We might surmise that releasing a pheremone into a pest population has the effect of increasing θ. The insects' chemical communication system becomes disrupted with high concentrations of pheremone in the air, making it difficult for mates to locate each other. Specific quantitative relationships between θ and pheremone concentrations have not been studied, however.

We note that any other probability density specified as the form for $f(\beta)$ in (9) yields a mating probability curve shaped similar to (10) as a function of n. In particular, (9) will not be sigmoid unless $f(\beta)$ depends on n. This stems from the complete monotone property of Laplace transforms (Feller, 1966, p. 415).

The hyperbolic mating function (10) curiously arises in an entirely different context as a model of the sterile male method of pest control. In this method, large numbers of males are reared and sterilized with radiation or chemicals. They are then released and maintained at a density of ξ in the pest population. If the density of wild males is νn, the probability that a female mates with a fertile (wild) male is $n/[(\xi/\nu) + n]$. The argument assumes the sterile males are as vigorous in mating as fertile males, the sex ratio remains constant, and the organisms have no trouble finding each other to mate. This sterile male model originated with Knipling (1955), though Kostitzin (1940) proposed a vaguely similar chance mechanism for using the hyperbola as a fertilization probability.

3. POPULATION GROWTH MODELS

3.1 Stochastic vs. Deterministic. Single species growth models in the ecological literature are customarily deterministic and continuous. The growth rate of a population is given by

$$dn/dt = \lambda(n) - \mu(n), \tag{11}$$

where n = population density (a continuous function of time), and $\lambda(n)$ and $\mu(n)$ are the instantaneous natality and mortality rates, respectively, in the population.

A stochastic approach, however, seems more appropriate for sparse populations, affording a variety of possible outcomes from a given initial population density. A stochastic model "corresponding" to the deterministic model (11) might be defined as the Markov birth-death process with birth rate $\lambda(n)$ and death rate $\mu(n)$. There are usually numerous stochastic "versions" of any deterministic model; a birth-death process has the advantage of treating n as a discrete variable. The fact that organisms come in integer packages is critical for endangered species. Furthermore, the probability of ultimate extinction, a quantity of obvious interest, is easily computed for a birth-death process.

Three forms for $\lambda(n)$ are considered:

$$\lambda(n) = \lambda n; \quad \lambda(n) = \lambda \varepsilon n^2; \quad \lambda(n) = \lambda n^2/(\theta + n). \qquad (12)\text{-}(14)$$

The first is a simple linear birth rate for a population not experiencing a mating shortage. The second, the bimolecular collisions model suggested by Volterra (1938), is the simple linear rate multiplied by a factor proportional to the expected per capita mating frequency (7). The third is the simple linear rate multiplied by the expected proportion of organisms that find mates under the hyperbolic mating function (10). A fourth birth rate based on the negative exponential function (5), $\lambda(n) = \lambda n(1 - e^{-\beta n})$, behaves quite similarly to (14) in dynamical growth models, but tends to be less tractable.

3.2 Pure Birth Models and Waiting Times. Pure birth models set $\mu(n) = 0$ for all n. This assumption is reasonable when losses from the population are insignificant. Though decline or extinction is not possible, the effect of a mating shortage is to increase greatly the waiting time necessary for the population to reach a certain size. This is seen by examining such waiting times in both the deterministic and stochastic birth models.

The deterministic waiting time, t, required for a population of initial size m to reach a given size n is found explicitly by integrating (11):

$$t = \int_m^n [1/\lambda(u)]du. \qquad (15)$$

For the three birth rates (12)-(14), we have, respectively,

$$t = (1/\lambda)[\log n - \log m], \qquad (16)$$

$$t = [1/(\lambda\varepsilon)][(1/m) - (1/n)], \qquad (17)$$

$$t = (1/\lambda)[\log n - \log m] + (\theta/\lambda)[(1/m) - (1/n)]. \qquad (18)$$

Rearranged, (16) gives the more familiar form $n = me^{\lambda t}$. In (17), $t \to 1/(m\lambda\varepsilon)$ as $n \to \infty$, showing that n becomes infinite in a finite time under this model. A population's growth could thus be approximated by this model only for short time periods and low initial densities. The reciprocal of birth rate (14) in (15) is $1/\lambda(n) = 1/(\lambda n) + \theta(\lambda n^2)$. The waiting time (18) is thus the sum of (16) and a component resembling (17). Growth in the hyperbolic mating-limited population is delayed over (16) by an amount related to the "collisions" between the sexes.

In stochastic pure birth processes, the waiting time to reach size n from initial size m is a continuous random variable, denoted T. The expected value of T is similar to (15), except that it is found by summing, rather than integrating, the reciprocal birth rates.

$$E[T] = \sum_{k=m}^{n-1} 1/\lambda(k). \tag{19}$$

With the birth rates (12), (13), and (14), we have

$$E[T] = (1/\lambda)[\Psi(n) - \Psi(m)], \tag{20}$$

$$E[T] = [1/(\lambda\varepsilon)][\Psi'(m) - \Psi(n)], \tag{21}$$

$$E[T] = (1/\lambda)[\Psi(n) - \Psi(m)] + (\theta/\lambda)[\Psi'(m) - \Psi'(n)] \tag{22}$$

Here $\Psi(\cdot)$ is the digamma function (Abramowitz and Stegun, 1965, p. 258). Expression (20) is always slightly greater than its deterministic counterpart (16) for the same values of λ, n, and m. Noting that $d(\log n)/dn = 1/n$, we see that (21) bears the same relationship to (20) as does (17) to (16). The birth process resulting from (13) is stochastically explosive, in that there is a positive probability that the population becomes infinite in a finite time (see Feller, 1968, p. 453). Using the more realistic birth rate (14), (22) is again the sum of a simple linear component (20) and a collisions component, as was the case for (18). However, (22) is always somewhat greater than (18).

3.3 Birth-Death Models and Extinction. Models allowing the possibilities of decline and extinction are appropriate for populations experiencing significant mortality losses. For simplicity, the form of the loss rate is hereafter assumed to be $\mu(n) = \mu n$, where μ is a constant less than λ.

Population increase occurs for all initial densities under the deterministic model (11) with the simple linear birth rate (12). For either of the mating birth rates (13) and (14), an equilibrium, denoted $\bar{n}$, typically exists where $\lambda(\bar{n}) = \mu\bar{n}$. The equilibrium is unstable, that is, the population increases if $m > \bar{n}$. This equilibrium is termed the *critical density* in a loose analogy to the critical mass of atomic fission. Indeed, the solution trajectories of (11) using (13) as a birth rate are explosive when $m > \bar{n}$. The critical density is $\bar{n} = \mu/(\lambda\varepsilon)$ using (13) and is $\bar{n} = \theta\mu/(\lambda-\mu)$ using (14).

Stochastic birth-death processes incorporating (12), (13), or (14) more reasonably allow possibilities of increase or decrease from a given initial population size. This property is particularly important for low population densities. As $\lambda(0) = 0$, zero is an absorbing state; extinction is a possible outcome of these processes.

The chance of extinction is a quantity of special interest for preserving endangered species or eradicating injurious ones. It is a well-known result from birth-death processes (e.g., Karlin and Taylor, 1975, p. 149) that the probability of extinction from an initial size m, denoted $\alpha(m)$, is

$$\alpha(m) = [\sum_{x=m}^{\infty} \rho(x)]/[1 + \sum_{x=1}^{\infty} \rho(x)], \tag{23}$$

where

$$\rho(x) = [\mu(1)\mu(2)\cdots\mu(x)]/[\lambda(1)\lambda(2)\cdots\lambda(x)], \tag{24}$$

under the condition that $\sum \rho(x)$ converges. Extinction is certain if the sum does not converge.

Observe that a discrete probability distribution can be defined by

$$\Pr[X = x] = \begin{cases} 1/[1 + \sum_{k=1}^{\infty} \rho(k)], & x = 0; \\ \rho(x)/[1 + \sum_{k=1}^{\infty} \rho(k)], & x = 1,2,\cdots . \end{cases} \tag{25}$$

The probability of extinction is seen to be the tail of this distribution:

$$\alpha(m) = \Pr[X \geqslant m] = \sum_{x=m}^{\infty} \Pr[X = x]. \tag{26}$$

Ecologists have adopted the simple linear birth-death process incorporating (12) as a stochastic model of a species colonizing a new environment such as an island (MacArthur and Wilson, 1967; Crowell, 1973). The birth-death process incorporating (14) is a slightly modified version of the pest control model given by Costello and Taylor (1975). The probabilities of extinction for the three birth-death processes incorporating (12), (13), and (14), are, respectively, the tails of geometric, Poisson, and negative binomial distributions:

$$\alpha(m) = \sum_{x=m}^{\infty} (1-\mu/\lambda)(\mu/\lambda)^{x} = (\mu/\lambda)^{m}, \tag{27}$$

$$\alpha(m) = \sum_{x=m}^{\infty} e^{-(\mu/\lambda\varepsilon)}(\mu/\lambda\varepsilon)^x/x! = \gamma(m, \mu/\lambda\varepsilon)/\Gamma(m), \tag{28}$$

$$\alpha(m) = \sum_{x=m}^{\infty} \binom{\theta + x}{x}(1 - \mu/\lambda)^{\theta+1}(\mu/\lambda)^x. \tag{29}$$

In (28), $\gamma(\cdot,\cdot)$ represents the incomplete gamma function (Gradshetyn and Ryzhik, 1965, p. 940).

In (27), $\alpha(m)$ as a function of m decreases in the characteristic geometric fashion. If $\lambda\varepsilon > \mu$, (28) as a function of m resembles (27) in shape. If $\lambda\varepsilon < \mu$, (28) acquires a declining sigmoid shape with the inflection point, $\bar{m}$ say, occurring at $\bar{m} \simeq \mu/\lambda\varepsilon$. This quantity corresponds to the mode of the Poisson probabilities in (28) and is also the population density at which $\lambda(\bar{m}) = \mu\bar{m}$. The function (29) also displays the declining sigmoid shape for values of $\theta > (\lambda-\mu)/\mu$; the extinction probabilities for this mating model are greatly increased over those in (27) when θ represents a sizable fraction of m. The inflection point of (29) occurs at $\bar{m} \simeq \theta\mu/(\lambda-\mu)$, corresponding to the mode of the negative binomial probabilities and to the point where $\lambda(\bar{m}) = \mu\bar{m}$. The quantity $\bar{m}$ for both (28) and (29) is the stochastic counterpart to the critical density of the deterministic models.

Setting $\theta = 0$ in birth rate (14) of course recovers birth rate (12). The geometric distribution in (27) follows as a special case of the negative binomial in (29). Additionally, (29) is equivalent to the left tail of a binomial distribution when θ is an integer:

$$\alpha(m) = \sum_{x=0}^{\theta} \binom{m + \theta}{x} (1 - \mu/\lambda)^x(\mu/\lambda)^{m+\theta-x}. \tag{30}$$

3.4 Relationship to Harvesting Models. Curiously, the mating models incorporating (13) and (14) are identical to population harvesting models with respect to their extinction probabilities. Consider a population with a simple linear birth rate (12) that is harvested at a constant (stochastic) rate μ/ε. Thus, $\lambda(n) = \lambda n$ and $\mu(n) = \mu/\varepsilon$. The probability of extinction for this process is exactly (28). Also, consider a population growing according to a simple linear birth-death process that is harvested constantly at rate $\mu\theta$. Thus, $\lambda(n) = \lambda n$ and $\mu(n) = \mu n + \mu\theta$. The probability of extinction for this second process is exactly (29). Failure to mate essentially represents removal of population members from the reproductive process.

It is interesting to further consider the effect of harvesting on a mating-limited population. Letting $\lambda(n) = \lambda n^2/(\theta + n)$ and $\mu(n) = \mu n + \gamma$, where γ is a constant stochastic harvesting rate, we obtain the tail of a probability distribution based on a hypergeometric power series:

$$\alpha(m) = \sum_{x=m}^{\infty} \frac{(\gamma\theta/\mu)\binom{\gamma/\mu + x}{x}\binom{\theta + x}{x}(\mu/\lambda)^{x+1}}{F(\theta, \gamma/\mu, 1, \mu/\lambda) - 1}. \qquad (31)$$

Here $F(\cdot,\cdot,\cdot)$ is the hypergeometric function (Gradshteyn and Ryzhik, 1965, p. 1039). This probability distribution is a truncated, translated version of a generalized hypergeometric distribution given by Kemp (1971).

4. DISTRIBUTIONS AND EXTINCTION PROBABILITIES

The negative binomial in (29), the Poisson in (28), and the generalized hypergeometric in (31) are all weighted versions of the geometric distribution in (27). A generalized weighted distribution (Patil and Rao, 1978) takes the form

$$f^w(x) = w(x)f(x)/E[w(X)], \qquad (32)$$

where $f(x)$ is a probability density, and $w(x)$ represents the weight associated with each value. Here $f(x)$ is the geometric distribution in (27), $w(x) = (1/\varepsilon)^x/x!$ in (28), $w(x) = (\theta + 1)(\theta + 2)\cdots(\theta + x)/x!$ in (29), and

$$w(x) = (\theta+1)(\theta+2)\cdots(\theta+x)(\gamma/\mu+1)(\gamma/\mu+2)\cdots(\gamma/\mu+x)/(x!)^2$$

in (31). The latter two weight functions are increasing in x, giving to their corresponding probability densities heavier tails than the geometric, and hence, higher extinction probabilities for all values of m. By contrast, the extinction probability (28) is much smaller than (27) for high m values due to the unrealistically high birth rate (13).

In other applications of stochastic birth-death processes, many traditional probability distributions may emerge as forms for $\Pr[X = x]$ in (25) provided the birth and death rates are suitably chosen. For instance, the recursion relationship for such a distribution is given by

$$\Pr[X = x]/\Pr[X = x - 1] = \mu(x)/\lambda(x). \qquad (33)$$

Also, $\rho(x)$ (24) frequently takes the form $\rho(x) = h(x)r^x$, where r is a constant, and $h(0) = 1$. The probability of extinction is then the tail of a power series distribution (Patil, 1962; Ord, 1972):

$$\Pr[X = x] = h(x)r^x / \sum_{x=0}^{\infty} h(x)r^x, \quad x = 0,1,2,\cdots . \tag{34}$$

Examples seen in this paper have all been power series. An additional, unrelated example yielding the log series can be catalogued: let $\lambda(n) = \lambda\cdot(n + 1)$, $n = 1,2,\cdots$; $\lambda(0) = 0$; and $\mu(n) = \mu n$; then

$$\alpha(m) = \sum_{x=m}^{\infty} -(\mu/\lambda)^{x+1}/[(x+1)\log(1 - \mu/\lambda)]. \tag{35}$$

Finally, it is worthwhile to note a relationship between the generating functions for $\alpha(m)$ and $\Pr[X = x]$. Define the following:

$$\tau(s) = \sum_{x=0}^{\infty} s^x \Pr[X = x]; \ \phi(s) = \sum_{x=0}^{\infty} s^x \alpha(x) \tag{36}$$

One obtains $\phi(s)$ from the familiar recursion relation for $\alpha(m)$ (see Karlin and Taylor, 1975, p. 140).

$$\alpha(m) = \alpha(m+1)\lambda(m)/[\lambda(m)+\mu(m)] + \alpha(m-1)\mu(m)/[\lambda(m)+\mu(m)], \tag{37}$$

using the conventions $\alpha(0) = 1$ and $\alpha(m) = 0$, $m < 0$. One might then utilize the relation (see Feller, 1968, p. 265)

$$\phi(s) = [1 - s\tau(s)]/(1 - s). \tag{38}$$

REFERENCES

Abramowitz, M. and Stegun, I. A., eds. (1965). *Handbook of Mathematical Functions*. Dover, New York.

Allee, W. C. (1938). *The Social Life of Animals*. W. W. Norton, New York.

Andrewartha, H. G. and Birch, L. C. (1954). *The Distribution and Abundance of Animals*. University of Chicago Press, Chicago.

Bartlett, M. S. (1978). *An Introduction to Stochastic Processes with Special Reference to Methods and Applications* (third edition). Cambridge University Press, Cambridge.

Boswell, M. T., Ord, J. K., and Patil, G. P. (1979). Chance mechanisms underlying univariate distributions. In *Statistical Distributions in Ecological Work*, J. K. Ord, G. P. Patil, and C. Taillie, eds. International Co-operative Publishing House, Fairland, Maryland.

Braumhover, A. H., Graham, A. J., Bitter, B. A., Hopkins, D. E., New, W. D., Dudley, F. H., and Bushland, R. C. (1955). Screwworm control through release of sterlilized flies. *Journal of Economic Entomology*, 48, 462-466.

Costello, W. G., and Taylor, H. M. (1975). Mathematical models of the sterile male technique of insect pest control. In *Mathematical Analysis of Decision Problems in Ecology*, A. Charnes and W. R. Lynn, eds. Springer-Verlag, Berlin.

Crowell, K. L. (1973). Experimental zoogeography: introduction of mice to small islands. *American Naturalist*, 107, 535-558.

Feller, W. (1966). *An Introduction to Probability Theory and Its Applications, Vol. 2.* Wiley, New York.

Feller, W. (1968). *An Introduction to Probability Theory and Its Applications, Vol. 1.* (third edition). Wiley, New York.

Gradshteyn, I. S. and Ryzhik, I. M. (1965). *Table of Integrals, Series, and Products*. Academic Press, New York.

Karlin, S. and Taylor, H. M. (1975). *A First Course in Stochastic Processes* (second edition). Academic Press, New York.

Kemp, C. D.(1971). Properties of some discrete ecological distributions. In *Statistical Ecology, Volume 1*, G. P. Patil, E. C. Pielou, and W. E. Waters, eds. The Pennsylvania State University Press, University Park, Pennsylvania.

Knipling, E. F. (1955). Possibilities of insect control or eradication through the use of sexually sterile males. *Journal of Economic Entomology*, 48, 459-462.

Kostitzin, V. A. (1940). Sur la loi logistique et sus generalisations. *Acta Biotheoretica*, 5, 155-159.

MacArthur, R. H. and Wilson, E. O. (1967). *The Theory of Island Biogeography*. Princeton University Press, Princeton, New Jersey

Mosimann, J. E. (1958). The evolutionary significance of rare matings in animal populations. *Evolution*, 12, 246-261.

Ord, J. K. (1972). *Families of Frequency Distributions*. Griffin, New York.

Patil, G. P. (1962). Certain properties of the generalized power series distributions. *Annals of the Institute of Statistical Mathematics*, 14, 179-182.

Patil, G. P. and Rao, C. R. (1978). Weighted distributions and size-biased sampling with applications to wildlife populations and human families. *Biometrics*, 34, 179-189.

Patil, G. P. and Stiteler, W. M. (1974). Concepts of aggregation and their quantification: a critical review with some new results and applications. *Researches on Population Ecology*, 15, 238-254.

Philip. J. R. (1957). Sociality and sparse populations. *Ecology*, 38, 107-111.

Prout, T. (1978). The joint effects of the release of sterile males and immigration of fertilized females on a density regulated population. *Theoretical Population Biology*, 13, 40-71.

Rade, L. (1972). On the use of generating functions and Laplace transforms in applied probability theory. *International Journal of Mathematical Education in Science and Technology*, 3, 25-33.

Richerson, J. V., Brown, E. A., and Cameron, E. A. (1976). Premating sexual activity of gypsy moth males in small plot field tests (*Lymanthria* (*=Porthetria*) *dispar:* Lymantriidae). *Canadian Entomologist*, 108, 439-448.

Sower, L. L. and Whitmer, G. P. (1977). Population growth and mating success of Indian meal moths and almond moths in the presence of synthetic sex pheremone. *Environmental Entomology*, 6, 17-20.

Steiner, L. F., Hart, W. C., Harris, E.J., Cunningham, R. T., Ohinata, K., and Kamakahi, D. C. (1970). Eradication of the oriental fruit fly from the Mariana Islands by the methods of male annihilation and sterile insect release. *Journal of Economic Entomology*, 63, 131-135.

Volterra, V. (1938). Population growth, equilibria, and extinction under specified breeding conditions: a development and extension of the logistic curve. *Human Biology*, 3, 1-11.

Williams, G. C. (1975). *Sex and Evolution*. Princeton University Press, Princeton, New Jersey.

[*Received May* 1980. *Revised October* 1980]

THE POISSON LOGNORMAL DISTRIBUTION AND ITS USE AS A MODEL OF PLANKTON AGGREGATION

D. D. REID

CSIRO Division of Mathematics & Statistics
P. O. Box 21
Cronulla N.S.W. 2230 Australia

SUMMARY. In an attempt to derive functional models or obtain convenient descriptions of the data in plankton ecology studies, biologists have utilized a number of univariate frequency distributions. This paper reviews the statistical properties of the Poisson lognormal distribution, compares maximum likelihood and moment estimators, and considers results of fitting the distribution to several data sets in the field of plankton ecology. The role of frequency distribution models in describing the spatial aggregation of plankton is discussed.

KEY WORDS. Poisson lognormal distribution, moment estimators, maximum likelihood, plankton aggregation.

1. INTRODUCTION

Interest in the Poisson lognormal (PsLN) distribution, as with a number of other contagious distributions, has stemmed from a practical need for a distribution which is consistent with certain biological assumptions or mechanisms. The PsLN distribution has been used to model two different aspects of biological systems:

(i) The distribution has been fitted to abundance data for a single species in a series of samples. Cassie (1962) discussed the application of the PsLN distribution to data from plankton ecology studies and postulated a (log-linear) mechanism whereby the aggregation of organisms is determined

C. Taillie et al. (eds.), Statistical Distributions in Scientific Work, Vol. 6, 303–316.

by the action of a series of environmental factors, assumed to be normally distributed.

(ii) The distribution has been fitted to data for the abundances of the different species in a multi-species community. In this case, it is assumed that the log abundances constitute a random sample from a normal distribution (Bulmer, 1974; Kempton and Taylor, 1975).

In the present paper, only applications of type (i) above are considered.

2. STATISTICAL PROPERTIES

2.1 Probability Function. The PsLN distribution may be defined as the mixture of Poisson(λ) distributions, where λ follows a lognormal (μ,σ) distribution. Thus the probability function for the PsLN distribution is given by

$$p_r = C \int_0^\infty e^{-\lambda} \lambda^{r-1} \exp[-(\ell n\ \lambda - \mu)^2/2\sigma^2]d\lambda \qquad r = 0,1,2,\cdots \tag{1}$$

where $C = 1/(\sqrt{2\pi}\sigma r!)$. Equation (1) may be expressed in several alternative forms. Two such are given by Brown and Holgate (1971):

$$p_r = C \int_{-\infty}^{\infty} \exp[\lambda r - e^{\lambda} - (\lambda-\mu)^2/2\sigma^2]d\lambda$$

$$p_r = C\sigma \int_{-\infty}^{\infty} \exp[r(\mu+\sigma\lambda) - e^{\mu+\sigma\lambda} - \lambda^2/2]d\lambda$$

These integrals cannot be evaluated analytically, so numerical integration or some other form of approximation must be used. Grundy (1951) tabulated p_r for $r = 0$ and 1 over a grid of values for the mean and variance (M,V) while Brown and Holgate (1971) listed p_r, $r = 0,1,\cdots$, for various values of M and V.

Bulmer (1974) provides an approximation for the evaluation of p_r, $r \geqslant 10$, for parameter values "likely to be encountered in practice," but for the applications discussed later in this paper, the approximation is not generally usable as the sample means are too small.

An approximation for p_r using the first two terms of the Gram-Charlier Type B Series was unsuccessfully attempted by Brown and Holgate (1971). In fact, using a result established by Kullback (1947), it can be shown that the above series expansion does not converge for any parameter pair (μ,σ). It is true that the only real test of an expansion of this type is whether it is sufficiently accurate for the purpose at hand, but the above observation may point to why it seems to fail in this particular case.

2.2 Moments. Heyde (1963) points out that the lognormal distribution is not determined by its moment sequence (also see Feller, 1971, p. 227). Likewise, then, the PsLN distribution is not determined by its moments.

We may obtain an expression for the factorial moments of the Poisson lognormal by noting that the *kth* factorial moment of PsLN (μ,σ^2) equals the *kth* central moment of lognormal (μ,σ^2), that is,

$$\mu'_{(k)} = \exp(k\mu + k^2\sigma^2/2), \quad k = 1,2,\cdots.$$

In particular, the mean and variance are

$$M = \exp(\mu + \sigma^2/2), \; V = M + M^2[\exp(\sigma^2) - 1].$$

Notice that the variance to mean ratio is greater than unity (provided σ is non-zero).

2.3 Skewness and Kurtosis. The measures of skewness and kurtosis used in this paper are $\surd\beta_1$ and β_2-3, where β_1 and β_2 are Pearson's coefficients. Formulae for skewness and kurtosis are:

$$\surd\beta_1 = [\delta^2(B+2) + 3\delta+1]/(\delta+1)^{3/2}\surd M$$

$$\beta_2-3 = M[\delta^3(B^3 + 3B^2 + 6B + 6) + 6\delta^2(B+2) + 7\delta + 1]/V^2$$

where $B = \exp(\sigma^2)$, $\delta = V/M - 1 = M(B - 1)$.

Values of the skewness and kurtosis statistics were calculated over a grid of means and variance/mean ratios. For means less than 5, kurtosis increases very rapidly with increasing variance/mean ratio. The skewness increases rapidly for means of less than about 1.

2.4 Modality. Cassie (1962, p. 86) had conjectured that the PsLN may have properties similar to those of the Thomas and Neyman distributions, *inter alia,* polymodality. However, as noted by Brown and Holgate (1971), the unimodality property of the PsLN was first conjectured by Anscombe (1950), and proven as part of a general result in Holgate (1970), which gives a condition under which compound Poisson distributions (Poisson mixtures) are unimodal.

2.5 Parameter Estimation.

(a) Method of Moments. The moment estimators of the parameters μ and σ are obtained by equating the sample mean and variance (M and V) to the corresponding population moments. This gives

$$\mu_m = \ln[M^2/(V+M^2-M)^{\frac{1}{2}}],\ \sigma_m = \{\ln[V+M^2-M)/M^2]\}^{\frac{1}{2}}.$$

The asymptotic (in 1/n) covariance matrix of (μ_m, σ_m) is

$$\frac{1}{n}\begin{bmatrix} \frac{\partial\mu_1'}{\partial\mu} & \frac{\partial\mu_1'}{\partial\sigma} \\ \frac{\partial\mu_2'}{\partial\mu} & \frac{\partial\mu_2'}{\partial\sigma} \end{bmatrix} \begin{bmatrix} \mu_2'-(\mu_1')^2 & \mu_3'-\mu_2'\mu_1' \\ \mu_3'-\mu_2'\mu_1' & \mu_4'-(\mu_2')^2 \end{bmatrix} \begin{bmatrix} \frac{\partial\mu_1'}{\partial\mu} & \frac{\partial\mu_2'}{\partial\mu} \\ \frac{\partial\mu_1'}{\partial\sigma} & \frac{\partial\mu_2'}{\partial\sigma} \end{bmatrix}$$

where μ_k' is the k*th* power moment about origin of the PsLN distribution. Thus

$$V(\mu_m) = (ae + bf)a + (af + bg)b$$

$$C(\mu_m,\sigma_m) = (ae + bf)c + (af + bg)d$$

$$V(\sigma_m) = (ce + bf)c + (cf + dg)d$$

where $a = (4BM+1)/2BM^2\ ,\ b = -1/2BM^2$

$$c = -(MB+1)/2BM^2\sigma\ ,\ d = 1/2\ BM^2\sigma$$

$$e = M(1+\delta)/n\ ,\ f = M[M^2B(B^2-1) + M(3B-1) + 1]/n$$

$$g = M[M^3B^2(B^4-1) + M^2B(6B^2-2) + M(7B-1) + 1]/n$$

(b) Maximum Likelihood Estimation. Bulmer (1974) discussed maximum likelihood estimation for the zero-truncated PsLN distribution. The likelihood equations for the non-truncated case are:

$$\sum_{r=0}^{\infty} p_r[r^2 - (2r+1)\ q_r + q_r] = 0$$

$$\sum_{r=0}^{\infty} r\, p_r - q_r f_r = 0$$

where f_r are sample frequencies and $q_r = (r+1)p_{r+1}/p_r$. The p_r, and hence the likelihood equations, must be evaluated by numerical integration; obtaining the maximum likelihood estimates requires an optimizing computer routine used in conjunction with numerical integration. The MLP program (Ross, 1978) was used for the applications discussed below in Section 3.

(c) Relative Efficiency of Moment Estimators. The efficiency of moment estimators relative to ML estimators was measured by $E = |A|/|W|$ where A is the asymptotic variance-covariance matrix (inverse of expected Fisher information) for the ML estimators and W is the covariance matrix for moment estimators. The parameter values used in the efficiency calculations were set equal to the ML estimates. One of the problems in estimating asymptotic efficiency in this case is that the variance estimates sometimes change very markedly between values of the parameter estimate found by the two methods. Another relevant point is that first order asymptotics may be misleading when used in variance calculations for finite samples (Shenton and Bowman, 1977). For data sets used in the present study, relative efficiency of the moment estimates was low, ranging from 1% to 39%.

3. APPLICATIONS

3.1 Fitting Frequency Distributions. Fitting the PsLN model to counts of individuals in a series of "quadrats" appears to have been limited to the applications discussed by Cassie (1962). Other plankton workers have noted the possibility of using the model, but dismissed it on the grounds of fitting difficulties.

In discussing the interpretation of frequency distribution models fitted to plankton data, Cassie (1962) considers two alternative basic models for generating the distributions: (a) The aggregation model (also described as generalized or stopped distribution model), based on random aggregations with the number of individuals per aggregation distributed in a specified manner; and (b) the fluctuating mean model, which refers to a mixture of Poisson distributions, the means of which are distributed in a specified manner (lognormal mixing distribution results in PsLN, gamma mixing distribution gives NB, Poisson gives Neyman type A (NY)). The mechanism favored by Cassie for the plankton data is the fluctuating mean model, and the parameter estimates are not used directly as ecological parameters. Cassie mentions the use of the NB parameter $(C = 1/\kappa)$ as a measure of aggregation, and points out that the equivalent measure for the PsLN is $C' = (\sigma^2-\mu)/\mu^2$, but his main emphasis is on the use of the PsLN in the form of the fluctuating mean model. Taylor *et al.*, (1979) have pointed out the severe limitations of κ as an ecological parameter. In fitting the PsLN, NY and NB models to the data sets of the present study, no attempt has been made to interpret parameters in terms of the aggregation (generalized or stopped distribution) model. A fairly extensive range of data sets was analyzed by Evans (1953) in an attempt to give some guidelines as to the applicability of the NY and NB models.

3.2 Cassie's Data Sets. The series of plankton data fitted graphically to the PsLN by Cassie (1962, p. 75) was re-analyzed using the models and estimators described in the present paper. The data show the number of individuals of a particular zooplankton species in each of 100 samples collected using a plankton pump. The sample quadrats are cylindrical parcels of water approximately 100m long and 15cm diameter. Results of fitting the negative binomial (NB) and Poisson lognormal (PsLN) distributions are shown in Table 1. The NB results are based on a maximum likelihood (ML) fit, while the PsLN distribution was fitted using both ML and moments methods. Deviance of the model as given in the tables is $-2F(\hat{\theta})$ where $F(\theta)$ is the support given by the data for the parameter set θ (Nelder and Wedderburn, 1973). It is apparent that PsLN gives a fit to the data set similar to that for the NB model. Parameter estimates obtained by the method of moments vary somewhat from the maximum likelihood estimates, and the relative efficiency of moment estimates is about 8% in this case. The PsLN fit obtained by graphical methods in Cassie (1962) is very close to the maximum likelihood fit.

Table 2 shows results of fitting the PsLN and NB distributions to another series from Cassie (1962). While Cassie did not fit this series to the PsLN, he noted that a distribution with skewness between that of the negative binomial and lognormal distributions should provide a satisfactory fit to the data. It is evident from Table 2 that the PsLN performs better than the NB in the tail of this series, but both models give a poor fit for r less than about 15. Results of fitting the Neyman type A distribution are not quoted here, but the fit was a very poor one.

3.3 Other Plankton Data Series. The results of fitting the PsLN, NB and NY distributions of two of the data sets described in Barnes and Marshall (1951) are given in Tables 3 and 4.

TABLE 1: Frequency distribution of Elminius modestus *nauplii (from Cassie,* 1962)

r	Observed f_r (n=100)	NB (ML)	Expected PsLN (Moments) Method	PsLN (ML)
0	4	4.6	.8	2.6
1	4	5.2	2.2	4.8
2	4	5.3	3.6	5.9
3	2	5.2	4.8	6.3
4	12	5.1	5.5	6.2
5	9	4.9	5.9	6.0
6	7	4.6	6.0	5.5
7	2	4.4	5.9	5.1
8	5	4.2	5.6	4.6
9	5	3.9	5.3	4.2
10	1	3.7	4.9	3.8
11	4	3.5	4.5	3.5
12	0	3.2	4.1	3.1
13+	7	8.5	10.3	7.8
16+	5	6.9	7.6	5.9
19+	10	7.1	7.1	5.8
23+	7	9.2	7.9	7.2
31+	6	5.4	4.3	4.4
40+	6	5.2	3.7	7.3
Deviance χ^2 (d.f.)		28.0(16)	36.4(15)	26.9(16)

	$\hat{\kappa}$	μ_m	σ_m	$\hat{\mu}$	$\hat{\sigma}$
Parameter estimate	1.22	2.35	.74	2.22	.99
Standard error	.19	.10	.12	.11	.09
Correlation (r)		-.56		-.14	

$\hat{}$ ML estimate   m method of moments estimate

TABLE 2: Frequency distribution of Podon polyphemoides *(from Cassie,* 1962*)*

r	Observed f_r (n=100)	Expected NB (ML)	PsLN (Moments)	PsLN (ML)
0	7	6.3	.5	3.4
1	4	5.0	1.3	4.8
2	3	4.4	2.2	5.2
3	3	4.0	2.9	5.1
4	3	3.7	3.5	4.8
5	2	3.5	3.9	4.5
6	6	3.3	4.1	4.1
7	1	3.1	4.2	3.8
8	1	2.9	4.2	3.5
9	8	2.7	4.1	3.2
10	4	2.6	4.0	2.9
11	1	2.5	3.8	2.7
12	6	2.3	3.6	2.5
13	3	2.2	3.4	2.3
14	7	2.1	3.2	2.2
15	1	2.0	3.1	2.0
16	1	1.9	2.9	1.9
17	2	1.8	2.7	1.7
18	1	1.8	2.5	1.6
19	2	1.7	2.4	1.5
20	2	1.6	2.2	1.4
21+	3	5.8	7.5	4.9
25+	8	7.0	8.1	5.6
31+	3	8.4	8.1	6.3
41+	8	9.3	7.1	7.0
60+	10	7.9	4.5	10.9
Deviance χ^2 (d.f.)		35.8(23)	46.2(21)	37.2(23)

	$\hat{\kappa}$	μ_m	σ_m	$\hat{\mu}$	$\hat{\sigma}$
Parameter estimate	.83	2.70	.83	2.51	1.29
Standard error	.12	.14	.16	.14	.12
Correlation (r)			-.71		-.09

TABLE 3: Frequency distribution of Oithona similis *nauplii (from Barnes and Marshall, 1951)*

r	Observed f_r (n=120)	Expected PsLN	NB	NY
0	23	23.3	21.0	21.4
1	28	34.0	33.1	32.7
2	34	27.9	29.2	28.9
3	17	17.3	18.9	18.9
4	8	9.7	10.1	10.2
5	7	4.6	4.7	4.7
6	3	2.1	1.9	2.0
7+	0	1.1	1.1	1.1
Deviance χ^2 (d.f.)		6.23(5)	6.07(5)	6.24(5)

	$\hat{\mu}$	$\hat{\sigma}$	μ_m	σ_m	$\hat{\kappa}$	$\hat{M}_1$	$\hat{M}_2$
Parameter estimate	.56	.41	.61	.32	8.62	8.19	.24
Standard error	.08	.05	.08	.13	6.54	5.78	.17
Correlation (r)	.10		-.12			-.99	

TABLE 4: Frequency distribution of Lamellibranch Larvae (from Barnes and Marshall, 1951)

r	Observed f_r (n=120)	Expected PsLN	NB	NY
0	11	11.9	11.3	11.8
1	24	22.5	21.5	20.8
2	20	24.5	24.4	23.8
3	23	20.5	21.4	21.4
4	17	15.3	16.0	16.3
5	11	10.0	10.7	11.0
6	5	6.3	6.6	6.8
7	4	3.8	3.8	3.9
8	3	2.2	2.1	2.1
9+	2	2.5	2.1	1.9
13+	0	.5	.1	.1
Deviance χ^2 (d.f.)		3.21(8)	2.34(8)	2.36(8)

	$\hat{\mu}$	$\hat{\sigma}$	μ_m	σ_m	$\hat{\kappa}$	$\hat{M}_1$	$\hat{M}_2$
Parameter estimate	.98	.47	1.00	.43	5.31	5.67	.53
Standard error	.07	.05	.07	.08	2.04	1.99	.19
Correlation (r)	-.18		-.21			-.98	

The three fitted distributions show very similar deviances of the data from fitted model for the first series, only minor variations in the shapes of the fitted models being apparent. For the second series the NB and NY distributions show slightly smaller deviances than the PsLN, but each of the three models provides an adequate fit to the data. Parameter estimates using the method of moments are close to the maximum likelihood estimates, and relative efficiencies for the moment estimates are 17% and 39% respectively.

Wiebe (1968), in a detailed study of the structure of zooplankton patchiness, collected contiguous plankton samples using a Longhurst-Hardy continuous plankton recorder along a 2-dimensional grid. Wiebe was interested in using the spatial covariance structure of the samples, thus he did not fit his data to univariate frequency distribution models. Two of Wiebe's data sets are examined here for the purpose of illustrating the results of fitting some univariate models to data which are based on different methods of collection, volumes of water sampled, and a sample environment different from the other series considered above. The two series quoted were chosen arbitrarily and are not meant to be representatives of the 400 samples enumerated in Wiebe (1968).

From Table 5, the NY distribution provides a better fit to the first series. Bimodality in the series results in the PsLN and NB distributions giving a poor fit, the fitted values being similar for the two models. Table 6 also shows a polymodal sample distribution, but in this case all three distributions show a poor fit to the data.

4. DISCUSSION

The six data sets analyzed in the present study obviously cannot be representative of the totality of such series, collected over many years by a large number of number of plankton workers. The aim of presenting these series was primarily to deal with some of the questions posed in Cassie (1962), and by subsequent workers in this field, who have referred to the PsLN model, but haven't used it because of the difficulties of fitting the model and estimating parameters. The data sets chosen seem to be reasonably typical of the field, viz. often highly aggregated and with a large variation between samples. While the fit of a range of contagious distributions is quite poor with some data sets, other data series give satisfactory fits to a number of models.

Cassie's interest in proposing the application of the PsLN to data of this type was twofold: firstly, the property of the

TABLE 5: Frequency distribution of Euphausia furcilia *(from Wiebe, 1968).*

r	Observed f_r (n=44)	Expected PsLN	NB	NY
0	20	15.8	17.6	21.1
1	1	11.3	9.7	4.9
2	10	6.6	5.9	5.4
3	8	3.8	3.8	4.4
4	2	2.2	2.4	3.0
5	1	1.4	1.6	2.0
6	0	.9	1.0	1.2
7	1	.6	.7	.8
8+	1	1.5	1.4	1.1
Deviance χ^2(d.f.)		22.4(5)	19.8(5)	11.8(5)

	$\hat{\mu}$	$\hat{\sigma}$	$\hat{\kappa}$	$\hat{M}_1$	$\hat{M}_2$
Parameter estimate	.99	1.00	.81	.85	1.98
Standard error	.25	.25	.35	.21	.44
Correlation (r)	-.57			-.64	

TABLE 6: Frequency distribution of Stylocheiron furcilia *(from Wiebe, 1968)*

r	Observed f_r (n=44)	Expected PsLN	NB	NY
0	9	7.6	8.4	10.5
1	3	9.2	8.4	5.7
2	11	7.7	7.0	6.4
3	12	5.7	5.5	5.7
4	1	4.0	4.1	4.6
5	3	2.8	3.0	3.5
6	1	1.9	2.2	2.5
7+	2	4.1	4.7	4.9
13+	2	.9	.6	.3
Deviance χ^2(d.f.)		18.4(6)	20.6(6)	21.5(6)

	$\hat{\mu}$	$\hat{\sigma}$	$\hat{\kappa}$	$\hat{M}_1$	$\hat{M}_2$
Parameter estimate	.80	.80	1.48	1.76	1.70
Standard error	.17	.16	.53	.47	.44
Correlation (r)	-.34			-.85	

PsLN having a skewness between the negative binomial and lognormal distributions seemed to be an empirical requirement of at least some of these series; secondly, the PsLN distribution corresponded to a plausible biological or physical model - the number of organisms present in a sample being determined by an exponential function of physical factors such as temperature and salinity. The use of frequency distributions thus fulfills the function of: (1) providing an empirical description of the data, to help assess differences between sites or time periods, etc; (2) providing a mathematical model of a physical process.

The results from the present study indicate that fitting the PsLN model to this type of data may sometimes succeed in the first function stated above, but, as has been indicated previously, the fit of the model will often be close to that of the negative binomial distribution. If a computer program such as MLP (the Rothamsted Maximum Likelihood Program) is available very little effort is required to fit the PsLN model. Using moment estimators will often result in low relative efficiency, and the expected frequencies must still be evaluated by numerical integration, unless parameter values are within the range of the tables of Brown and Holgate (1971).

Alternative approaches to the analysis of aggregation of plankton have been reviewed by Fasham (1978), who suggests the use of spectral methods as the most suitable method for analyzing plankton spatial pattern. If analysis of spatial pattern is the primary objective, then, as recognized by Cassie (1962, p. 89), use of frequency distributions is not the most appropriate approach, since it fails to take into account the spatial structure in the data.

Cassie's contention that the PsLN is a useful and realistic model in plankton distribution studies, can only be properly confirmed or refuted by analysis of a much more extensive range of data sets, but Wiebe (1980, personal communication) has observed that a large body of recent plankton data indicates that models such as the PsLN often do not adequately fit the data, and thus do not represent the (fundamental model) generalization conjectured by Cassie.

ACKNOWLEDGEMENTS

The author gratefully acknowledges the guidance and encouragement of J. B. Douglas. David Griffiths and the referees made a number of helpful comments. Thanks are also due to P. H. Wiebe for permission to use data from his unpublished thesis, and for useful criticisms and advice.

REFERENCES

Anscombe, F. J. (1950). Sampling theory of the negative binomial and logarithmic series distribution. *Biometrika*, 37, 358-382.

Barnes, H. and Marshall, S. M. (1951). On the variability of replicate plankton samples and some applications of contagious series to the statistical distribution of catches over restricted periods. *Journal of the Marine Biological Association of U.K.*, 30, 233-263.

Brown, S. and Holgate, P. (1971). Table of the Poisson lognormal distribution. *Sankhya, Series B*, 33, 235-248.

Bulmer, M. G. (1974). On fitting the Poisson lognormal distribution to species abundance data. *Biometrics*, 30, 101-110.

Cassie, R. M. (1962). Frequency distribution models in the ecology of plankton and other organisms. *Journal of Animal Ecology*, 31, 65-91.

Evans, D. A. (1953). Experimental evidence concerning contagious distributions in ecology. *Biometrika*, 40, 186-211.

Fasham, M. J. (1978). The statistical and mathematical analysis of plankton patchiness. *Oceanography and Marine Biology: Annual Review*, 16, 43-79.

Feller, W. (1971). *An Introduction to Probability Theory and its Applications, Vol. II* (second edition). Wiley, New York.

Grundy, R. M. (1951). The expected frequencies in a sample of an animal population in which the abundances of species are lognormally distributed. *Biometrika*, 38, 427-434.

Heyde, C. C. (1963). On a property of the lognormal distribution. *Journal of the Royal Statistical Society, Series B*, 2, 392-393.

Holgate, P. (1970). The modality of some compound Poisson distributions. *Biometrika*, 57, 666-667.

Kempton, R. A.(1975). A generalized form of Fisher's logarithmic series. *Biometrika*, 62, 29-38.

Kempton, R. A. and Taylor, L. R. (1974). Log-series and lognormal parameters as diversity discriminants for the Lepidoptera. *Journal of Animal Ecology*, 43, 381-399.

Kullback, S. (1947). On the Charlier type B series. *Annals of Mathematical Statistics*, 18, 574-581.

Nelder, J. A. and Wedderburn, R. W. M. (1973). Generalized linear models. *Journal of the Royal Statistical Society, Series A*, 135, 370-384.

Ross, G. J. S. (1978). Curve fitting using the Rothamsted maximum likelihood program. In *Numerical Software - Needs and Availability*, D. Jacobs, ed. Academic Press.

Shenton, L. R. and Bowman, K. O. (1977). *Maximum Likelihood Estimation in Small Samples*. Charles Griffin, London.

Taylor, L. R., Woiwod, I. P. and Perry, J. N. (1979). The negative binomial as a dynamic ecological model for aggregation, and the density dependence of k. *Journal of Animal Ecology*, 48, 289-304.

Wiebe, P. H. (1968). *Zooplankton patchiness and its effects on sampling error: a field and computer model study*. Ph.D. Thesis, University of California, San Diego.

[*Received May* 1980. *Revised November* 1980]

SOME APPLICATIONS OF STATISTICAL DISTRIBUTION THEORY TO BIOLOGY AND MEDICINE

ALAN J. GROSS and
M. CLINTON MILLER III

Department of Biometry
Medical University of South Carolina
Charleston, South Carolina 29403 USA

SUMMARY. Often, a particular statistical distribution will fit collected medical or biological data quite well. When this occurs there is usually an underlying reason. For example, if an individual dies due to some random phenomenon such as an accident where the probability of occurrence is proportional to the length of the interval under observation and whose rate of occurrence is constant over time and if there is no other cause of death, e.g. aging, the statistical distribution that characterizes survival in this instance is well known to be the negative exponential distribution. Other statistical distributions such as the Poisson and gamma distributions are also known to fit certain types of biological and medical data quite well for similar reasons.

In this article, specific statistical distributions are proposed for fitting biological and medical data arising from other mechanisms. In particular the sum of two exponentials with different failure rates is proposed as a model for kidney failure data. The Poisson distribution with the zero class truncated is proposed as a model for estimating a potential population of a health clinic and the number of exposed families during an epidemic in which no member of the family incurred the symptoms of the disease.

Reasons why specific models may not fit biological or medical data are discussed. In these instances, more complicated models are suggested such as the compound Poisson distribution. However, there are instances where no specific statistical distribution may provide a good fit to biological data because

C. Taillie et al. (eds.), Statistical Distributions in Scientific Work, Vol. 6, 317–336.

of its complicated genesis. For example, survival in a human population with all of its ramifications is difficult to describe by any single statistical distribution.

Repeated measurements is a pervasive problem in biomedical research. Examples include measurements, such as systolic blood pressures, recorded on the same patient within a sample of patients at different times. Statistical procedures are proposed for examining such patient data.

There are biological and medical data that are multivariate in nature to which a multivariate exponential rather than a multivariate normal distribution would provide an appropriate fit. These distributions are discussed, briefly.

Finally, there is an interesting extension of the geometric distribution that has applications in clinical trial studies. This distribution is introduced.

KEY WORDS. Time-response distributions, zero class estimation, concomitant variables, competing risks, Poisson distribution, compound Poisson distribution.

1. INTRODUCTION

Medical and biological data occur and are observed in many forms. Specifically, however, medical and biological data are either discrete (such as the number of accidents seen weekly at an emergency clinic) or continuous (such as the diastolic and systolic blood pressures of hypertensive patients). This of course is true of all collected data regardless of the source of the data. Furthermore, just as with other statistical data, biological and medical data can be univariate or multivariate in nature. Although concern in this article is primarily directed towards univariate data, an application of the multivariate ANOVA model (Potthoff and Roy, 1964) is discussed with regard to systolic and diastolic blook pressure data.

Many medical and biological data are time-response data in nature. That is, one measures the length of time to a specific response such as death. relapse, failure of a patient on a kidney machine, etc. Statistical distributions relating to a time-response data are treated in Section 2.

Biological and medical data also deal with counts; e.g. blood counts, counting number of patients entering a health facility, or counting the number of cases of a chronic disease such as cervical cancer in a given community. Often, when dealing with count data a particular category is not recorded. For example,

McKendrick (1926) examines a chlorea epidemic in India wherein he is able to ascertain the number of households containing one, two, three, or four cholera cases but not the number of infected households that have zero recorded cholera cases. Information was available on the total number of households with no cholera cases, but there is no way to partition those zero case households into infected and noninfected. Thus, estimating the "infected" zero case number becomes an important public health problem. Dahiya and Gross (1973) consider this problem employing the truncated Poisson distribution. Dahiya (1980) tests the fit of the truncated Poisson with the estimated "infected" zero case number to McKendrick's data. These problems are discussed in some detail in Section 3.

The problem of "repeated measures" is continually occurring in various biomedical settings. For example, hypertensive patients receiving different treatment regimens to lower their blood pressures have these blood pressures recorded at different times. Thus, a two way design is formed with "treatment" as one classification and "time" as the other classification. It is clear that the usual two-way ANOVA model is not appropriate for data analysis in this situation because of the intrapatient correlations that exist from time to time. However, the model proposed by Potthoff and Roy (1964) for analyzing growth curve data does provide a method for analyzing such "repeated measures" data. Unfortunately, the Potthoff-Ray model is not as frequently used as it should be in dealing with the "repeated measures" data, perhaps because the procedure is not well known. In Section 4 this model is put forth as one possibility in dealing with this problem. Other approaches to this problem have been discussed by Federer (1977), Hills and Armitage (1979), and Brown (1980).

Section 5 describes the application of non-normal types of multivariate distributions to biomedical data. In particular, multivariate exponential and exponential-type distributions have had application in response-time situations. Gross and Lam (1980) discuss an exponential-type model that is somewhat analogous to the paired t-test. Their distribution is based on the bivariate exponential distribution developed by Block and Basu (1974).

Finally, in Section 6, an extension of the geometric distribution is introduced that has application to clinical trials studies.

2. TIME-RESPONSE DISTRIBUTIONS

The most widely used time-response distribution is the exponential distribution. It is used to describe the time-response event, e.g. accidental death, whose occurrence rate is constant with respect to time.

It is well known (see, for example, Feller, 1966, p. 10) that the sum of exponential random variables each with the same occurrence rate has a gamma distribution. However, should the occurrence rates not be the same, the sum of exponential variables no longer has a gamma distribution. To pursue this idea further, consider the following situation. Suppose a patient with kidney disease enters a study and is observed until both kidneys fail. It is assumed that the failure rate of each kidney is constant and equals α while both kidneys are functioning. However, as soon as one kidney fails the failure rate of the remaining functioning kidney is also constant and equals $\beta > \alpha$. It is not difficult to show that the failure distribution for the patient is

$$f(t;\alpha,\beta) = 2\alpha\beta\{\exp(-2\alpha t)-\exp(-\beta t)\}/(\beta-2\alpha), \qquad (1)$$

$\alpha > 0$, $\beta > 0$; $t \geq 0$. A reparameterization of the model with $\gamma = 2\alpha$ is desirable. Thus,

$$f(t;\beta,\gamma) = \beta\gamma\{\exp(-\gamma t)-\exp(-\beta t)\}/(\beta-\gamma). \qquad (1')$$

This distribution is not well-known. However, a version of it is found in Feller (1966; p. 39, problem 6). Gross *et al.* (1971) consider this distribution and its properties. They examine the problem of estimating the parameters β and γ by the method of maximum likelihood (m.ℓ.). First, Monte-Carlo simulation techniques were used to generate samples from this distribution. The Newton-Raphson method and the Method of Scoring introduced by Rao (1952) were then used to obtain the m.ℓ. estimates from the generated samples. Large sample properties of the estimates were also examined. Thus, the distribution (1) which may be applied to measure the length of life of two-organ systems has been studied in the complete sample case; i.e. in the case where no observations have been censored.

A two-organ system in which the organs fail independently of each other demonstrates a different failure distribution. To see this more clearly and to describe this failure distribution, suppose that within a two-organ system the first organ fails independently of the second organ and their respective failure rates are α_1 and α_2. Let T_1 and T_2 be the respective failure times of the two organs. Then, $T = \max(T_1, T_2)$ is the failure time of the organ system. Thus,

$$\text{pr}\{T\leqslant t\} = \text{pr}\{T_1 \leqslant t;\ T_2 \leqslant t\} \quad . \tag{2}$$

Since T_1 and T_2 are independent the cumulative distribution is

$$F(t;\alpha_1,\alpha_2) = \text{pr}\{T\leqslant t\} = \{1-\exp(-\alpha_1 t)\}\{1-\exp(\alpha_2 t)\} \quad . \tag{3}$$

It then follows that this two-organ failure density is

$$f(t;\alpha_1,\alpha_2) = \alpha_1\exp(-\alpha_1 t) + \alpha_2\exp(-\alpha_2 t) -(\alpha_1+\alpha_2)\exp\{-(\alpha_1+\alpha_2)t\} \quad , \tag{4}$$

The moment generating function, mean, and variance of T are straightforward to obtain. Let $m_\theta(T)$, μ_T, and σ_T^2 represent them (respectively). Then

$$m_\theta(T) = (1-\theta\alpha_1^{-1})^{-1} + (1-\theta\alpha_2^{-1})^{-1} - \{1-\theta(\alpha_1+\alpha_2)^{-1}\}^{-1}, \tag{5}$$

$$\mu_T = \alpha_1^{-1} + \alpha_2^{-1} - (\alpha_1+\alpha_2)^{-1} \quad , \tag{6}$$

and $$\sigma_T^2 = (\alpha_1^{-1} - \alpha_2^{-1})^2 + 2(\alpha_1 + \alpha_2)^{-1}\mu_T - (\alpha_1 + \alpha_2)^{-2} . \tag{7}$$

Estimation of the parameters α_1 and α_2 is still under investigation.

A rather interesting generalization of the constant occurrence rate is given in Gross and Clark (1975, Ch. 1). Suppose the occurrence rate is constant except that at specific, known times the rate changes. That is, all changes in the occurrence rate are in jumps, i.e. the rate is constant except for changes at times $t_1, t_2, \cdots, t_k$. Put $t_0 = 0$, $t_{k+1} = +\infty$, and let α_j be the occurrence rate during the interval (t_{j-1}, t_j), $j=1,2,\cdots,k+1$. The resulting time-response pdf, $f(t;\alpha_1,\cdots,\alpha_{k+1})$, is zero for negative t and for $t_{j-1}\leqslant t<t_j$ is given by

$$f(t;\alpha_1,\cdots,\alpha_{k+1}) = \alpha_j\exp\{-\alpha_j(t-t_{j-1}) - \sum_{i=1}^{j-1}\alpha_i(t_i-t_{i-1})\}, \tag{8}$$

$$j = 1,2,\cdots,k.$$

This model is known as the piecewise exponential distribution. It is useful in assessing population life tables, because the failure rate in these tables is generally calculated for every one-year interval from birth to over 110 years. Aroian and Robinson (1966) discuss an application of sequential analysis to this distribution. Calculation of the moment generating function, mean, and variance are straight forward but rather tedious. If the parameters $\alpha_1,\cdots,\alpha_k$ are mathematically independent, the m.ℓ. estimators and their large sample properties present no particular difficulties. If, however, these parameters are functionally dependent, then consideration of the m.ℓ. estimators and their large sample properties becomes somewhat complicated. This problem is currently under investigation.

Often response times are related to exogenous variables. This is particularly true in biomedical applications. One of the earliest biomedical studies of response times in the presence of concomitant information was conducted by Feigel and Zelen (1965). In their study they considered survival times of leukemia patients along with each patient's initial white blood count. White blood counts were considered concomitant observations so that on each of n patients the observations $(x_1, y_1),\cdots,(x_n, y_n)$ were collected. The variable x_i is the logarithm to the base 10 of the white blood count at the time of diagnosis of the *ith* patient and y_i is the patient's survival time; $i=1, 2,\cdots,n$. The statistical distribution Feigl and Zelen (1965) hypothesized for patient survival was exponential with failure rate λ_i for the *ith* patient, where $\lambda_1^{-1} = a+bx_i$, $i=1,2,\cdots,n$. Thus, the mean survival time was hypothesized to be a linear function of the base 10 logarithm of the white blood count in the patient population. Estimation of the parameters a and b was then carried out using m.ℓ. estimation techniques. The large sample properties of the estimates were also examined. Actual leukemia patient data were used in two examples that were presented by Feigel and Zelen. It should not be surprising that the estimates of b in these examples were negative, indicating that higher initial white blood counts indicate shorter survival times for leukemia patients.

Zippin and Armitage (1966) extended the Feigel-Zelen model to the case where the survival data underwent multiple censoring. Thus, for each patient with acute leukemia the white blood count at diagnosis was available. In addition it was noted whether the patient was dead or alive at the end of the study. Those patients who were alive constituted the censoring in the sample. Furthermore, since patients entered the study at different times, patient censoring was multiple, Zippin and Armitage (1966) again used m.ℓ. techniques to estimate the parameters a and b and examined their large sample properties. The estimate of b was still negative, an encouraging sign. Even with the censored observations the lower initial white blood counts led to longer survival among patients.

The ideas of concomitant variables and censored response times were dealt with in a very elegant manner by Cox (1972). Cox's model allows elimination of the restrictive assumption of a constant failure rate. In fact, the Cox model is non-parametric in nature. There are, however, some drawbacks to the model. The m.ℓ. estimators of the regression parameters are based on obtaining a partial likelihood function rather than a full likelihood function on the set $\{t_{(j)}\}$, $t_{(1)} \leqslant \cdots \leqslant t_{(d)}$, of times when the deaths or failures occur. (See Cox (1975) for a discussion of partial likelihood functions.) The other drawback to Cox's model is that the data are singly censored; i.e., the only knowledge available on the survivors is that each survivor has a failure time that is at least as large at $t_{(d)}$, the largest failure time. In general, however, Cox's model has enjoyed wide usage in survival studies, particularly clinical trials, since its inception. For example, several studies of the survival of heart transplant patients at Stanford University have used Cox's model as well as other response time (survival) distributions in analyzing the survival patterns of these patients; see Turnbull *et al.* (1964), Mantel and Byar (1974), and Crowley and Hu (1977).

The last topic concerning time-response distributions that is covered in this survey is the competing risk problem. It is assumed that response that is considered is death of an individual and there may be k risks that "compete" for his life.

If one uses the approach by Moeschberger and David (1971), the following model may be appropriate. Let Y_j be the survival time of someone who is subject to the risk R_j only. If all risks of death are included, the length of life, T, of the individual is given as $T = \min_j Y_j$. The probability density function of T is then

$$f(t) = p^{-1} \int_t^{\infty} \cdots \int_t^{\infty} g(y_1, \cdots, y_k) \big|_{y_i = t} \prod_{\substack{\ell=1 \\ \ell \neq 1}}^{k} dy_\ell, \tag{9}$$

where $p = \mathrm{pr}\{Y_i = \min_j Y_i\}$, i.e. the probability that death is due to the *ith* risk, $g(y_1, \cdots, y_k)$ is the joint density function of the times to death from all k risks, and t is the actual time of death, the realization of T.

Suppose each individual in a sample of size n in a population dies as a consequence of one of the k risks. If n_i individuals die according to the *ith* risk and $n = \Sigma_i n_i$, the lifetime of the *jth* individual dying from risk R_i is t_{ij}. If the k risks are mutually independent, then the likelihood function of the sample is

$$L = \frac{n!}{\Pi_i n_i!} \prod_{i=1}^{k} \prod_{\ell=1}^{k} \prod_{j=1}^{n_i} \lambda_i(t_{ij})(1-F_\ell(t_{ij})) \tag{10}$$

where $F_\ell(t)$ and $\lambda_\ell(t)$ are the cumulative distribution function and failure (occurrence) rate, respectively, for an individual whose risk of death is R_ℓ, $\ell=1,2,\cdots,k$.

Moeschberger and David (1971) investigate the likelihood function (10) in specific instances when, for example, each underlying failure distribution is exponential. Estimators of the parameters are obtained and their large sample properties are considered. A dependent risk model is also considered by Moeschberger (1974) which is based on a multivariate Weibull distribution that was proposed by Marshall and Olkin (1967).

3. COUNT DATA WITH A MISSING CATEGORY

In biomedical applications as in other applications of statistics cound data abound. Simple examples of biomedical count data include estimating the number of seizures of epileptic patients, determining, within specified limits, the number of pets in a certain community immunized against rabies, and determining whether the incidences of certain diseases such as lung cancer and hypertension are increasing or decreasing.

Count data are often fitted by well-known statistical distributions such as the binomial, multinomial, and Poisson. In general terms, however, if X is a count or discrete random variable whose domain is the set of natural integers pluz zero, or a subset thereof, the probability function $f(x:\underset{\sim}{\theta})$ is by definition,

$$f(x;\underset{\sim}{\theta}) = \underset{\underset{\sim}{\theta}}{pr}\{X = x\}, \tag{11}$$

$x=0,1,\cdots$, and $\underset{\sim}{\theta}$ is a vector parameter (univariate or multivariate). If a sequence of independent observations is available from (11), $x_1, x_2, \cdots, x_n$, (say) a common method of estimating $\underset{\sim}{\theta}$ as with the response-time distributions is the m.ℓ. method.

Often a certain value of values of X are unavailable. In applications the most frequently missing value of X is the value $X = 0$. For example, a repairman is interested in estimating the number of a certain item, e.g. color television sets, in his service area. The repairman observes only those items that require at least one repair during the year. Thus, those items not requiring repair are not seen and hence the repairman does not know how many items there are in his service area. As a second example, consider the question of how many people are served by a medical facility such as outpatient clinic. The number of people who used the clinic during a given time period is of course known, as is the number of times each person used the clinic. However, the number of people who should have visited the clinic but did not is unknown. A final example, in this context, involves estimating the number of households that were infected with cholera during a cholera epidemic in a village in India; see McKendrick (1926). The available information is the number of households with at least one active case of cholera. The missing information is the number of exposed households with no clinically active cholera cases. Thus, the total number of infected households is unknown.

Consider now the problem of estimating the parameter θ from the probability function

$$f(x;\theta) = \underset{\theta}{\mathrm{pr}}\{X = x\}$$

when the value of $X = 0$ is missing or truncated. Then, the conditional probability function given $X > 0$ is

$$f^*(x;\theta) = f(x;\theta)/\{1-f(0;\theta)\}, \tag{12}$$

$x=1,2,\cdots$. Estimation of θ can be achieved by m.ℓ. estimation. Thus, if $x_1,\cdots,x_n$ is a random sample of size n from a population with probability function (11), then m.ℓ. estimator, $\hat{\theta}$, is the value of θ that satisfies the equation

$$\sum_i \frac{\partial \ln f(x_i;\theta)}{\partial\theta} - n\,\frac{\partial \ln\{1-f(0;\theta)\}}{\partial\theta} = 0 \tag{13}$$

and depends on the data set $x_1,\cdots,x_n$. It should also be noted that this maximization procedure is valid as long as the regularity conditions concerning θ hold. These conditions are discussed by Rao (1952, pp. 157-8) and are generally satisfied for the distributions that are most frequently applied to count data; e.g. the Poisson, binomial, and negative binomial distributions.

Suppose now that

$$f(x;\theta) = \exp(-\theta)\ \theta^x/[x!\{1-\exp(-\theta)\}] \quad (14)$$

$x = 1,2,\cdots$. That is, the Poisson with the zero class truncated is the distribution under consideration. Estimation of θ, is principally m.ℓ. estimation, has been considered by a number of authors including Cohen (1960), David and Johnson (1952), Irwin (1959), Johnson and Kotz (1969), and McKendrick (1926). The m.ℓ. estimator $\hat{\theta}$ of θ is the unique solution of the equation

$$\hat{\theta}/\{1-\exp(-\hat{\theta})\} = \sum_{i=1}^{n} x_i/n = \bar{x}, \quad (15)$$

where $x_1,\cdots,x_n$ is the random sample from the population whose probability function is (14). Cohen (1960) provides tables for obtaining $\hat{\theta}$ when the value of $\bar{x}$ is given. He also considers the large sample properties of $\hat{\theta}$.

The problem that still remains is the estimation of N, the total Poisson sample, or, equivalently, the estimation of n_o, the number of zero class observations that cannot be observed. This problem is considered in two related papers: Dahiya and Gross (1973) and Blumenthal *et al.* (1978). Since in applications the two procedures are quite similar and since the method Gross and Dahiya (1973) is somewhat easier to apply, this method is reviewed. Their estimator $\hat{N}$ is, however, conditional on the value of $\hat{\theta}$.

Dahiya and Gross show that $\hat{N} = [n/\{1-\exp(-\hat{\theta})\}]$ where $[a]$ is the largest integer less than or equal to a. Furthermore, a $(1-\alpha)100$ percent confidence interval for N is given as $\hat{N} \pm z_\alpha\, \hat{\sigma}\, \sqrt{\hat{N}}$ where z_α is such that $\text{pr}\{|Z| > z_\alpha\} = \alpha$ when Z has a standard normal distribution and

$$\sigma^2 = \exp(-\hat{\theta})/[1-\{\exp(-\hat{\theta})\}(1+\hat{\theta})\}]. \quad (16)$$

Again to obtain integer lower and upper limits the appropriate integer values are easily obtained.

Finally, in this context, Dahiya (1980) has considered the problem of how well a hypothesized discrete distribution fits the observed data in each class using the estimated zero class value. Specifically, if $n_1, \cdots, n_{k-1}$ are observed values in the $(k-1)$ mutually exclusive classes, where n_{k-1} is the number of ones, n_{k-2} the number of twos, $\cdots n_1$ the number of integers at least size $(k-1)$, then the statistic W, which is defined as

$$W \equiv \sum_{1}^{k-1} (n_i - \hat{N}\hat{p}_i)^2 / \hat{N}\hat{p}_i, \tag{17}$$

is asymptotically $\chi^2_{k-3} + \lambda\chi^2_1$ where χ^2_{k-3} and χ^2_1 are independent chi-square variables with $k-3$ and one degrees of freedom, respectively, and $\lambda = \theta/\{\exp(\theta)-1\}$. Hence $0 < \lambda < 1$, $\hat{N} = [n/\{1-\exp(-\hat{\theta})\}]$, $\hat{p}_i = \hat{\theta}^{k-i} \exp(-\hat{\theta})/(k-i)!$, $i = 2, \cdots, k-1$, and $\hat{p}_1 = 1 - \sum_{1}^{k-2} \hat{p}_i$. Since θ is unknown one can apply Dahiya's procedure by noting that for any constant c

$$\mathrm{pr}\{\chi^2_{k-3} > c\} \leqslant \mathrm{pr}\{\chi^2_{k-3} + \lambda\chi^2_1 > c\} \leqslant \mathrm{pr}\{\chi^2_{k-2} > c\}.$$

Thus, ordinary chi-square tables are sufficient in most cases.

In other biomedical applications the Poisson distribution is not appropriate as a model for count data. For example, suppose an estimate is required of the total sample of individuals who visit a medical facility in a given period of time. Counts are available on those individuals who visit the facility at least once. These counts may not be independent, however, because an individual may visit the facility more than once for the same medical problem. Thus, a more complex distribution, e.g. a compound Poisson distribution, may be required to fit this type of count data. It is well known that the negative binomial distribution arises as a compound Poisson distribution, e.g. see Sherbrooke (1966). Thus, there may be occasion for fitting a negative binomial distribution, with $X = 0$ truncated, to counts at certain medical facilities. Again the problem of estimating the number of individuals who should have visited the clinic but didn't arises. That is, estimation of the missing zero class sample size for the negative binomial distribution that is truncated at $X = 0$ would be appropriate and necessary. Generally, the methodology pioneered by Blumenthal (1979) and Sanathanan (1972) is useful in approaching these problems.

A final application of estimating the size of the missing zero class occurs in Phase IV trials of newly marketed drugs. These trials occur after a drug has been approved and marketed. The purpose of the trials is to test efficacy and record the side effects (both types and numbers of patients complaining of each type) that occur with each drug. If no side effect occurs, then no report is made. Since the drug is on the market, it is difficult to determine the number of individuals who are taking it. The problem is currently under study by Cantor and Miller (1980). Suppose there are k independent side effects or clusters of side effects associated with quasi-independent physiological systems. In general $S_0, S_1, \cdots, S_r$, $r=2^k-1$, are the 2^k possible subsets of the integers $1,2,\cdots,k$. For example, if $k=2$ systems for which side effects have been reported, are gastro-intestinal and drowsiness, then S_0 is the set of persons with no side effects, S_1 is the set of persons with gastro-intestinal side effects only, S_2 is the set of persons with drowsiness as the side effect, and S_3 is the set of persons with both side effects.

Let p_j be the probability an individual has side effect j, $j=1,2,\cdots,k$, and let n_i be the number of people in subset S_i, $i=1,2,\cdots,r$. The quantity n_0 is unknown. First of all, if $f(s_i)$ is the probability a person who receives medication has the side effects described by S_i, $i=0,1,\cdots,r$, then

$$f(s_i) = \prod_{j \varepsilon S_i} p_j \prod_{j \notin S_i} (1-p_j),$$

$i=1,\cdots,r$. Note that $f(s_0) = \prod_{j=1}^{k} (1-p_j)$. Thus, the conditional probability of incurring the side effects in S_i, given $i > 0$, is

$$f^*(s_i) = f(s_i)/\{1-f(s_0)\}, \tag{18}$$

$i=1,2,\cdots,r$, and, hence, the conditional likelihood function L^* is given as

$$L^* = \prod_{i=1}^{r} [f(s_i)/\{1-f(s_0)\}]^{n_i} . \tag{19}$$

Estimators of $p_1, \cdots, p_k$ are then obtained by the m.ℓ. method. Thus, by methods similar to Dahiya and Gross (1973) an estimator of n_0, $\hat{n}_0$, is found as $\hat{N}-n$, where $n = \sum_{i=1}^{r} n_i$, and $\hat{N} = [n/\{1-f(s_0)\}]$. This procedure and its properties are still under investigation.

4. REPEATED MEASURES

One of the important applications of statistics to biomedical data is with regard to repeated measures. In this connection, an important and common problem that occurs in hypertension research is when one (or more) group(s) of patients is (are) given medication to reduce patient blood pressures over a specified period of time, and, in the course of the study, a sequence of blood pressure readings is recorded for each patient.

Often, not much attention is paid to the intrapatient correlations that exist within patients as their blook pressures are measured at different times. Thus, one may be tempted to apply the standard two-way analysis of variance model to these readings in which type of medication is one effect and time of measurement is the other effect. Unfortunately, such an analysis can and often does lead to incorrect interpretations.

An approach to this problem was proposed by Potthoff and Roy (1964) in which they were to analyze dental measurements on 11 girls and 16 boys where measurements were recorded on each of the 27 children at ages 8, 10, 12, and 14. They obtain estimates of the growth curves of each group of children using the points 8, 10, 12, and 14 years. Confidence intervals for each growth curve are obtained and a confidence interval for their difference is also calculated.

More generally, the Potthoff-Roy model is applicable to the following situations:

(i) Suppose there are n experimental units all observed under identical conditions. Each unit is observed at times $t_1, \cdots, t_r$. The observations $X(t_1), \cdots, X(t_r)$ (say) on each experimental unit are assumed to follow a multivariate normal distribution with unknown variance-covariance matrix Σ ($\Sigma \neq I\sigma^2$) and

$$E\{X(t)\} = a_0 + a_1 t + \cdots + a_{u-1} t^{u-1} . \tag{20}$$

Experimental units are assumed to be stochastically independent.

(ii) The experimental situation in (i) may be generalized to include m groups of experimental units in which the *ith* group is of size n_i, $i=1,2,\cdots,m$. Each group may be represented by a different growth curve.

(iii) Other generalizations of the model are also possible, including measurement of more than a single characteristic that varies with time as a polynomial, a two-way classification model, and/or a combination of these.

Extension of the work of Potthoff and Roy (1964) have been investigated by Rao (1965), Khatri (1966), and Grissle and Allen (1969). These extensions deal primarily with how one estimates Σ, a problem not fully investigated by Potthoff and Roy. There is also an allowance for the addition of co-variables into the model.

Finally, it has been observed in many applications that a polynomial fit of the time points yields inadequate results. Hence other more sophisticated methods have been and are being developed including such methods as stochastic differential equations. A good discussion of these methods is found in Sandland and McGilchrist (1979).

Another repeated measure situation occurs when different medications are administered to the same group of patients in sequence. If all medications were to be given in the same order then the effects of each treatment depend on the order in which these treatments are given. That is, if k treatments are given to a patient the only effect that can be observed from the last treatment in the sequence is its direct effect. The next-to-last treatment will not only have a direct effect but also a one-period residual effect as will all other treatments back to the first treatment that will have a direct effect, a one-period residual effect, a two-period residual effect, $\cdots$, a (k-1)-period residual effect. Federer (1977) discusses designs that are applicable in measuring not only the direct effects of the treatments administered sequentially, but also their various residual effects. Application of these designs in such a treatment sequencing may prove very useful.

The cross-over designs have been used frequently as a method to mitigate the residual treatment effects. For example, the two-period cross-over design examines the effects of two treatments administered sequentially in which the patient sample is divided randomly into two groups. One group of patients receives the treatments A and B (say) in the order A before B. The other patient group receives the same treatments in the order B before A. These designs have been quite useful in clinical trial situations. If one goes beyond two periods and two treatments then intrapatient correlations as well as residual treatment effects come into play. To our knowledge not much research has been carried out to apply the Potthoff-Roy-Grizzle-Allen multivariate methods to multiperiod cross-over designs. Some of the important articles concerning cross-over designs are by Grizzle (1965), Hills and Armitage (1979), and Brown (1980).

5. MULTIVARIATE EXPONENTIAL DISTRIBUTIONS

Multivariate distributions, as in other disciplines, have applications in biology and medicine. Applications of the multivariate normal and related distributions such as Hotelling's T^2 distribution are common. For example, heights and weights of school age children are often fitted by a bivariate normal distribution when these data are to be analyzed in tandem. Methodology on repeated measures presented in Section 4 depends on multivariate normal assumptions.

There are situations involving multivariate data for which the multivariate normal distribution is not the best candidate for fitting the distribution. A more appropriate multivariate density, especially when dealing with response-time situations, is likely to be a multivariate extension of the exponential or a related density such as the Weibull or gamma density. To illustrate these ideas, the bivariate situation is considered here.

One of the earliest bivariate exponential distributions was introduced by Gumbel (1960). He proposed the joint cumulative density function of X, Y to be

$$F(x,y) = 1 - \exp(-x) - \exp(-y) + \exp(-x-y-\sigma xy) \quad (21)$$

where $x \geqslant 0$, $y \geqslant 0$, and the parameter $\sigma \varepsilon [0,1]$. It is easily shown that both marginals $F(x)$ and $G(y)$ are exponential.

Freund (1960) developed a bivariate extension of the exponential distribution that is applicable to a two component system (see Section 2). However, Freund's bivariate density $f(x,y)$ given as

$$f(x,y) = \begin{cases} \alpha\beta'\exp[-\{\beta'y + (\alpha+\beta-\beta')x\}], & 0 < x < y \\ \beta\alpha'\exp[-\{\alpha'x + (\alpha+\beta-\alpha')y\}], & 0 < y < x \end{cases} \quad (22)$$

does not have marginal distributions that are exponential.

Marshall and Olkin (1967) developed a multivariate exponential distribution that is based on preserving the lack of memory property of the exponential. In the bivariate case

$$F(x,y) = 1 - \exp\{-\alpha x - \beta y - \gamma \max(x,y)\}, \tag{25}$$

$x \geqslant 0$, $y \geqslant 0$, $\alpha > 0$, $\beta > 0$, $\gamma > 0$. It is not difficult to show for this distribution that

$$\text{pr}\{X>x+t,\ Y>y+t \mid X>x,\ Y>y\} = \text{pr}\{X>t,\ Y>t\},$$

i.e. the lack of memory property is satisfied.

All these multivariate distributions are precursors to the distribution developed by Block and Basu (1974). The Block-Basu distribution is quite similar to Freund's distribution and is given as

$$f(x,y) = \begin{cases} \dfrac{\alpha_1\alpha(\alpha_2+\alpha_{12})}{\alpha_1+\alpha_2} \exp[-\{\alpha_1 x + (\alpha_2+\alpha_{12})y\}], & y > x > 0 \\ \\ \dfrac{\alpha_2\alpha(\alpha_1+\alpha_{12})}{(\alpha_1+\alpha_2)} \exp[-\{(\alpha_1+\alpha_{12})\, x + \alpha_2 y\}], & 0 < y < x, \end{cases} \tag{24}$$

where $\alpha_1>0$, $\alpha_2>0$, $\alpha=\alpha_1 + \alpha_2 + \alpha_{12}$, and $\alpha_{12}>\max\{-\alpha_1,-\alpha_2\}$. The Block-Basu distribution satisfies the lack of memory property and is absolutely continuous. However, the marginal distributions are not exponential.

Gross and Lam (1980) apply the Block-Basu distribution when response times for two different treatments are recorded in the same patient sample; i.e. each patient has two recorded response times, one for each treatment.

Here, it should be noted that applications of non-normal distributions to biological and medical data are not as common as they should be. In this section the multivariate exponential distribution, in particular the bivariate exponential distribution, is discussed as a specific non-normal distribution that can be applied to biological and medical data.

6. A GENERALIZED GEOMETRIC DISTRIBUTION

In clinical trial studies treatments that patients receive often produce dichotomous results. That is, the treatment is classified only as effective or noneffective. Furthermore, when a patient receives the treatment he may suffer a side effect or he may not suffer it. Also, a sequence of clinical trials may be required to develop an effective treatment regimen for a particular disease entity. An example of this is the current treatment sequence for childhood leukemia, a disease that in many instances can now be controlled if not cured. Thus, a question may be asked: How many clinical trials are required to develop an effective treatment for a particular disease process? The purpose of this section is to propose a model that may be useful in at least resolving this question, partially.

Suppose q_x is the probability that an event does not occur on the *xth* trial (e.g. q_x is the probability that the *xth* clinical trial does not produce an effective treatment for a given disease). Let $p_x = 1-q_x$ be the complementary porbability; $x=1,2,\cdots$. (In the special case of the geometric distribution $p_x=p$ for all x.) Define $Q_x \equiv \Pi_1^x q_y$ and $Q_0 \equiv 1$. Clearly, the probability the event occurs for the first time on the *xth* trial is

$$f(x) = \begin{cases} Q_{x-1}\, p_x, & x=1,2,\cdots \\ 0, & \text{elsewhere.} \end{cases} \tag{25}$$

Let us assume that $f(x)$ sums to unity so that equation (25) describes a discrete probability distribution whose support is the set of positive integers. [Using the Borel-Cantelli lemma, a necessary and sufficient condition for the $f(x)$ to sum to unity is that either (i) $p_x = 1$ for some x or (ii) $\Sigma p_x = \infty$. A sufficient condition is that $\limsup \{q_x\} < 1$.]

As an example, suppose $q_x = q/x$, $x \geqslant 1$, where $0 < q < 1$. It then follows that

$$f(x) = \begin{cases} (1-q/x)q^{x-1}/(x-1)!, & x=1,2,\cdots \\ 0, & \text{elsewhere.} \end{cases} \tag{26}$$

The characteristic function C_t of (26) is easily seen to be

$$C_t = \exp(e^{it}q)\ (e^{it}-1) + 1.$$

The mean and variance of (26) are thus e^q and $e^q\ (2+q-e^q)$, respectively. Other properties of (26) including estimation of q are currently under investigation. The fitting of (26) to clinical trials data is also under consideration.

This model also has application in the following situation. Suppose milk cows undergo antibiotic treatment for an infection that is discovered in the herd. As long as this substance remains in measurable quantities milk extracted from a given cow may not be sold for consumption. Blood samples are taken daily to determine whether the cow's milk may be sold. Each day is assumed to represent a new trial and on the *xth* day, the probability the milk cannot be sold is q/x, $0 < q/x < 1$, $x=1,2,\cdots$.

ACKNOWLEDGEMENT

The authors thank Dr. Charles Taillie for his valuable comments on an earlier manuscript leading to substantial improvement in the content of this article.

REFERENCES

Aroian, L. A. and Robinson, D. E. (1966). Sequential life tests for the exponential distribution with changing parameter. *Technometrics*, 8, 217-227.

Block, H. W. and Basu, A. P. (1974). A continuous bivariate exponential extension. *Journal of the American Statistical Association*, 69, 1031-1037.

Blumenthal, S. (1979). *Stochastic Expansions for Point Estimation From Complete, Censored and Truncated Samples.* (In Press.)

Blumenthal, S., Dahiya, R. C., and Gross, A. J. (1978). Estimating the complete sample size from an incomplete Poisson sample. *Journal of the American Statistical Association*, 70, 182-187.

Brown, B. Wm., Jr. (1980). The cross-over experiment for clinical trials. *Biometrics*, 36, 69-79.

Cantor, A. B. and Miller, M. C., III (1980). Estimating the patient sample size in Phase IV clinical trials based on recorded side effects. (Unpublished manuscript.)

Cohen, A. C. (1960). Estimating the parameter in a conditional Poisson distribution. *Biometrics*, 16, 203-211.

Cox, D. R. (1972). Regression models and life tables. *Journal of the Royal Statistical Society*, B34, 187-220 (with disucssion).

Cox, D. R. (1975). Partial likelihood. *Biometrika*, 62, 269-276.

Crowley, J. and Hu, M. (1977). Covariance analysis of heart transplant data. *Journal of the American Statistical Association*, 72, 27-36.

Dahiya, R. C. (1980). Pearson goodness-of-fit test when the sample size is unknown. (Submitted for publication).

Dahiya, R. C. and Gross, A. J. (1973). Estimating the zero class from a truncated Poisson sample. *Journal of the American Statistical Association*, 68, 731-733.

David, F. N. and Johnson, N. L. (1952). The truncated Poisson. *Biometrics*, 8, 275-285.

Federer, W. T. (1977). Applications and concepts of repeated measures designs when residual effects are present. (Unpublished manuscript.)

Feigel, P. and Zelen, M. (1965). Estimation of exponential survival probabilities with concomitant information. *Biometrics*, 21, 826-838.

Feller, W. (1966). *An Introduction to Probability Theory and Its Applications, Vol. II*. Wiley, New York.

Freund, J. E. (1961). A bivariate extension of the exponential distribution. *Journal of the American Statistical Association*, 56, 971-977.

Grizzle, J. E. (1965). The two-period change-over design and its use in clinical trials. *Biometrics*, 21, 467-480. (Corrigenda in *Biometrics*, 30, 727).

Grizzle, J. E. and Allen, D. M. (1969). Analysis of growth and dose response curves. *Biometrics*, 25, 357-381.

Gross, A. J. and Clark, V. A. (1975). *Survival Distributions: Reliability Applications in the Biomedical Sciences*. Wiley, New York.

Gross, A. J. and Lam, C. F. (1980). Paired observations from a survival distribution. *Biometrics* (to appear).

Gross, A. J., Clark, V. A., and Liu, V. (1971). Estimation of survival parameters when one of two organs must function for survival. *Biometrics*, 27, 369-377.

Gumbel, E. J. (1960). Bivariate exponential distributions. *Journal of the American Statistical Association*, 55, 698-707.

Hills, M. and Armitage, P. (1979). The two-period cross-over clinical trial. *British Journal of Clinical Pharmacy*, 8, 7-20.

Irwin, J. O. (1959). On the estimation of the mean of a Poisson distribution from a sample with the zero class missing. *Biometrics*, 15, 324-326.

Johnson, N. L. and Kotz, S. (1969). *Discrete Distributions*. Houghton Mifflin, Boston.

Khatri, C. G. (1966). A note on a MANOVA model applied to problems in growth curves. *Annals in the Institute of Statistical Mathematics*, 18, 75-86.

Mantel, N. and Byar, D. P. (1974). Evaluation of response-time data involving transient states: An illustration using heart transplant data. *Journal of the American Statistical Association*, 69, 81-86.

Marshall, A. W. and Olkin, I. (1967). A multivariate exponential distribution. *Journal of the American Statistical Association*, 32, 30-44.

McKendrick, A. G. (1926). Application of mathematics to medical problems. *Proceedings of the Edinburg Mathematical Society*, 44, 98-103.

Moeschberger, M. (1974). Life tests under dependent causes of failure. *Technometrics*, 16, 39-47.

Moeschberger, M. and David, H. A. (1971). Life-tests under competing causes of failure and the theory of competing risks. *Biometrics*, 27, 909-933.

Potthoff, R. F. and Roy, S. N. (1964). A generalized multivariate analysis of variance model useful especially for growth curve problems. *Biometrics*, 51, 313-326.

Rao, C. R. (1952). *Advanced Statistical Methods in Biometric Research*. Wiley, New York.

Rao, C. R. (1965). The theory of least squares when the parameters are stochastic and its application to the analysis of growth curves. *Biometrika*, 52, 447-458.

Sandland, R. L. and McGilchrist, C. A. (1979). Stochastic growth curve analysis. *Biometrics*, 35, 255-271.

Sanathanan, L. (1972). Estimating the size of the multinomial population. *Annals of Mathematical Statistics*, 43, 142-152.

Sherbrooke, C. C. (1966). *Discrete Compound Poisson Processes and Tables of the Geometric Poisson Distribution*. RM-4831-PR, RAND Corporation, Santa Monica, CA.

Turnbull, B. W., Brown, B. Wm., Jr., and Hu, M. (1974). Survivorship analysis of heart transplant data. *Journal of the American Statistical Association*, 69, 74-80.

Zippin, C. and Armitage, P. (1966). Use of concomitant variables and incomplete survival with estimation survival parameter. *Biometrics*, 22, 655-672.

[*Received May* 1980. *Revised September* 1980]

EXTREME VALUE THEORY WITH APPLICATION TO HYDROLOGY

R. V. CANFIELD

Department of Applied Statistics
Utah State University
Logan, Utah

D. R. OLSEN

Department of Statistics
Texas A & M University
College Station, Texas

T. L. CHEN

Northrop Aviation
Hawthorne, California

SUMMARY. There is little to guide the researcher in the selection of a continuous distribution to describe natural phenomena. Very often empirical fit is the sole cirterion used to choose the best distribution. Problems associated with the empirical fit are described.

Whenever possible, theoretical considerations should be used to help determine the best distribution. Extreme value theory is one of the few guides with easily recognized hypotheses. Extreme value theory is reviewed and applied to stream flow data.

Application of the classical theory is limited due to the common occurance of mixture random variables in natural phenomena. Mixture random variables can considerably reduce the rate at which the extreme value of a sequence converges to its asymptotic form. An extension of extreme value theory which specifically considers mixture random variables is given. The resulting distributions are products of extreme value distributions rather than the standard mixture form. Estimation and application of these forms to river flow data is illustrated.

C. Taillie et al. (eds.), Statistical Distributions in Scientific Work, Vol. 6, 337–350.

KEY WORDS. Extreme value theory, mixture random variables, peak annual river flow.

1. INTRODUCTION

The choice of probability function for discrete phenomena is greatly aided by the many probability models which determine a unique function. For example, the Poisson assumptions, binomial and negative binomial models are usually easily associated with the underlying structure of observed phenomena. Although there is a vast number of continuous distributions, there are relatively few which possess a probability structure easily associated with observed circumstances. In most applications empirical fit is used to direct the choice of distribution with little or no theoretical justification. Blind use of empirical fit as the sole criterion for selecting a distribution can be dangerous as illustrated in Section 2 of this paper. Normal theory and extreme value theory are two examples of theory providing continuous probability models which may be associated with observed phenomena.

Application of extreme value theory is discussed in this paper. The assumptions of and distributions resulting from extreme value theory are reviewed in Section 3. Application of the theory to the distribution of annual flood peaks is also given in Section 3.

Mixed distributions are common in natural processes. They occur in river flow data for example, because storms may originate from more than one source (e.g., hurricane, nonhurricane storms). River flows are also affected by the amount of snow pack and rate of melt in some areas. Thus, when such influences are present, the result may be a random variable which is a mixture with several components. The random variable, time to failure, for devices with more than one mode of failure can be a mixture (Canfield and Borgman, 1975).

Mixture random variables present a special problem for extreme value theory. In effect, they can greatly increase the sample size (i.e., the number of observations from which the extreme is selected) which is necessary for the asymptotic theory to be effective. In most applications, this sample size is fiexed by nature (e.g., 365 days/year), and can not be controlled by the experimenter. The nature of this problem and methods which can be used to recognize it are discussed in Section 4. The last solution deals with an extension of extreme value theory (Canfield and Borgman, 1975) to mixture random variables. A more effective extreme value form for the mixture case is given and applied to riverflow data.

2. EMPIRICAL FIT

A common practice used to find a distribution for existing data is to select a few parametric families (which seem reasonable), estimate the parameters for each family and then pick the best according to some appropriate goodness-of-fit criterion. This procedure was used for example by Benson (1968) to develop a uniform technique to determine flood frequency. The log-Pearson type III distribution was selected from among 6 candidate distributions as providing the best overall fit to a group of long term annual flood peaks. The results of this study have been challenged (Bobee and Robitaille, 1977) wherein a distribution not considered previously (the Pearons Type 3 distribution) was also evaluated. This kind of reevaluation is to be expected and is encouraged by Benson (1968). However, a simple plot easily illustrates the fallacy of determining a distribution with "best overall fit" to many data sets. Figure 1 shows annual flood peak data from the Headingly River, Canada, plotted $\ln(b-X_i)$ *vs* $\ln[-\ln(N-i+1)/(N+1))]$ where X_i is the *ith* order statistic of N observations and b is an estimated upper bound on flows. If the data is from a Pearson type 3 or a log-Pearson type 3 distribution, a gentle arc is expected. The "S" shape of the plot indicates otherwise. It is not hard to imagine the generous error in extrapolating the magnitude of low probability flood events from this population if either of these distributions is used as a model. Yet, this data was among the sets used to show the Pearson type 3 distribution is better overall than the log Pearson Type 3 (Bobee and Robitaille, 1977). It is clear that neither should be used and would result in serious error if used to estimate the probability of rare events.

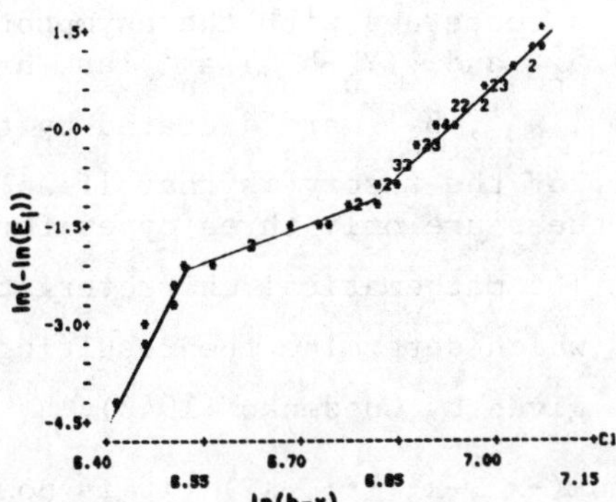

FIG. 1: Annual Flood Peaks, Headingly River, Assinibione, Canada. 1912-1970.

It may be argued that better goodness-of-fit criteria needs to be used. However, Monte Carlo experiments (Canfield *et al.*, 1980) have shown that the wrong distribution can provide a better fit at least occasionally when compared with the true distribution using an empirical fit. It seems reasonable that this could happen with any criteria adopted. It has also been suggested that the best fitting distribution cannot be too different from the true distribution. This may be true in the regions of the random variable which have high density. However as pointed out in the previous paragraph, it need not hold in the tails.

Thus, when available, theoretical considerations should play an important role in the selection of a distribution. Application of extreme value theory as a tool in the selection of a distribution is discussed in the remainder of this paper.

3. EXTREME VALUE DISTRIBUTIONS

The most authoritative work on extreme value theory is provided by Gnedenko (1943). The assumptions and distributions resulting from the theory are reviewed in the first part of this section. Application is illustrated in the second part.

3.1 Review. Extreme value random variables are defined as follows. Let $x_1, x_2, \cdots, x_n$ be a sequence of independently, identically distributed, continuous random variables. Let

$$Z_n = \max(x_1, x_2, \cdots, x_n) \quad \text{and} \quad Y_n = \min(x_1, x_2, \cdots, x_n).$$

Extreme value theory is concerned with the asymptotic distribution of sequences $(Z_n - b_n)/a_n$ and $(Y_n - b_n')/a_n'$ as $n=1,2,\cdots$. The norming values a_n, b_n, a_n', b_n' are dictated by the theory. The interesting result of the theory is that if an asymptotic distribution exists, there are only three types for Z_n and three types for Y_n. The mathematical characteristics for the random variables x_1 which determine the resulting distribution for Z_n and Y_n are given by Gnedenko (1943).

Because $-Y_n = \max(-x_1, -x_2, \cdots, -x_n)$ it is possible to restrict attention to distributions of the maximum extremes. Thus, the distributions of Z_n only will be considered here. The three possible distributions of Z_n are (Gnedenko, 1943)

$$F_1(x) = \exp -\exp - \frac{x-b}{c} \quad -\infty<x<\infty,\ c>0 \tag{1}$$

$$F_2(x) = \begin{cases} 0 & x<b \\ \exp\left\{-\left(\frac{b-x}{c}\right)^{-a}\right\} & x\geqslant b,\ c>0,\ a>0. \end{cases} \tag{2}$$

$$F_3(x) = \begin{cases} 1 & x>b \\ \exp\left\{-\left(\frac{b-x}{c}\right)^{a}\right\} & x\leqslant b,\ c>0,\ a>0. \end{cases} \tag{3}$$

Qualitative characteristics of these distributions are discussed in the next section. The assumption of independence of the $x_1, x_2, \cdots, x_n$ random variables is violated in many applications. However, Watson (1952) has shown that independence is not a necessary assumption. If the randomized sequence of x_i's satisfies the assumption for all n, the theory holds.

The advantage of the theory is that once an extreme value situation is recognized one can legitimately confine the search for best fit to three extreme value distributions. The mathematical characteristics of the three distributions are very different, thus it is relatively easy to determine the correct one for a given set of data. A graphical procedure is given below for use in identifying which of the extreme value distributions should be used with a given set of data.

Distributions (1), (2), and (3) have some easily observed characteristics. The function $F_3(x)$ is limited to some maximum value b (i.e., $F_3(x) = 1$ for $x \geqslant b$), thus random variables which have an upper limit have extreme value form $F_3(x)$. The converse of this statement is not necessarily true, however, and variables which are not limited may also have this form (Gnedenko, 1943).

The form $F_2(x)$ is referred to as a "Cauchy type" because the extreme values for the Cauchy distribution follow distribution (5). Cauchy type distributions are "heavy tailed" and seldom occur in nature. Thus, distribution (5) has limited usefulness compared with the other type types. There is, however, reference to its use in Gumbel (1954). The form $F_1(x)$ is the one most widely used and generally the only one explained in textbooks.

Three simple plots constitute the easiest method of determining which extreme value distribution is appropriate. Let $x_{(1)}, x_{(2)}, \cdots, x_{(n)}$ represent the ordered extreme value data for the observed maximums. For any random variable, the expected value of its distribution function evaluated at the *ith* order statistic is $i/(n+1)$ where the sample size is n (i.e., $E(F(x_{(i)})) = i/(n+1)$) (Lindgren, 1976). Define $E_i = i/(n+1)$. Note that from equation (3)

$$\ln(-\ln F_1(x_{(i)})) = -x_{(i)}/c + b/c. \tag{4}$$

The relationship in equation (4) is linear in $x_{(i)}$. Substituting E_i for $F(x_{(i)})$ in equation (4) and plotting $X_{(i)}$ *vs.* $\ln(-\ln F(x_{(i)}))$ identifies data from a population with distribution function $F_1(x)$. If equation (1) is appropriate the plot will be a straight line. If the data are from any other distribution the plot will not be a straight line.

The plot which identifies data from an $F_2(x)$ population is similar. From equation (2) it follows that

$$-\ln F_2(x_{(i)}) = -a \ln(x-b) + a \ln c. \tag{5}$$

Thus, if data are from a population with distribution $F_2(x)$, the plot of $\ln(x(i)-b)$ *vs.* $\ln(-\ln E_i)$ will be a straight line with negative slope. The plot is relatively insensitive to b. Thus, although more precise estimation procedures are available, b can be crudely estimated in 1.4 or 2 times the largest value.

The third plot which identifies $F_3(x)$ is motivated from equation (3) in the same manner, i.e., the plot of $\ln(b-x_{(i)})$ *vs.* $\ln(-\ln E_i)$ is a straight line with positive slope.

3.2 Application. The application of extreme value theory to peak annual floods is illustrated in this section.

As discussed by Bobee and Robitaille (1977), the physical limitations of meteorological phenomena and basin characteristics which control river flow suggest that flows are bounded by an upper limit. Thus it seems that the most logical distribution for the statistical description of flood peaks is $F_3(x)$. Figure 2 verifies this choice for the Kymijoki River in Finland. It is

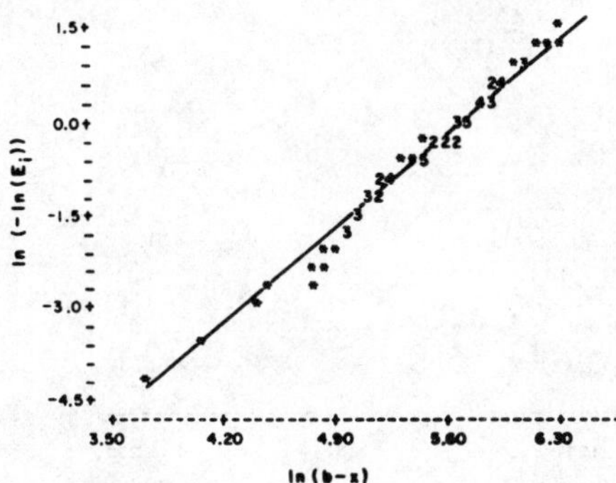

FIG. 2: Annual Flood Peaks, Kymijoki River, Peruoo, Finland. 1900-1968.

very evident from a glance that the data plot is linear in this case.

The distribution $F_3(x)$ is transformed Weibull, i.e., if the $F_3(x)$ is transformed by $y = -x$ the distribution of y is Weibull with the same parameters as $F_3(x)$ (b is negative). Therefore a program available for maximum likelihood (ML) estimation of Weibull parameters (Harter and Moore, 1965) can be used to estimate the parameters by transforming the original data.

The previous discussion would seem to establish $F_3(x)$ as the distribution to use with peak annual floods. Unfortunately few of the rivers plotted produced straight line plots as in Figure 2. Figure 1 is such an example. It is shown in the next section that the presence of mixture random variables can easily produce plots of the type in Figure 1.

4. EXTREME OF MIXTURE RANDOM VARIABLES

4.1 The Problem. One of the problems associated with the application of extreme value theory to mixture random variables is illustrated in Figures 3-6. These figures represent plots of Monte Carlo data which simulates a very natural sampling situation. Suppose the period of sampling (e.g., 365 days/year) is fixed. Observations occur from either of two populations, however, the second population is rarely observed. When it does occur, it is usually the maximum event during the observation

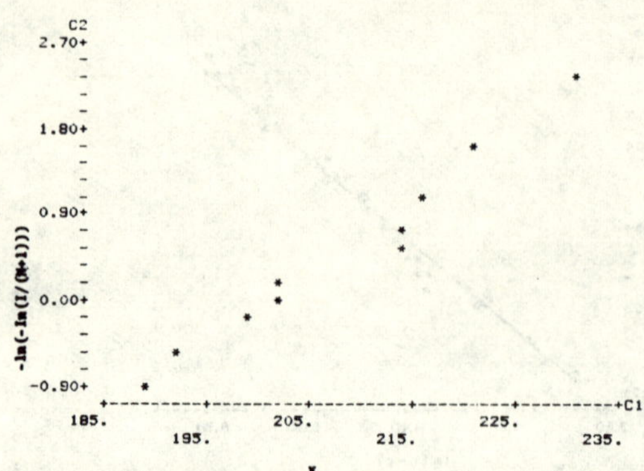

FIG. 3: Extremes of 10 *Random Samples, Each of Size* 50.

period. This is the case for example with river flows in some regions where infrequeny hurricanes strike the area. In the Monte Carlo simulation, a two component mixture random variable $Y = p_1X_1 + p_2X_2$ was sampled. The components X_1 and X_2 are each normal with means $\mu_1=235$, $\mu_2=350$ and standard deviations $\sigma_1=40$, $\sigma_2=30$, respectively. The mixture probabilities are $p_1=.9992$ and $p_2 = .0008$.

Figure 3 shows the maximums from the first 10 sets of 50 random observations of Y. Since $F_1(x)$ (equation 1) is the appropriate extreme value distribution, the plot is ln(-ln(i/ (10+i) *vs.* y_i, $i=1,2,\cdots,10$ where y_i is the *ith* order statistic of the 10 extremes. The linear nature of the plot is evident, indicating appropriate application of extreme value theory. Figure 4 shows the same plot with 20 more maximums added. In this case one of the rare events (X_2) has occurred and appears as an outlier. The same plot appears in Figure 5 with 20 additional observations and again in Figure 5 with a total of 100 observations. Two straight line plots are now evident.

A problem associated with extremes of mixture random variables can now be recognized. No matter how many maximums are observed, due to the sample size (50) from which each maximum is chosen, about 1 in 25 extremes will be from X_2 with the remaining extremes from X_1. Thus two straight lines will occur. One solution is to increase the sample size from which the extreme is taken. However, for the random variable simulated here,

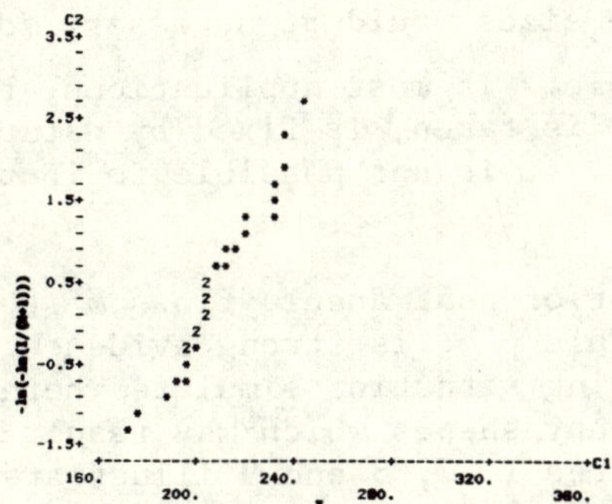

FIG. 4: Extremes of 30 *Random Samples, Each of Size* 50.

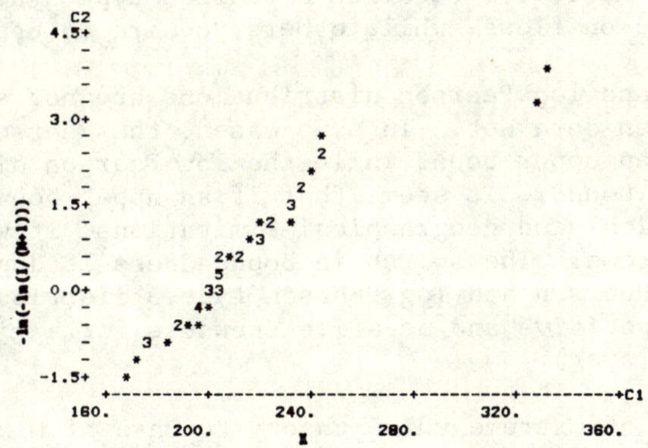

FIG. 5: Extremes of 50 *Random Samples, Each of Size* 50.

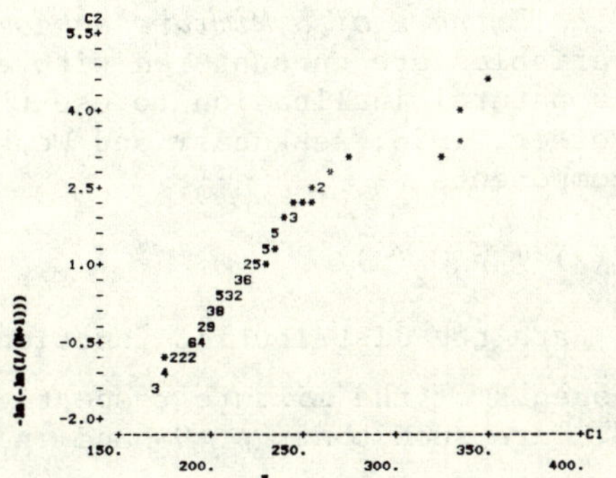

FIG. 6: Extremes of 100 *Random Samples, Each of Size* 50.

extremely large sample sizes would be necessary for the X_2 to determine each sample. In most applications, the sample size from which the maximum is taken, is fixed by nature (e.g., 365 days/year). Therefore, it is not possible to increase the sample size.

Figure 7 is a plot of peak annula flows of the Vuoski River at Imatra, Finland. This plot is strong evidence of the reality in nature of the sampling situation simulated here. It is not difficult to imagine many shapes which may result from mixture random variable. Figures 1, 7, 8 and 9 illustrate this variety for the rivers indicated.

The variety of possible shapes is an explanation for a rather perplexing problem encountered by Bobee and Robitaille (1977). They noted that some rivers required a Pearson type 3 distribution with a lower bound on flows, while others require an upper bound.

The Pearson and log Pearson distributions are not even consistent for a given data set. In some cases, the Pearson distribution calls for an upper bound while the log Pearson distribution calls for a lower bound. It seems that if an upper bound is valid due to meteorological and geograohical limitations, it would be valid for all systems. The switch in boundedness is due to the inability of the Pearson and log Pearson type 3 distributions to accommodate both positive and negative skewness for a given bound (upper and lower).

An extension of extreme value theory to case of mixture random variables (Canfield and Borgman, 1971) is a useful tool for describing the distributions encountered here. Application of this extension is given in the next section.

4.2 Distribution of the Extreme of a Mixture Random Variable. When mixture random variables are encountered with extreme value situations, there is a natural inclination to use the standard mixture form (e.g., Potter, 1958; Ashkanasy and Weeks, 1975). For the case of two components

$$F(x) = p_1G_1(x) + p_2G_2(x) \tag{6}$$

where $G_i(x)$, $i=1,2$, are the distribution functions of the first and second components of the mixture respectively. The parameters p_i, $i=1,2$, are such that $p_i>0$ and $p_1+p_2 = 1$. Estimation of the parameters in equation (6) is very difficult because p_1 and p_2 must be estimated in addition to all of the parameters of both $G_1(x)$ and $G_2(x)$. A more adequate approxi-

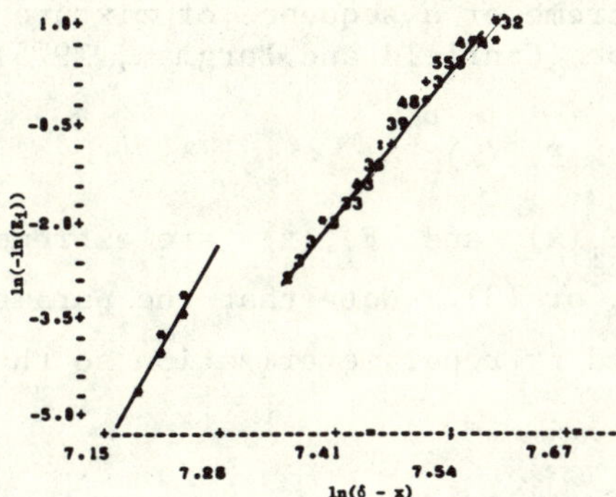

FIG. 7: Annual Flood Peaks, Vuoksi River, Imatra, Finland. 1847-1968.

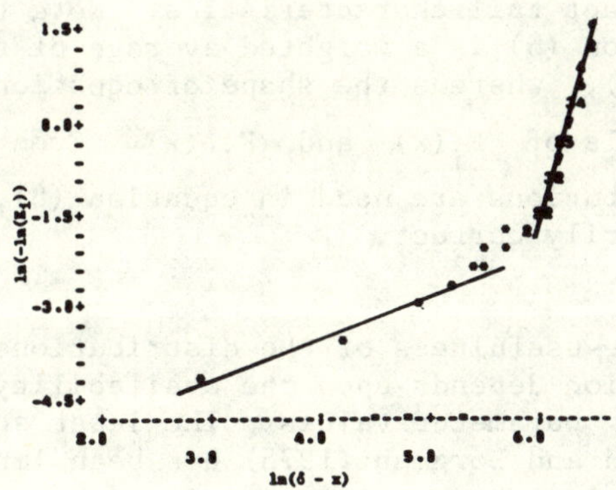

FIG. 8: Annual Flood Peaks, Assiniboine River, Brandon, Canada. 1902-1980.

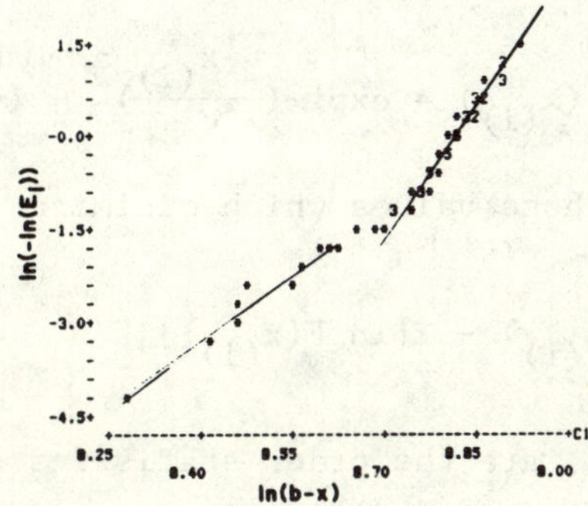

FIG. 9: Annual Flood Peaks, Saskatoon River, South Saskatchewan, Canada. 1912-1970.

mating form for the extreme of a sequence of mixture random variables is shown to be (Canfield and Borgman, 1975)

$$F(x) = F_i(x)^{p_1} \; F_{i'}(x)^{p_2} \tag{7}$$

where the components $F_i(x)$ and $F_{i'}(x)$ are extreme value distributions (1), (2), or (3). Note that the parameters p_1 and p_2 can be absorbed by reparameterization so that equation (7) can be rewritten,

$$F(x) = F_i(x)F_{i'}(x) \tag{8}$$

thereby reducing the number of parameters in the distribution. Because of its theoretical basis, a distribution of this form should have the correct tail characteristics. Note that the tail shape in equation (8) is a weighted average of the tails of $G_1(x)$ and $G_2(x)$, whereas the shape of equation (8) is a product of the tails of $F_i(x)$ and $F_{i'}(x)$. Even if two extreme value distributions are used in equation (8), the tail shape is not necessarily correct.

4.3 Estimation. The usefulness of the distributions described in the previous section depends upon the availability of techniques for estimating parameter values. The least squares technique in Canfield and Borgman (1975) has been improved and applied to several peak annual flood records (Canfield *et al.*, 1980) with excellent results. The method is illustrated here for (8) with each component of the type $F_3(x) = \exp\{-((b-x)/c)^a\}$, $x \leq b$. The upper bound b is assumed to be the same for both components. For this case (8) becomes

$$F(x_{(i)}) = F_3(x_{(i)})\; F_{3'}(x_{(i)}) = \exp\{-(\frac{b-x_{(i)}}{c})^{a} - (\frac{b-x_{(i)}}{c'})^{a'}\}.$$

Parameter estimates are those values which minimize

$$Y = \sum_{i=1}^{n} \{\ln F(x_{(i)}) - E[\ln F(x_{(i)})]\}^2 \tag{9}$$

where $x_{(1)}, x_{(2)}, \cdots, x_{(n)}$ are the order statistics of a random sample of size n from a population with distribution $F(x)$, and W_i^{-1} is the variance of $\ln F(x_{(i)})$. Because the distribution of $F(x_{(i)})$ is $\beta(i,n-1+1)$ (Lindgren, 1976), the constants

W_i and $E[\ln F_3(x_{(i)})]$, $i=1,2,\cdots,n$ are nonparametric. Therefore, the values are useful for estimation no matter what distribution is used. Gaussian quadrature may be used to compute these constants. For convenience of notation let

$$\theta' = (c,c'), \quad \gamma = (a,a')$$

$$E' = (E \ln F(x_{(1)}), E \ln F(x_{(2)}), \cdots, E \ln F(x_{(n)})$$

$$D = \mathrm{Diag}(\ln x_{(1)}, \ln x_{(2)}, \cdots, \ln x_{(n)})$$

and

$$A = \begin{pmatrix} (b_{(1)}-x)^{a_1}, & (b_{(2)}-x)^{a_2}. & .(b_{(n)}-x)^{a_1} \\ (b_{(1)}-x)^{a_2}, & (b_{(2)}-x)^{a_2}, & ,(b_{(n)}-x)^{a_2} \end{pmatrix}$$

Then minimization of (9) occurs when

$$\theta = (A\,A')^{-1}AE \quad \text{and} \quad ADA'(A\,A')^{-1}AL = ADL.$$

The second of these equations is a function of γ and b and requires numerical methods for solution. The subroutine ZSYSTM in the IMSL Library (1977) has been used for this purpose (Canfield *et al.*, 1980).

5. CONCLUSIONS

Always graph the data.

ACKNOWLEDGEMENTS

The authors wish to acknowledge the cooperation extended by B. B. Bobee and R. Robitaille in providing the river flow data for this study. We also express appreciation to the Office of Water Research and Technology of the U. S. Department of the Interior for partial support of this research.

REFERENCES

Ashkanasy, N. M. and Weeks, W. D. (1975). Flood frequency distribution in a catchment subject to two storm rainfall producing mechanisms. Hydrology symposium, Armidale, NSW, Australia, May 18-21, Institute of Engineering, Sydney, Australia, p. 153-157.

Benson, M. A. (1968). Uniform flood-frequency estimating methods for federal agencies. *Water Resources Research*, 891-908.

Bobee, B. B. and Robitaille, R. (1977). The use of the Pearson type 3 and log Pearson type 3 distributions revisited. *Water Resources Research*, 13, 427-443.

Canfield, R. V. and Borgman, L. E. (1975). Some distributions of time to failure for reliability applications. *Technometrics*, 17, 263-268.

Canfield, R. V., Olsen, D. R., Hawkins, R. H., and Chen, T. L. (1980). Use of Extreme Value Theory in Estimating Flood Peaks from Mixed Populations. Hydrolics and Hydrology Series, UWRL/H-80/01, Utah Water Research Laboratory, Utah State University, Logan, Utah.

Gnedenko, B. (1943). Sur la distribution limite du terms maximum d'une serie aleatorie. *Annals of Mathematics*, 44, 423-453.

Gumbel, E. J. (1958). *Statistics of Extremes*. Columbia University Press, New York.

Harter, H. L. and Moore, A. H. (1965). Maximum likelihood estimation of Gamma and Weibull populations from complete and censored samples. *Technometrics*, 7, 639-643.

IMSL. (1977). International Mathematical and Statistical Libraries, Inc., Library 3, Edition 6, Vol. 2, Houston, Texas.

Lindgren, B. W. (1976). *Statistical Theory*. Macmillan, New York.

Potter, W. D. (1958). Upper and lower frequency curves for peak rates of runoff. *Transactions of the American Geophysical Union*, 39, 100-105.

U. S. Water Resources Council. (1976). Guidelines for determining flood flow frequency. Bulletin No. 17 of the Hydrology Committee, Washington, D. C.

Watson, G. S. (1952). Extreme value theory for m dependent stationary stochastic processes. *Annals of Mathematical Statistics*, 25, 798-803.

[*Received May* 1980. *Revised September* 1980]

PROPERTIES OF EXTREME ORDER STATISTICS AND THEIR APPLICATION TO FIRE LOSSES AND EARTHQUAKE MAGNITUDES

G. RAMACHANDRAN

Operational Research and
Systems Studies Section
Fire Research Station
Borehamwood, Hertfordshire
WD6 2BL, ENGLAND

SUMMARY. In accident studies where data are available only for large events there are certain advantages in arranging the observations for damages in decreasing order of magnitude and applying the theory of extreme values to these ranked variables. This has been illustrated in this paper with reference to fire protection and insurance problems and prediction of earthquake magnitudes. "Parent" probability distributions considered in these studies belong to the "exponential type" which include the normal, log-normal, gamma, chi-square and logistic as well as the exponential distribution.

KEY WORDS. exponential type distributions, extreme order statistics, fire loss, fire protection, insurance, earthquake magnitudes.

1. INTRODUCTION

In some accident studies it will be time consuming and expensive to collect data on all incidents and hence figures for damage caused may be available only for a few large events of economic importance. For assessing the economic value of fire protection measures, for example, loss data are generally available only for large fires which constitute less than 10 percent of the total number of fires in many groups of buildings but account for more than 50 percent of the total loss. In such cases the large observations could be arranged in decreasing order of magnitude and the asymptotic theory of extreme values applied to these

C. Taillie et al. (eds.), Statistical Distributions in Scientific Work, Vol. 6, 351–366.

ranked variables which would be more useful and simpler than other techniques normally employed for analyzing censored data. This approach has been illustrated in this paper with reference to fire protection and insurance problems and prediction of earthquake magnitudes.

The asymptotic theory discussed pertains to "parent" probability distributions which are classified by Gumbel (1958) as "exponential type" with reference to their limiting behavior (or in the domain of attraction of Gumbel's double exponential law $\Lambda(x) = \exp(-\exp(-x))$ as used by other authors on extreme value theory). This type includes the normal, lognormal, gamma, chi-square and logistic as well as the exponential distribution. The ranked observations at the upper tail (larger values) of the parent distribution may be defined as extreme order statistics with the largest (first rank) as a particular case among them.

2. FIRE PROTECTION

The essential first step in any application of extreme value theory is to identify the nature of the parent probability distribution which expresses mathematically the probabilities with which the random variable considered could attain various values. According to studies carried out by Ramachandran (1974, 1975a), Shpilberg (1974) and other authors mentioned in these papers, fire loss distribution is skewed and in general the logarithm of loss has a probability distribution belonging to the "exponential type". Among the distributions of this type, lognormal is recommended widely for modelling fire insurance claims. Exponential for log loss or Pareto for loss is another distribution considered by some actuaries.

If, during any period, n fires occur in a group of buildings with identical fire risks the losses in these fires could be regarded as a sample of observations generated by the parent probability distribution $F(z)$ where z is the logarithm of loss. Logarithms of large losses in this sample arranged in decreasing order of magnitudes and denoted by $z_{(m)n}$ $(m = 1,2,\cdots)$ are referred to as extreme order statistics. The probability density function of $z_{(m)n}$ would be well approximated, for large n and small m, by

$$\frac{m^m a_{(m)n}}{(m-1)!} \exp[-ma_{(m)n}(z-b_{(m)n}) - m\exp\{-a_{(m)n}(z-b_{(m)n})\}] , \quad -\infty<z<\infty.$$

This function is the basis for studying the statistical behavior of each large loss of given rank. The parameter $b_{(m)n}$ is the modal (most probable) value of $z_{(m)n}$ and $a_{(m)n}$ the "failure rate" of z at $b_{(m)n}$ as given by $f(b_{(m)n})/[1-F(b_{(m)n})]$. For the largest value the suffix m takes the value $m = 1$.

Large losses are of economic importance and hence it would be useful to predict the behavior of top, say, r losses ($m = 1$ to r) over a period of time. For this purpose it is necessary to estimate the values of $a_{(m)n}$ and $b_{(m)n}$ ($m = 1$ to r). This is a simple problem if the value of n and the exact form of $F(z)$ is known, e.g. normal with known mean and standard deviation. However, in most practical situations the exact nature of $F(z)$ is generally unknown but can be assumed to be of exponential type. In such cases if values of $z_{(m)n}$ are available for, say, N samples each of size n, $a_{(m)n}$ and $b_{(m)n}$ can be estimated by any one of the methods described by Ramachandran (1975b).

Consider, for example, the analysis described by Ramachandran (1974) using the top 17 fire losses ($m = 1$ to 17) in the UK textile industry during the 21 year period ($N = 21$) from 1947 to 1967. The linear method was applied for estimating the parameters of the extreme order distributions. Since the number of fires per year increased over the period considered the following approximate correction for variation in sample size was included in the estimation process.

$$z_{(m)n_j} = b_{(m)n} + [y_{(m)j} + \log_e(n_j/n)]/a_{(m)n} \qquad (1)$$

approximately where $z_{(m)n_j}$ came from a sample ie year of n_j fires. The parameters $a_{(m)n}$ and $b_{(m)n}$ referred to $n(=465)$ fires in 1947. Equation (1) is based on the relationship

$$b_{(m)n_j} = b_{(m)n} + (1/a_{(m)n}) \log_e(n_j/n) \qquad (2)$$

where $b_{(m)n_j}$ is the modal value of $z_{(m)}$ for samples each of size n_j. Before estimating the parameters $a_{(m)n}$ and $b_{(m)n}$, the figures for losses (in units of £'000) were corrected for inflation expressing them at 1947 values with the aid of indices of retail prices. Base e was used for calculating the logarithms of losses. The results reproduced in Table 1 indicated an increasing failure rate for large values of the variable z, the logarithm of loss.

Equation (2) is exactly true for a parent of exponential form. Ramachandran (1975b) has studied the errors in this approximation for gamma and standard normal distributions for sample sizes from 450 to 1000 with $n = 450$. The errors are not serious for this range but cause some concern for samples of size less than 50 from a normal population.

Figure 1 is based on equations (1) and (2) and the probability points of $\tau_{(m)} = a_{(m)n}(Z_{(m)n} - b_{(m)n})$. It is on a log scale showing the relationship between the annual frequency of fires in the textile industry and the probable size, at 1947 prices, of the largest $(m = 1)$, *7th* $(m = 7)$ and *16th* $(m = 16)$ fire in a year. For each of these three ranks, the modal sizes of the losses are shown with confidence bands. For an estimated number of fires in any year, an ordinate erected at the corresponding point on the x axis would intersect the upper and lower confidence lines at points giving the corresponding confidence limits. The probability of exceeding the upper or falling short of the lower is 0.1. As an example, if the number of fires expected in a year in the textile industry is 1000, the most probable value of the largest loss would be £260,000 with upper and lower confidence limits of £700,000 and £180,000; all these figures are at 1947 money values.

The confidence lines represent a control chart based on the current trend. The increase in the frequency n of fires may be partly due to the inadequacy of fire prevention measures. In addition, if some or all of the actual large losses corrected for inflation exceeded the corresponding upper limits it may be concluded that general changes in fire fighting and fire protection methods or in the industrial processes are taking place to alter the picture for the worse. If the losses are less than the lower limits, then the changes are for the better. These arguments and the data on losses for the period 1968 and 1978 suggest that protection measures are coping well with fire outbreaks in the UK textile industry.

For assessing the value of fire protection devices at the industry level it is necessary to estimate the average and total loss in all fires for each industry and for each important class of risk within an industry. The problem is to estimate the average loss making the best use of available data which is restricted to large values. For this purpose generalized Least Square and Maximum Likelihood models have been developed by Ramachandran (1974, 1975a,b) for estimating the parameters of the parent distribution.

TABLE 1: Textile industry, UK.

Extreme Order (m)	$a_{(m)n}$	$b_{(m)n}$	Extreme Order (m)	$a_{(m)n}$	$b_{(m)n}$
1	2.247	5.214	10	1.034	3.259
2	1.785	4.829	11	0.973	3.137
3	1.626	4.534	12	0.925	2.972
4	1.460	4.327	13	0.886	2.832
5	1.387	4.113	14	0.924	2.749
6	1.424	3.988	15	0.937	2.680
7	1.239	3.749	16	0.950	2.583
8	1.163	3.564	17	1.002	2.537
9	1.212	3.448			

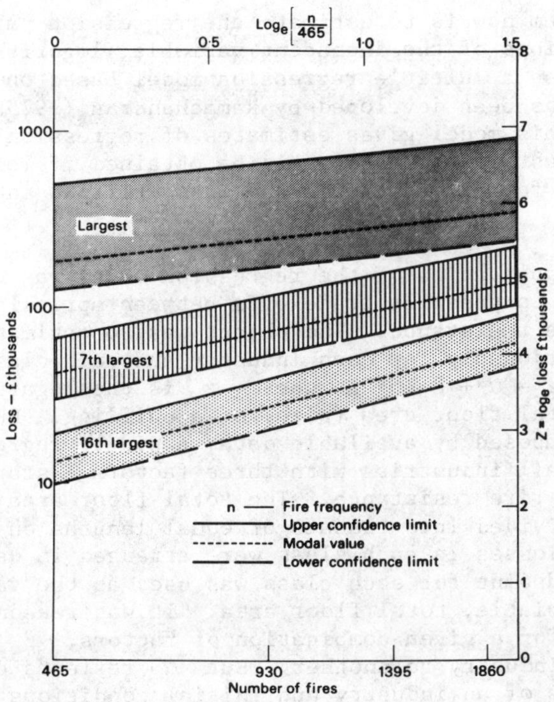

FIG. 1: Fire frequency and large losses.

Rogers (1977) has applied the least square method to large losses in a number of industries and trades. He assumed specifically that fire loss has a lognormal probability distribution. Table 2 contains some interesting results obtained by Rogers for average loss in all fires (estimated from large losses). Figures in this table relate to fires that survived "infant mortality"; very small fires were excluded from the total sample size since their inclusion would distort the shape of the loss distribution particularly at the upper tail which is of economic importance. It is apparent that sprinklers reduce the loss expected in multi-storey buildings to a considerable extent.

Results of the kind shown in Table 2 are helpful to estimate the effectiveness of a single protection measure such as sprinklers. Other devices like detectors and structural fire resistance also reduce the fire damage. Some factors could enhance the damage, e.g. delay in discovering the fire or extinguishing it. A full assessment of fire risk ought to consider all the relevant factors and their interactions and evaluate their independent contributions to the damage by performing a multiple regression analysis.

The problem now is to estimate the regression parameters using large observations of the dependent variable, logarithm of loss. For this purpose a multiple regression model based on extreme value theory has been developed by Ramachandran (1975b,c). Using large losses this model gives estimates of regression parameters equivalent to estimates which would be obtained if loss figures were available for all the fires and were utilized in the calculations.

A pilot application of the regression model was carried out with reference to possible trade-offs between sprinklers and structural fire resistance. The object was to estimate the following relationship for each industry and each class within an industry: $z = \alpha + \beta \log A$ where z is the logarithm of loss and A the total floor area (size) of a building. Due to restrictions imposed by available data, a single regression was performed for all industries with three factors - storeys, sprinklers and fire resistance. The total floor area of the building was divided into classes of equal lengths on a log scale and the losses in each class were arranged in decreasing order. The midpoint for each class was used as the value of the independent variable, total floor area. It was reasonably assumed that, for a given combination of factors, β will not vary from one industry to another. But α reflects certain characteristics of an industry and initial conditions at the time of ignition. Hence an estimate of α was obtained for each class of each industry by estimating the average loss as described earlier and the mean and standard deviation of logarithm of floor area using a large sample of all fires.

TABLE 2: Average loss per fire at 1966 *prices* (£000).

	Sprinklered single storey	Sprinklered multi storey	Nonsprinklered single storey	Nonsprinklered multi storey	Overall
Textiles	2.9	3.5	6.6	25.2	9.3
Timber and furniture	1.2	3.2	2.4	6.5	3.8
Paper, printing and publishing	5.2	5.0	7.1	16.2	10.9
Chemical and allied	3.6	4.3	4.3	8.2	6.4
Wholesale distributive trades	–	4.7	3.8	9.4	7.0
Retail distributive trades	–	1.4	0.4	2.4	1.9

TABLE 3: Textile industry, multistorey buildings. Annual loss (£) *and total fire cost* (£) *at* 1978 *prices.*

Building Category		Annual loss		Total fire cost	
		Floor area (sq. ft.)		Floor area (sq. ft.)	
		100,000	1,000,000	100,000	1,000,000
Sprinklered	High fire resistance	735	2,375	37,900	322,800
	Low fire resistance	1,163	9,460	29,600	268,600
Non Sprinklered	High fire resistance	3,583	15,686	48,300	281,900
	Low fire resistance	5,672	62,489	56,700	624,900

The annual probability of fire starting (or the frequency of fires per year) also has a "power relationship" with the size of building. For a building of given size the annual ignition probability revealed by another study was multiplied by the expected loss in a fire estimated by the regression to provide an estimate of the annual loss expected in the building. As an example results for two hypothetical multistorey buildings engaged in textile manufacture are given in Table 3.

It is apparent that sprinklers reduce the annual fire loss to a considerable extent whether the building has low or high fire resistance. Such gains are only probable and are spread over the life of a building, say, 40 years. Assuming a discount rate of 10 percent per annum the net present value (NPV) of the gain is approximately 10 times the annual value. The costs of installing sprinklers and providing fire resistant structures are incurred once and for all at the beginning and hence they may be taken as their NPV. Annual maintenance costs of protection systems are comparatively negligible. The total cost of fire for any category is $C = C_f + C_p$ where C_f = NPV of fire loss (10 times annual loss) and C_p = cost of fire protection. The total cost C for the four categories and two buildings considered would be as in Table 3.

If, for some reason, it is decided not to install sprinklers in a large multistorey building it would be more economical (less total cost) to make the building highly fire resistant. However, if sprinklers are to be installed it is not economical to spend money to increase the fire resistance of the structure as well. It appears that requirements for structural fire resistance specified in building regulations or codes could be relaxed if a multistorey industrial building is provided with sprinklers. Such concessions already exist in some countires, e.g. USA, Australia, and have recently been introduced for Greater London area but are yet to be clearly defined in the building regulations for other regions of United Kingdom.

3. FIRE INSURANCE

Only from recent times statistical techniques are being seriously considered for tackling actuarial problems in fire insurance. A few papers have been published in the Transactions of the Actuarial Congress held once in four years and the ASTIN Bulletin containing papers presented at ASTIN Colloquia. (ASTIN is the non-life branch of the International Actuarial Association.) Problems where extreme value theory could be fruitfully applied are discussed in this section.

Falling at the upper tail of the claim distribution, large claims exercise a critical effect on the financial performance of an insurance company. In order to reduce the seriousness of this effect the insurer usually arranges a reinsurance contract for each policy dividing a claim between himself and the reinsurer. According to an "excess of loss" treaty the reinsurer takes responsibility for compensating a loss in excess of an agreed level L. For calculating the premium it has to collect a reinsurance firm has, generally, information only on large claims incurred in the past. For this purpose, Ramachandran (1974, 1976) has proposed models based on extreme value theory. In one of these methods the exact nature of the "parent" distribution, e.g. lognormal is regarded as unknown but assumed to be of the "exponential type". As described earlier large claims over a period of years are analyzed to estimate the parameters $a_{(m)n}$ and $b_{(m)n}$. In this case, with any rank m, the net premium per claim for an excess of loss cover above L is approximately given by

$$P(L) = \{mL/n(a_{(m)n}-1)\}\exp(-y_L)$$

where $y_L = a_{(m)n}[\text{Log}_e L - b_{(m)n}]$. Depending on the data on large claims, P(L) would have different values for different rank m.

For purposes of illustrating the formula mentioned above Ramachandran (1976) used the results for the textile industry (Table 1) and obtained estimates of the function P(L)/L which are reproduced in Table 4. In the last line of the table estimates based on actual large losses during 1968-73 are also shown. For the range of losses considered the formula has given premium rates lower than the actual for $m \leqslant 3$ and higher rates for $m \geqslant 4$. Rank 10 has given very high rates due to the fact that $a_{(10)n}$ is very close to unity. Determining the optimum value of m is a useful statistical problem for a future study.

The problem of "maximum retention" consists in determining for a risk the portion which it is to the insurer's advantage to retain itself and, as complement, the remaining portion which should be passed off (ceded) to the reinsurer. The insurer bears the entire amounts of all claims up to a retention level M; also for any claim L greater than M he will pay only M, the reinsurer taking the responsibility for paying the balance, (L - M). For assessing the insurer's liability it is necessary first to determine the distribution function of a claim amount. This can be achieved by applying the theory of extreme order statistics to large claims and estimating the parameters for a number of exponential type distributions for logarithms of loss.

TABLE 4: Textile industry, UK. (n = 465; L *in units of* £ 000).

Extremes (m)	$a_{(m)n}$	$b_{(m)n}$	P(L)/L: L = 20.086, $Log_e L = 3$	P(L)/L: L = 54.598, $Log_e L = 4$	P(L)/L: L = 148.413, $Log_e L = 5$
1	2.247	5.214	0.2496	0.0264	0.0028
2	1.785	4.829	0.1434	0.0241	0.0040
3	1.626	4.534	0.1248	0.0246	0.0048
4	1.460	4.327	0.1298	0.0301	0.0070
5	1.387	4.113	0.1301	0.0325	0.0081
6	1.424	3.988	0.1243	0.0299	0.0072
7	1.239	3.749	0.1593	0.0462	0.0134
8	1.163	3.564	0.2034	0.0637	0.0199
9	1.212	3.448	0.1571	0.0468	0.0139
10	1.034	3.259	0.8268	0.2940	0.1045
Actual experience* (1968-73)			0.1280**	0.0280	0.0065

*At 1947 prices; 8000 fires ** Estimated

TABLE 5: Return periods of Canadian earthquakes.

$n_0 = 134,\ \mu = 1.212,\ \theta = 1$

Magnitude (M)	n_M	n_M/n_0	Return periods (years)
3.5	20.697	0.154	0.598
4.5	69.547	0.519	2.016
5.6	263.802	1.969	7.650
5.8	336.164	2.509	9.747
6.3	616.217	4.599	17.867
7.05	1529.350	11.413	44.340
7.5	2638.590	19.691	76.500

Among these one should finally select the distribution that will give the closest fit to the observed large claims though this might lead to large deviations at the unknown lower tail (small claims). Large claims are of economic importance and hence it is necessary to obtain a good representation of the probability curve in the region spanned by these values.

For a maximum retention of M, the insurer can expect to be responsible for a sum of

$$x_M = \int_0^M xv(x) + M(1 - V(M))$$

where $v(x)$ and $V(x)$ are the density and distribution functions of loss x. If the parent is judged to be log normal with μ and σ as the mean and standard deviation for logarithm (base e) of loss, value of x_M is easily seen to be

$$G(k - \sigma)\exp(\mu + \sigma^2/2) + M\{1 - G(k)\}$$

where $k = (\log_e M - \mu)/\sigma$ and $G(t) = \{1/\sqrt{2\Pi}\} \int_{-\infty}^{t} \exp(-u^2/2)$
With a maximum retention of M and n claims the total liability of the insurer is $n \cdot x_M$. For each claim under an excess of loss treaty the reinsurer is responsible for the balance R_M given by

$$[1 - G(k - \sigma)]\exp(\mu + \sigma^2/2) - M\{1 - G(k)\}$$

since $R_M = \int_M^\infty (x - M)v(x)$. The reinsurer is hence likely to be liable for a total sum of $n \cdot R_M$. It may be verified that

$$x_M + R_M = \exp(\mu + \sigma^2/2)$$

which is the expected value of x in the entire unlimited range. The values x_M, R_M and $P(L)$ are net amounts per claim to which appropriate "loading" will be added to cover administrative costs of the insurance company concerned.

A "deductible" can be defined as the participation of the insured in a loss up to a certain limit agreed on in advance. When a deductible is introduced in an insurance contract, it is hoped that the insured will show greater interest in adopting loss prevention and reduction measures, since he will have to bear part of the financial burden himself whenever a loss occurs. As an inducement the insured is given the advantage of reduced premium for obtaining insurance coverage for his property. On the other hand when a deductible is applied the insurer will not have to pay and settle small losses which obviously relieve him of considerable amount of work.

The insured need not report to his insurance company any losses less than the deductible amount. Hence an insurance firm's account of reported losses (claims) will typically include only those losses that are larger than the deductible carried by the particular insured. Different insureds carry different deductibles some of which could be large. Hence an insurance firm generally has data providing only a truncated distribution of claim amount for each insured in any risk category; this complicates the construction of an overall claim distribution for the category. This problem can be resolved by arranging in decreasing order of magnitude all losses greater than the largest deductible for the category and applying extreme order theory to these ranked observations. Two courses of action are suggested as follows.

All the policyholders may be asked to report the *number* of losses incurred below the deductible levels; information about the *amounts* of losses are not required. This procedure will provide an estimate of the total number n of observations. Then, applying extreme value theory, one could try several distributions and select the best one as judged from large losses. If it is difficult to obtain information about n large claims may be classified according to years and arranged in decreasing order for each year. The parameters $a_{(m)n}$ and $b_{(m)n}$ may then be estimated for a number of ranks m as in Table 1; this will provide some description of the claim distribution. In this method it has to be assumed, however, that the number of claims per year is large and does not vary significantly from year to year. This approach also requires the use of several years of data in order to have a large sample size. This requirement introduces some problems especially if the nature of insurance portfolio whose claims are being analyzed has changed significantly in the observed period of time.

The insured has to bear the entire amount of any loss up to the deductible level D. For a loss L greater than D, his liability is D since he will receive the difference $(L - D)$ from his insurer. The insured thus "retains" a maximum amount of D and "reinsures" with the insurer losses exceeding D. Hence the expected value of the net amount per claim which the insured has to provide in his budget is x_D obtained by setting $M = D$ in the statistic x_M discussed earlier. The insurer thus saves an amount of x_D per claim against which he normally offers the insured a reduction in the premium.

4. EARTHQUAKE MAGNITUDES

For calculating insurance premium rates and designing the construction of buildings to withstand earthquake tremors it is necessary to determine the probability distribution of magnitudes of earthquakes. In a recent paper, Ramachandran (1980) has studied this problem in detail reviewing the works of other scientists and proposing the application of extreme order theory.

"Intensity" is an indication of the severity of earthquake at a specific location and depends on the focal depth, subsoil condition, building construction and other such factors; the effects are usually measured in Modified Mercalli grades. "Magnitude" is the term used to express the total amount of energy released by an earthquake; it is a measure of the absolute size of an earthquake and does not refer to the effect at any particular location. The Richter Scale gives the numerical value of the magnitude (M) which has the following relationship with energy (E) expressed in ergs: $\log_{10}E = 11.4 + 1.5M$. Although magnitude and intensity need not be closely correlated these two variables might have some relationship for a small geographic region.

The relationship between magnitude and frequency of occurrence takes the form $\log_{10}N = a - bM$ where N is the number of earthquakes of magnitude M. From this it may be easily deduced that magnitude has the exponential probability density

$$f(M) = \mu \exp[-\mu(M-\theta)]$$

where θ is the magnitude of the smallest earthquake that could be detected or for which data are available. It has theoretically no upper bound since there is a small but finite probability that magnitude in a future earthquake could exceed any observed maximum value. An estimate of μ is given by $\{1/(\bar{M} - \theta\}$ where $\bar{M}$ is the average magnitude in a large number of earthquakes. For earthquakes in a small area of Western Canada, with $\theta = 1$, μ was estimated to have the value 1.212 from data having the relationship $\log_{10}N = 4.10 - 0.67M$.

Consider now a number of periods in time with an average of n earthquakes during each period. Following extreme value theory the largest earthquake during each period whose magnitude may be denoted by $M_{(1)}$ will have a probability distribution with cumulative distribution function $\exp[-\exp(-y_{(1)})]$ where $y_{(1)} = \alpha_{(1)}[M'_{(1)} - U_{(1)}]$, $M'_{(1)} = M_{(1)} - \theta$. For the exponential parent pertaining to earthquake magnitudes,

$$\alpha_{(1)} = \mu \quad \text{and} \quad U_{(1)} = (1/\mu)\log_e n$$

since $F(U_{(1)}) = 1 - (1/n) = 1 - \exp(-\mu U_{(1)})$. The double exponential distribution for the largest magnitudes mentioned above can be derived from accepted seismological relationships. Several authors have used this distribution for predicting the magnitudes of largest earthquakes by fitting to the data the following straight line

$$M'_{(1)} = U_{(1)} + \{y_{(1)}/\alpha_{(1)}\}.$$

A value of $\theta = 0$ is implicit in these studies and this could introduce serious errors in the analysis of some data; for, $U_{(1)}$ is the modal value of $M'_{(1)}$ and not of $M_{(1)}$.

Using only the largest magnitudes is wasteful of information. Hence all the available large magnitudes during each period should be utilized and the general theory of extreme order statistics applied as in the case of fire losses.

If the magnitudes of n earthquakes during each period are arranged in decreasing order and $M_{(i)}$ is the *ith* magnitude in that arrangement, the probability density function of $M_{(i)}$, for large n, is approximately

$$\{i^i/(i-1)!\}\exp[-iy_{(i)} - i\exp(-y_{(i)})]\,\alpha_{(i)}$$

where $y_{(i)} = \alpha_{(i)}(M'_{(i)} - U_{(i)})$, $M'_{(i)} = M_{(i)} - \theta$.

Methods described by Ramachandran (1975b) can be used for estimating $\alpha_{(i)}$ and $U_{(i)}$ for a number of ranks without reference to the parent distribution. If the parent is exponential as one might expect, $\alpha_{(i)}$ will be constant $(=\mu)$ and $U_{(i)} = (1/\mu)\ \log_e(n/i)$ approximately. After ascertaining the structure of the parent distribution, its parameters can be estimated as illustrated by Ramachandran (1980) with reference to the Canadian data mentioned earlier. Using estimated values of 28 large magnitudes ($i = 1$ to 28) in a hypothetical sample of 10,000 earthquakes and applying the maximum likelihood method he obtained $\mu = 1.3854$ and $\theta = 0.75$ assuming that the latter parameter was unknown. The value of μ was quite close to the estimate 1.212 obtained from the complete sample with unity as the lower bound for θ.

"Return period" is defined as the average time which lapses between two unrelated earthquakes of the same magnitude M in the same area. Equivalently it is also the period during which only one earthquake is expected to exceed magnitude M. The probability of exceeding $M'(= M - \theta)$ is $\{1 - F(M')\}$. If $n_M = \{1 - F(M')\}^{-1}$, we can regard M' as the most probable largest magnitude in repeated periods with an average of n_M earthquakes during each period. For $F(M') = 1 - (1/n_M)$. If earthquakes occur at the rate of n_0 per year the parameter (n_M/n_0) may be termed as the "reference period" of M' (in years). As discussed earlier the reduced variable $y_{(1)} = \alpha_{(1)} (M - U_{(1)})$ where $U_{(1)}$ is the modal largest value has the cumulative frequency $\exp[-\exp(-y_{(1)})]$. Since M' has been defined as the mode $U_{(1)} = M'$ so that $y_{(1)} = \alpha_{(1)}(M - M') = \theta\ \alpha_{(1)}$. Hence a large magnitude M has the cumulative frequency $\exp[-\exp(-\theta\ \alpha_{(1)})]$ and return period (in years).

$$(n_M/n_0)[1 - \exp\{-\exp(-\theta\ \alpha_{(1)})\}]^{-1}.$$

If $\theta = 0$ and M is the modal value of the largest annual magnitudes so that $n_M = n_0$ the return period may be seen to reduce to 1.582, the quantity generally used in extreme value applications.

For the exponential distribution discussed earlier $\alpha_{(1)} = \mu$ and $n_M = \exp\{\mu(M - \theta)\}$. Using these particular values the return periods of Canadian earthquakes for selected magnitudes were calculated as shown in Table 5.

ACKNOWLEDGEMENT

The paper is Crown Copyright, reproduced by permission of the Controller, HM Stationery Office. It is contributed by permission of the Director, Building Research Establishment, Department of the Environment.

REFERENCES

Gumbel, E. J. (1958). *Statistics of Extremes*. Columbia University Press, New York.

Ramachandran, G. (1974). Extreme value theory and large fire losses. *ASTIN Bulletin*, VII, 293-310.

Ramachandran, G. (1975a). Extreme order statistics in large samples from exponential type distributions and their application to fire loss. In *Statistical Distributions in Scientific Work, Vol. 2*, G. P. Patil, S. Kotz and J. K. Ord, eds. Reidel, Dordrecht-Holland. Pages 355-367.

Ramachandran, G. (1975b). *Extreme order statistics from exponential type distributions with applications to fire protection and insurance*. Ph.D. thesis, University of London.

Ramachandran, G. (1975c). Factors affecting fire loss - Multiple regression models with extreme values. *ASTIN Bulletin*, VIII, 229-241.

Ramachandran, G. (1976). Extreme value theory and fire insurance. *Transactions of the 20th International Congress of Actuaries*, Tokyo, Oct-Nov 1976, Vol. 4, 695-707 and Vol. 5, 224-226.

Ramachandran, G. (1980). Extreme value theory and earthquake insurance. *Transactions of the 21st International Congress of Actuaries*, Zurich and Lausanne, June 1980, Vol. 1, 337-353.

Rogers, F. E. (1977). Fire losses and the effect of sprinkler protection of buildings in a variety of industries and trades. Building Research Establishment, Current Paper CP 9/77.

Shpilberg, D. C. (1974). Risk insurance and fire protection; a systems approach, Part 1; Modelling the probability distribution of fire loss amount. Factory Mutual Research Corporation, Norwood, Massachusetts, Technical Report No. 22431.

[*Received May* 1980. *Revised October* 1980]

STATISTICAL CHOICE OF UNIVARIATE EXTREME MODELS

J. TIAGO DE OLIVEIRA

Faculty of Sciences of Lisbon
Center of Statistics and Applications (INIC)
58, Rua da Escola Politécnica
1294 Lisboa Codex - Portugal

SUMMARY. The paper presents a method of (asymptotical) statistical choice between Weibull, Gumbel or Fréchet models to fit large samples of maxima (IID). The solution of the statistical trilemma, centered around Gumbel distribution, is asymptotically fair (equal probabilities of deciding Fréchet or Weibull distribution if Gumbel distribution is the right one) and consistent. If the 'significance level' is allowed to converge to zero, at a convenient rate, it is shown that the probabilities of wrong decisions converge to zero. Some applications are made. Unfortunately the method seems to be useful only for large samples ($n \geqslant 400$) as shown in applications at the end.

KEY WORDS. distribution of maxima, Weibull model, Gumbel model, Fréchet model, asymptotic statistical decision.

1. INTRODUCTION

As it is well known, the three univariate extreme models (for maxima), in reduced form, have the expressions:

$$\Lambda(z) = \exp(-e^{-z}),\ -\infty<z<+\infty, \qquad \text{Gumbel model;}$$

$$\Phi_\alpha(z) = 0 \text{ if } z<0$$

$$= \exp(-z^{-\alpha}) \text{ if } 0<z<+\infty,\ \alpha>0 \qquad \text{Fréchet model;}$$

C. Taillie et al. (eds.), Statistical Distributions in Scientific Work, Vol. 6, 367–387.

$$\Psi_\alpha(z) = \exp(-(-z)^\alpha) \text{ if } -\infty<z<0,\ \alpha>0$$
$$= 1 \text{ if } 0<z<+\infty, \qquad \text{Weibull model;}$$

see Gumbel (1958) for details.

Those models, with a convenient change of location and dispersion, can be put under the unified form, given in Jenkinson (1955):

$$G(z|k) = \exp(-(1+kz)^{-1/k}) \text{ for } 1+kz>0,\ -\infty<k<+\infty,\ k\neq 0,$$

with the usual continuation at $k=0$; when $k=0^+$ or $k=0^-$ we obtain

$$G(z|k) = \Lambda(z); \text{ for } k<0 \text{ we have } G(z|k) = \Psi_{-1/k}(-1-kz)$$

and for $k>0$ we obtain $G(z|k) = \Phi_{1/k}(1+kz)$.

Thus we will derive statistical choice between the statistical extreme models lead by the test about k ($k<0$, $k=0$, $k>0$). Statistical choice will, thus, be a statistical trilemma, in which Gumbel model ($k=0$) will play a central position. It must be noted that, here, we do not have a case of separation, as defined in Cox (1961) and (1962), because Gumbel model ($k=0$) can be obtained as a limit of Fréchet ($k>0$) and Weibull ($k<0$) cases, as shown before.

Evidently, as we are dealing with distinction of models, and as large samples lead to good decisions, the emphasis will be on asymptotic behavior. We will begin by using the forms of $G(z|k)$, without location and dispersion parameters. Later, we will introduce them and compensate through use of the maximum likelihood estimators for Gumbel model in the choice statistic, curiously decreasing the randomness (variance). We could have used methods of quasi-linear decision, as in Tiago de Oliveira (1966), but the final result is not manageable, as it will be seen.

2. STATISTICAL TESTING OF GUMBEL VS. FRÉCHET OR WEIBULL REDUCED MODELS

As the density is

$$g(z|k) = (1+kz)^{-1/k-1} \exp(-(1+kz)^{-1/k}) \text{ if } 1+kz>0$$

$$= 0 \text{ if } 1+kz<0$$

the likelihood of an independent sample $(z_1,\ldots,z_n)$ is

$$L(z|k) = L(z_1,\ldots,z_n|k) = \prod_1^n g(z_i|k).$$

The test of k vs. k' is given by the Neyman-Pearson theorem as usual.

The L.M.P. test of $k=0$ vs. $k>0$ $(k<0)$ is given by the rejection region.

$$R: \left.\frac{\partial \log L(z|k)}{\partial k}\right|_{k=0} \geq c^+ (\leq c^-) \text{ or } R: \sum_1^n v(z_i) \geq c^+ (\leq c^-)$$

where $v(z) = \left.\frac{\partial \log g(z|k)}{\partial k}\right|_{k=0} = \frac{z^2}{2} - z - \frac{z^2}{2} e^{-z};$

$c^+ = c_n^+(\alpha)$ and $c^- = c_n^-(\alpha)$ are obtained, as usual, by the condition $\int_R L(z|0)\,dz = \alpha$, where α is the significance level of the test.

The two-sided L.M.P. test of $k=0$ vs. $k\neq 0$ is given by the rejection region

$$R: \left.\frac{\partial^2 \log L(z|k)}{\partial^2 k}\right|_{k=0} + \left(\left.\frac{\partial \log L(z|k)}{\partial k}\right|_{k=0}\right)^2 \geq$$

$$> a_n \left.\frac{\partial \log L(z|k)}{\partial k}\right|_{k=0} + b_n$$

where $\underline{a}_n$ and $\underline{b}_n$ are determined by the relations

$$\int_R L(z|0)\,dz = \alpha$$

$$\int_R \left.\frac{\partial L(z|k)}{\partial k}\right|_{k=0} dz = 0 .$$

As $\bar{v}(z) = \left.\frac{\partial^2 \log g(z|k)}{\partial k^2}\right|_{k=0} = z - \frac{2}{3} z^2 + \left(\frac{2}{3} z^3 - \frac{z^4}{4}\right) e^{-z}$

the region has the form

$$R: \sum_1^n \bar{v}(z_i) + \left(\sum_1^n v(z_i)\right)^2 \geqslant a_n \sum_1^n v(z_i) + b_n ;$$

asymptotically (as could be expected) the region is of the form

$$R': \left| \sum_1^n v(z_i) \right| \geqslant c'$$

where $c' = c_n'(\alpha) = c_n^+(\alpha/2) = -c_n^+(\alpha/2)$ asymptotically.

The tests are evidently consistent, i.e., Prob $\{rej|k=0\} \to \alpha$ and Prob $\{rej|k>0\}$, Prob $\{rej|k<0\}$ and Prob $\{rej|k\neq 0\}$ converge to 1; see Fraser (1957).

3. ASYMPTOTIC TESTING

Let us denote by $V_n(z) = V_n(z_1,\ldots,z_n) = \sum_1^n v(z_i)$. For $k=0$ (Gumbel distribution) we have

$$\mu(0) = M(v(z)) = \int_{-\infty}^{+\infty} v(z) \exp(-z)\exp(-e^{-z})\, dz = 0$$

$$\sigma^2(0) = \mathrm{Var}(v(z)) = \int_{-\infty}^{+\infty} v^2(z)\, e^{-z}\, e^{-e^{-z}}\, dz =$$

$$= \frac{\Gamma^{(4)}(1)}{4} + \Gamma^{(3)}(1) + \Gamma''(1) = 2.42361.$$

In general, we will introduce the notation

$$\mu(k) = M(v(z)|k) = \int_{-\infty}^{+\infty} v(z)\ d\ G(z|k) \quad \text{and}$$

$$\sigma^2(k) = \mathrm{Var}(v(z)|k) = M(v^2(z)|k) - \mu^2(k) = \int_{-\infty}^{+\infty} v^2(z) d\ G(z|k)$$

$$- \mu^2(k),$$

for the mean value and variance of $v(z)$ with respect to $G(z|k)$. In Appendix I, we will show that they exist only when $-1<k<1/4$. We have the relations $\mu(k) = 2.42361k + 4.95362k^2 + 0(k^3)$ and $\sigma^2(k) = 2.42361 + 25.26470k + 275.03597k^2 + 0(k^3)$.

In those conditions, we know, by the central limit theorem that

$$\mathrm{Prob}\ \{(V_n - n\mu(k))/\sqrt{n}\ \sigma(k) \leqslant x\} = \mathrm{Prob}\ \{V_n \leqslant n\mu(k) +$$

$$\sqrt{n}\ \sigma(k)x\} \to N(x)$$

where $N(x)$ is the standard normal distribution function. Let λ_α denote the solution of

$$N(x) = 1-\alpha;\ \lambda_\alpha + \lambda_{1-\alpha} = 0 \quad \text{and} \quad \lambda_\alpha > 0 \quad \text{if} \quad \alpha > 1/2.$$

The asymptotic test of $k=0$ vs. $k<0$ is given, thus, by

$$R'\!:\ V_n > \sqrt{n}\ \sigma(0)\ \lambda_\alpha$$

and the asymptotic test of $k=0$ vs. $k>0$ is given by

$$R'\!:\ V_n < -\sqrt{n}\ \sigma(0)\ \lambda_\alpha .$$

Consider now the two-sided test given by the rejection region

$$\sum_1^n \bar{v}(z_i) + V_n^2 \geqslant a_n \; V_n + b_n \quad .$$

Denoting by $Z_n = V_n/\sqrt{n}\,\sigma(0)$ and $\bar{Z}_n = \left(\sum_1^n \bar{v}(z_i) - n\,\bar{\mu}(0)\right)/\sqrt{n}\,\bar{\sigma}(0)$ where $\bar{\mu}(0)$ and $\bar{\sigma}(0)$ have obvious meanings, region R can be written as

$$\sigma^2(0)Z_n^2 \geqslant z_n\,\sigma(0)/\sqrt{n}\,Z_n + b_n/\sqrt{n} - \bar{\mu}(0) - \sigma(0)\bar{Z}_n/\sqrt{n}$$

which, when $n\to\infty$, leads to

$$\sigma^2(0)Z_n^2 \geqslant \bar{\alpha}\,\bar{Z}_n + \bar{\beta} \; .$$

Symmetry implies $\bar{\alpha} = 0$ so that, finally, the rejection region is asymptotically

$$R': |\, Z_n \,| \geqslant \lambda_{\alpha/2}$$

or

$$R': |\, V_n \,| \geqslant \sqrt{n}\,\sigma(0)\,\lambda_{\alpha/2} \; .$$

The asymptotic two-sided test is, thus, the 'intersection' of two one-sided tests of level $\alpha/2$, as could be expected.

4. THE ASYMPTOTIC POWER OF THE TESTS; CONSISTENCY

As it is of interest in the sequel we will only consider the behavior of the two-sided test

$$|\, V_n \,| \geqslant \sqrt{n}\,\sigma(0)\,\lambda_{\alpha/2} \quad \text{of } k=0 \text{ vs. } k\neq 0.$$

Then $\text{Prob}\{\text{rej}|k\} = 1 - \text{Prob}\{|V_n| < \sqrt{n}\,\sigma(0)\,\lambda_{\alpha/2}|k\} =$

$$= 1 - \text{Prob}\left\{\frac{-\sqrt{n}\,\sigma(0)\,\lambda_{\alpha/2} - n\,\mu(k)}{\sqrt{n}\,\sigma(k)} \leq \frac{V_n - n\mu(k)}{\sqrt{n}\,\sigma(k)} \leq \frac{\sqrt{n}\,\sigma(0)\,\lambda_{\alpha/2} - n\,\mu(k)}{\sqrt{n}\,\sigma(k)}\right\}$$

which if $k=0$, as $\mu(0)=0$, converges to $1 - \{N(\lambda_{\alpha/2}) - N(-\lambda_{\alpha/2})\} = \alpha$ and if $\mu(k)\neq 0$ converges to 1. It should be noted that $\mu'(0)=\sigma^2(0)$ so that, in a neighborhood of 0, $\mu(k)\neq 0$ and, also, $\mu(k)$ increases.

Let us consider, now, the effect of using the rejection region R'': $|V_n| > \sqrt{n}\,\sigma(0)d_n$, with d_n to be defined conveniently. Then $\text{Prob}\{\text{rej}|k\} = 1 - \text{Prob}\{-\sqrt{n}\,\sigma(0)d_n < V_n < \sqrt{n}\,\sigma(0)d_n\} = 1 - \{\text{Prob}\{V_n \leq \sqrt{n}\,\sigma(0)d_n|k\} - \text{Prob}\{V_n < -\sqrt{n}\,\sigma(0)d_n|k\}\}$. If $k=0$, using Appendix II.1, we see that as

$\lambda_n = 0$, $\delta_n = \sqrt{n}\,\sigma(0)$ we get $\text{Prob}\,\{\text{rej}|0\} \to 1 - \{N(\lim d_n) - N(-\lim d_n)$

which is equal to 0 if $d_n \to +\infty$. If $k\neq 0$ (and $\mu(k)\neq 0$) as $\lambda_n = n\,\mu(k)$ and $\delta_n = \sqrt{n}\,\sigma(k)$, we get $\text{Prob}\{\text{rej}|k\neq 0\} \to 1 - \{N(\lim(d_n\,\sigma(0)/\sigma(k) - \sqrt{n}\,\mu(k)/\sigma(k) - N(\lim(-d_n\,\sigma(0)/\sigma(k) - \sqrt{n}\,\mu(k)/\sigma(k))$ which is 1 if $\dfrac{d_n}{\sqrt{n}} \to 0$. In fact, the condition $\text{Prob}\{\text{rej}|k\neq 0\} \to 1$ is equivalent to

$$N(\lim(\sigma(0)d_n - \mu(k)\sqrt{n})/\sigma(k)) + N(\lim(\sigma(0)d_n + \mu(k)\sqrt{n})/\sigma(k)) = 1.$$

If $k<0$ with $\mu(k)<0$, as the first summand converges to 1, we see that we must have $d_n+\mu(k)/\sigma(0)\cdot\sqrt{n}\to-\infty$ so that $\lim d_n/\sqrt{n}\leq -\mu(k)/\sigma(0)$ and as $\mu(k)\to\mu(0)=0$ when $k\to 0$ we get $\lim d_n/\sqrt{n}\leq 0$; analogously for $k>0$, with $\mu(k)>0$, we obtain $\lim d_n/\sqrt{n}\geq 0$. With $d_n\to+\infty$ we must have $d_n/\sqrt{n}\to 0$. Thus with a sequence d_n such that $d_n\to\infty$, $d_n/\sqrt{n}\to 0$ we can get a test which has a correct decision with probability converging to 1; $d_n\to\infty$, $d_n/\sqrt{n}\to 0$ is, thus, the consistency condition.

We can now evaluate approximately the level of significance. We have

$$\text{Prob}\{\text{rej}|0\} \simeq 1 - (N(d_n) - N(-d_n)) = 2(1-N(d_n)).$$

Thus if $\alpha_n(\to 0)$ is to be such the approximate level of significance is $2\alpha_n$, we must have $1-N(d_n)=\alpha_n$. But for $x\to\infty$ we know that $1-N(x) \sim \frac{1}{\sqrt{2\pi}x}\exp(-x^2/2)$ so that we have d_n given asymptotically by

$$d_n = \sqrt{-2\log\alpha_n} - \frac{\log(-\log\alpha_n)+\log(4\pi)}{2\sqrt{-2\log\alpha_n}}, \text{i.e.,}\quad 1 - N(d_n)\sim\alpha_n.$$

If we take $\alpha_n=1/n$ we get

$$d_n = \sqrt{2\log n} - \frac{\log\log n+\log(4\pi)}{2\sqrt{2\log n}}$$

which is such that

$$d_n \to +\infty \text{ and } d_n/\sqrt{n}\to 0.$$

Other choices of d_n are evidently possible.

Recall that, in all the reasoning, that preceded, we always assumed that $\mu(k)$ and $\sigma^2(k)$ existed, i.e., that $-1<k<1/4$.

5. THE STATISTICAL CHOICE BETWEEN THE REDUCED MODELS

What precedes suggests the use of the decision rule:

if $V_n \leq -\sqrt{n}\,\sigma(0)\;c_n$ - choose the Weibull model;

if $|V_n| < \sqrt{n}\;(0)\;c_n$ - choose the Gumbel model;

if $V_n \leq \sqrt{n}\,\sigma(0)\;c_n$ - choose the Fréchet model.

Thus, the error probabilities are:

$$E_n\,(k<0) = \text{Prob}\{V_n > -\sqrt{n}\,\sigma(0)\;c_n | k<0\};$$

$$E_n\,(0) = 1 - \text{Prob}\{|V_n| \leq \sqrt{n}\,\sigma(0)\;c_n | k=0\};$$

$$E_n\,(k>0) = \text{Prob}\{V_n < \sqrt{n}\,\sigma(0)\;c_n | k>0\},$$

which have the approximations:

$$E_n(k<0) \simeq 1 - N\left(-\frac{\sigma(0)}{\sigma(k)}c_n - \sqrt{n}\,\frac{\mu(k)}{\sigma(k)}\right) = N\left(\frac{\sigma(0)}{\sigma(k)}\;c_n + \sqrt{n}\,\frac{\mu(k)}{\sigma(k)}\right)$$

with $k<0$.

$$E_n(0) \simeq 1 - \{N(c_n) - N(-c_n)\} = 2(1 - N(c_n)$$

$$E_n(k>0) \simeq N\left(\frac{\sigma(0)}{\sigma(k)}\;c_n - \sqrt{n}\,\frac{\mu(k)}{\sigma(k)}\right) \quad \text{with } k<0.$$

As for $\underline{k}$ small we have $\mu(k)<0$ for $k<0$ and $\mu(k)>0$ for $k>0$ we see that the first and third probabilities converge to zero if $c_n/\sqrt{n}\to 0$ and that the second one, also, converges to zero if $c_n\to\infty$, as could be expected from the discussion in Section 4.

A solution is to take evidently

$$c_n = \sqrt{2\log n} - \frac{\log(\log n) + \log(4\pi)}{2\sqrt{2\log n}} \quad .$$

We can, even, evaluate the principal parts of the probability errors. As $1 - N(x) = N(-x) \sim \frac{1}{\sqrt{2\pi}\, x} e^{-x^2/2}$ when $x\to\infty$ we have:

$$E_n(k<0) \sim N\left(\frac{\sigma(0)}{\sigma(k)} c_n + \sqrt{n}\,\frac{\mu(k)}{\sigma(k)}\right) \frac{\sigma(k)}{\sqrt{2\pi n}\,(-\mu(k))}\, e^{-\frac{1}{2}\left(\frac{\sigma(0)}{\sigma(k)} c_n + \sqrt{n}\,\frac{\mu(k)}{\sigma(k)}\right)^2}$$

because, as $\mu(k)<0$, the argument of N is <0 for large n; if we take $c_n/\sqrt{n} \leqslant \frac{-\mu(k)}{2\sigma(0)}$ the last result is approximately majored by

$$\frac{\sigma(k)}{\sqrt{2\pi\, n}\,(-\mu(k))} \exp\left(-\frac{\sigma^2(0)}{2\sigma^2(k)}\, c_n^2\right); \; E_n(0) \approx 2(1-N(c_n)) \sim 2/n;$$

$$E_n(k>0) \approx N\left(\frac{\sigma(0)}{\sigma(k)} c_n - \sqrt{n}\,\frac{\mu(k)}{\sigma(k)}\right) \sim \frac{\sigma(k)}{\sqrt{2\pi n}\,\mu(k)}\, e^{-\frac{1}{2}\left(\frac{\sigma(0)}{\sigma(k)} c_n + \sqrt{n}\,\frac{\mu(k)}{\sigma(k)}\right)^2}$$

approximately majored by $\frac{\sigma^2(k)}{\sqrt{2\pi n}\,\mu(k)} \exp\left(-\frac{\sigma^2(0)}{2\sigma^2(k)}\, c_n^2\right)$ when

$$c_n/\sqrt{n} \leqslant \frac{\mu(k)}{2\sigma(0)} \quad .$$

We can justify the symmetry condition $(b_n+a_n=0)$ in another way which is illuminating. Let us impose that, for $k=0$, the probabilities of wrong decision for Weibull and Fréchet models are equal (*fairness* of the decision rule) and converge to zero. The proabilities of wrong decision

$$\text{Prob}\{V_n \leqslant \sqrt{n}\, b_n\, \sigma(0)\} \quad \text{and} \quad 1 - \text{Prob}\{V_n \leqslant \sqrt{n}\, a_n\, \sigma(0)\}$$

are approximated by $N(b_n)$ and $1-N(a_n)=N(-a_n)$ which suggests to take $b_n=-a_n$. Consequently it is natural to take $a_n=c_n$, $b_n=-c_n$ as done before.

6. THE INTRODUCTION OF LOCATION AND SCALE PARAMETERS

Until now we did suppose that the location and scale parameters of $G(z|k)$ (λ and δ) were known, taken for the sake of commodity to be $\lambda=0$ and $\delta=1$. We will suppose now that the underlying distribution is $G\left(\frac{x-\lambda}{\delta}\middle|k\right)$. Remark that λ and δ are not the usual location and dispersion parameters of the Weibull and Fréchet distributions. In fact, as for $k<0$ we have $G(z|k) = \Psi_{-1/k}(-1-kz)$ we have $G\left(\frac{x-\lambda}{\delta}\middle|k\right) = \Psi_{-1/k}\left(\frac{x-(\lambda-\delta/k)}{\delta/(-k)}\right)$ so that, then, the location and dispersion parameters of the usual Weibull distribution are $\lambda-\delta/k$ (the right-end of the distribution) and $\delta/(-k)$. Analogously, for $k>0$, we have $G\left(\frac{x-\lambda}{\delta}\middle|k\right) = \Phi_{1/k}\left(\frac{x-(\lambda-\delta/k)}{\delta/k}\right)$ so that the usual location and dispersion parameters are $\lambda-\delta/k$ (the left-end of Fréchet distribution) and δ/k. We could try to use the quasi-linear method of decision, as in Tiago de Oliveira (1966). But the choice of $k=0$ vs. $k<0$ or $k>0$ would lead to use of the choice statistic

$$V_n'(x) = \frac{\phi_0'(x_1,\ldots,x_n)}{\phi_0(x_1,\ldots,x_n)}$$

where $\phi_k(x_1,\ldots,x_n)$ is the statistic, defined for the sample $(x_1,\ldots,x_n)$, by

$$\phi_k(x_1,\ldots,x_n) = \int_{-\infty}^{-\infty} d\lambda \int_{-\infty}^{+\infty} \delta^{n-2}\, L(\lambda+\delta\, x_i|k)\, d\delta$$

and $$\phi'_k(x_1,\ldots,x_n) = \frac{\partial\, \phi_k(x_1,\ldots,x_n)}{\partial k}.$$

The development of the expression of V'_n leads to

$$V'_n(x_1,\ldots,x_n) =$$

$$\frac{\int_{-\infty}^{+\infty} d\lambda \int_{-\infty}^{+\infty} d\delta\, \delta^{n-2} V_n(\lambda+\delta x_i) e^{-n(\lambda+\delta\bar{x})} \exp(-e^{-\lambda}\Sigma e^{-\delta x_i})}{\int_{-\infty}^{+\infty} d\lambda \int_{0}^{+\infty} d\delta\, \delta^{n-2}\, e^{-n(\lambda+\delta\bar{x})} \exp(-e^{-\lambda}\Sigma e^{-\delta x_i})}$$

which, although the integration on δ is easy, leads to a non-manageable expression.

We will use the statistic $V_n(z) = \sum_{i=1}^{n} v(z_i)$ substituting the reduced values z_i by the "estimated" reduced values $\bar{z}_i=(x_i-\hat{\lambda}(x_i))/\hat{\delta}(x_i)$ where $\hat{\lambda}$ and $\hat{\delta}$ are the maximum likelihood estimators of λ and δ for the Gumbel model, given by the well-known equations

$$\hat{\delta}(x_i) = \bar{x} - \sum_{i=1}^{n} x_i \exp(-x_i/\hat{\delta}(x_i))/\sum_{i=1}^{n} \exp(-x_i/\hat{\delta}(x_i))$$

$$\hat{\lambda}(x_i) = -\hat{\delta}(x_i) \log\left(\left[\sum_{i=1}^{n} \exp(-x_i/\hat{\delta}(x_i))\right]/n\right).$$

Thus, we will use the statistic,

$$\hat{V}_n(x) = \sum_{1}^{n} v\left(\frac{x_i-\hat{\lambda}(x_i)}{\hat{\delta}(x_i)}\right)$$

obtaining its asymptotic distribution by the δ-method, based in Cramer (1946), Fraser (1957) and Mann and Wald (1942): recall that, in Gumbel model, the random pair $(\hat{\lambda},\hat{\delta})$ is asymptotically binormal with mean values λ and δ, variances $(1+\frac{6}{\pi^2}(1-\gamma)^2)\delta^2/n$ and $6\delta^2/\pi^2 n$, covariance $6(1-\gamma)\delta^2/\pi^2 n$ and correlation coefficient $\rho = \left(1+\frac{\pi^2}{6(1-\gamma)^2}\right)^{-\frac{1}{2}}$.

If we recall the reasoning of Sections 3, 4, and 5 we see that the basic result was that $V_n(z)$ was asymptotically normal with mean value $n\mu(k)$ and variance $n\sigma^2(k)$; as $\mu(0)=0$ the rule was to choose Weibull, Gumbel or Fréchet models according V_n was in the regions

$$V_n < -\sqrt{n}\,\sigma(0)\,c_n,\quad |V_n| < \sqrt{n}\,\sigma(0)\,c_n \quad\text{or}\quad V_n > \sqrt{n}\,(0)\,c_n;$$

the consistency was proven utilizing the sign of $\mu(k)$ ($\mu(k)<0$ if $k<0$ and $\mu(k)>0$ if $k>0$) in a neighborhood of $k=0$.

We now show, similarly to what happened to

$V_n(z_i)$, that as $\frac{1}{n}\hat{V}_n(x_i) \overset{P}{\to} 0 = \hat{\mu}(0)$ and $\mathrm{Var}(\frac{1}{n}\hat{V}_n(x_i)) \sim \hat{\sigma}^2(0)/n$,

$$\hat{V}_n(x_i)/\sqrt{n}\,\hat{\sigma}(0)$$

is asymptotically standard normal, i.e., the "estimated" $\hat{V}_n(x_i)$ (shown to be LMP) has analogous behavior to the original one. We have,

$$\hat{V}_n(x_i) = \sum_1^n \left\{ \frac{(x_i-\hat{\lambda})^2}{2\hat{\delta}^2} - \frac{(x_i-\hat{\lambda})^2}{2\hat{\delta}^2}\, e^{-\frac{(x_i-\hat{\lambda})}{\hat{\delta}}} - \frac{x_i-\hat{\lambda}}{\hat{\delta}} \right\}$$

which, by simple algebra, takes the form

$$\hat{V}_n(x_i) = 1/2\hat{\delta}^2{}_0\{\sum_1^n x_i^2 - 2\hat{\delta}\sum_1^n x_i - n\sum_1^n x_i^2\, e^{-x_i/\hat{\delta}} \Big/ \sum_1^n e^{-x_i/\hat{\delta}}\}.$$

Let us now obtain the asymptotic distribution of $\hat{V}_n(x_i) = \sum_1^n v\left(\frac{x_i-\hat{\lambda}}{\hat{\delta}}\right)$ (*). We have $\frac{1}{n}\hat{V}_n(x_i) = \frac{1}{n}\sum_1^n v\left(\frac{x_i-\lambda}{\hat{\delta}}\right) -$
$\frac{\hat{\lambda}-\lambda}{\delta} \times \frac{1}{n}\sum_1^n v'\left(\frac{x_i-\lambda}{\hat{\delta}}\right) - \frac{\hat{\delta}-\delta}{\delta} \times \frac{1}{n}\sum_1^n \frac{x_i-\lambda}{\delta}\ v'\left(\frac{x_i-\lambda}{\sigma}\right) + o_p(1/\sqrt{n})$.

As $\sqrt{n}(\hat{\lambda}-\lambda)$ is asymptotically normal with mean value zero and finite variance we see that

$$\frac{\hat{\lambda}-\lambda}{\delta}\left(\frac{1}{n}\sum_1^n v'(z_i) - M(v'(z))\right) = o_p(1/\sqrt{n})$$

so we can substitute $\frac{1}{n}\sum_1^n v'(z_i)$ by $M(v'(z))$ and analogously we can substitute $\frac{1}{n}\sum_1^n z_i\ v'(z_i)$ by $M(z\ v'(z))$. We have

$M(v') = \frac{1}{2}\left(\frac{\pi^2}{6} + \gamma^2\right) - \gamma$ and $M(zv') = \dot{\gamma} - \Gamma'''(2)/_2$.

Consequently, we have

$$\frac{1}{n}\hat{V}_n(x_i) = \frac{1}{n}\sum_1^n v(z_i) - M(v'(z))\frac{\hat{\lambda}-\lambda}{\delta} - M(z\ v'(z))\frac{\hat{\delta}-\delta}{\delta} + o_p(1/\sqrt{n}).$$

Evidently we can expect the asymptotic behavior to be normal as $(\hat{\lambda}-\lambda)/\delta$ and $(\hat{\delta}-\delta)/\delta$ are asymptotically binormal. But the computation of the correlation is difficult and we will proceed away.

Denoting by $L(x|\lambda,\delta)$ the likelihood

$$L(x|\lambda,\delta) = \frac{1}{n}\prod_1^n \Lambda'\left(\frac{x_i-\lambda}{\delta}\right)$$

the maximum likelihood equation (for $\hat{\lambda},\hat{\delta}$), after the first order development, are

(*) This proof is simpler and shorter than the one presented at the Triestre meeting, that one already simpler than the first one; for details of the method see Tiago de Oliveira (1980).

$$0 = \frac{1}{n}\frac{\partial \log L(x|\hat{\lambda},\hat{\delta})}{\partial\hat{\lambda}} = \frac{1}{n}\frac{\log L(x|\lambda,\delta)}{\partial\lambda} + (\hat{\lambda}-\lambda)\,\frac{1}{n}\frac{\partial^2 \log L(x|\lambda,\delta)}{\partial\lambda^2}$$

$$+ (\hat{\delta}-\delta)\,\frac{1}{n}\frac{\partial^2 \log L(x|\lambda,\delta)}{\partial\lambda\partial\delta} + o_p(1/\sqrt{n})$$

$$0 = \frac{1}{n}\frac{\partial \log L(x|\hat{\lambda},\hat{\delta})}{\partial\hat{\delta}} = \frac{1}{n}\frac{\partial \log L(x|\lambda,\delta)}{\partial\delta} + (\hat{\lambda}-\lambda)\cdot\frac{1}{n}\frac{\partial^2 \log L(x|\lambda,\delta)}{\partial\lambda\partial\delta}$$

$$+ (\hat{\delta}-\delta)\cdot\frac{1}{n}\frac{\partial^2 \log L(x|\lambda,\delta)}{\partial\delta^2} + o_p(1/\sqrt{n}).$$

By the same technique we can substitute the second derivatives of $\log L(x|\lambda,\delta)$ by their mean values and, after rearrangement, we obtain

$$\frac{1}{n}M\left(\frac{\partial^2 \log L}{\partial\lambda^2}\right)(\hat{\lambda}-\lambda) + \frac{1}{n}M\left(\frac{\partial^2 \log L}{\partial\lambda\partial\delta}\right)(\hat{\delta}-\delta) = -\frac{1}{n}\frac{\partial \log L}{\partial\lambda} + o_p(1/\sqrt{n})$$

$$\frac{1}{n}M\left(\frac{\partial^2 \log L}{\partial\lambda\partial\delta}\right)(\hat{\lambda}-\lambda) + \frac{1}{n}M\left(\frac{\partial^2 \log L}{\partial\delta^2}\right)(\hat{\delta}-\delta) = -\frac{1}{n}\frac{\partial \log L}{\partial\delta} + o_p(1/\sqrt{n}) .$$

It is very easy to show that

$$\frac{1}{n}M\left(\frac{\partial^2 \log L}{\partial\lambda^2}\right) = \frac{1}{\delta^2}\int_{-\infty}^{+\infty}\left(\frac{\Lambda''(z)}{\Lambda'(z)}\right)'\Lambda'(z)dz = \frac{A}{\delta^2}$$

$$\frac{1}{n}M\left(\frac{\partial^2 \log L}{\partial\lambda\partial\delta}\right) = \frac{1}{\delta^2}\int_{-\infty}^{+\infty} z\left(\frac{\Lambda''(z)}{\Lambda'(z)}\right)'\Lambda'(z)dz = \frac{B}{\delta^2}$$

$$\frac{1}{n}M\left(\frac{\partial^2 \log L}{\partial\delta^2}\right) = \frac{1}{\delta^2}\left(\int_{-\infty}^{+\infty} z^2\left(\frac{\Lambda''(z)}{\Lambda'(z)}\right)'\Lambda'(z)dz - 1\right) = \frac{C}{\delta^2}$$

with $A = -1$, $B = 1-\gamma$, $C = -\left(\frac{\pi^2}{6} + (1-\gamma)^2\right)$ and $AC - B^2 = \frac{\pi^2}{6}$.
Thus we obtain

$$\frac{\hat{\lambda}-\lambda}{\delta} = \frac{\delta}{AC-B^2}\left(B\cdot\frac{1}{n}\frac{\partial \log L}{\partial\delta} - C\cdot\frac{1}{n}\frac{\partial \log L}{\partial\lambda}\right) + o_p(1/\sqrt{n})$$

$$\frac{\hat{\delta}-\delta}{\delta} = \frac{\delta}{AC-B^2}\left(B\cdot\frac{1}{n}\frac{\partial \log L}{\partial\lambda} - A\cdot\frac{\partial \log L}{\partial\delta}\right) + o_p(1/\sqrt{n})$$

and substituting the expression in $\frac{1}{n}\hat{V}_n(x_i) = \frac{1}{n}\sum_1^n v\left(\frac{x_i-\lambda}{\hat{\delta}}\right)$ we get

$$\frac{1}{n}\hat{V}_n(x_i) = \frac{1}{n}\sum_1^n v(z_i) - \frac{M(v')\delta}{AC-B^2}\left(B\cdot\frac{1}{n}\frac{\partial \log L}{\partial\delta} - C\cdot\frac{1}{n}\frac{\partial \log L}{\partial\lambda}\right) -$$

$$- \frac{M(z\ v')\delta}{AC-B^2}\left(B\cdot\frac{1}{n}\frac{\partial \log L}{\partial\lambda} - A\cdot\frac{1}{n}\frac{\partial \log L}{\partial\delta}\right) + o_p(1/\sqrt{n})$$

and, finally,

$$\frac{1}{n}\hat{V}_n(x_i) = \frac{1}{n}\sum_1^n (v(z_i) + v_\Delta(z_i)) + o_p(1/\sqrt{n})$$

where $$v_\Delta(z) = \frac{M(v')}{AC-B^2}\left\{B\ (1+\frac{\Lambda''}{\Lambda'}z) - C\,\frac{\Lambda''}{\Lambda'}\right\} +$$

$$+ \frac{M(zv')}{AC-B^2}\left\{B\,\frac{\Lambda''}{\Lambda'} - A(\Lambda + z\,\frac{\Lambda''}{\Lambda'})\right\}.$$

Note that $M(v_\Delta(z)) = 0$ as should be expected. Consequently the asymptotic behavior of $\frac{1}{n}\sum_1^n v\left(\frac{x_i-\hat{\lambda}}{\hat{\delta}}\right)$ is the same of $\frac{1}{n}\sum_1^n (v(z_i) + v_\Delta(z_i))$ where we have, for $v_\Delta(z)$, the expression

$$v_\Delta(z) = \frac{1}{2}\left\{\frac{\pi^2}{6} + \gamma^2 - 2\gamma + 2\gamma(1-\gamma)\frac{6}{\pi^2}(2\gamma - \Gamma'''(2)) + (1-\gamma)\Gamma''(2)\right\}$$

$$(e^{-z}-1) + \frac{3}{\pi^2}(2\gamma - \Gamma'''(2) + (1-\gamma)\,\Gamma''(2)(1-z+ze^{-z}) = 0.498086$$

$$(e^{-z} - 1) + 0.307974(1 - z + ze^{-z}).$$

After a simple, but long, computation, we get

$$\sigma^2(0) = \text{Var}\,(v(z) + v_\Delta(z))$$

$$= \sigma^2(0) = \frac{3}{2\pi^2}\,(2\gamma - \Gamma'''(2) + (1-\gamma)\,\Gamma''(2))^2 - \frac{1}{4}\,\Gamma''(2)^2 = 2.09748$$

$$< \sigma^2(0) = 2.43261.$$

Consequently $\hat{V}_n(x_i)/\sqrt{n}\,\hat{\sigma}(0)$ is also asymptotically standard normal and the decision is similar to the previous one. Note that $\hat{\mu}(k)$ and $\hat{\sigma}^2(k)$, the mean value and variance of $v(z)+v_\Delta(z)$ under $G(z|k)$, exist in the same region of $\mu(k)$ and $\sigma^2(k)$.

The reduction of the variance, of about 14%, is easy to explain: In fact for $z\to+\infty$ and $z\to-\infty$ we see that $v(z)/v_\Delta(z)\sim -z/0.61596$ so that the additive term $v_\Delta(z)$ to $v(z)$ has a counter-effect for (large) $z>0$ which has larger probability. In fact, the correlation between $v(z)$ and $v_\Delta(z)$ is -0.36655 as follows from $\text{Var}(v) = \sigma^2(0) = 2.43261$, $\text{Var}(v_\Delta(z)) = 0.32563$ and $\text{Cor}(v(z), v_\Delta(z) = -0.32563$.

From the practical point of view we will use $2\hat{V}_n$ and the regions will be defined using $\pm\, 2\sqrt{n}\hat{\sigma}(0)c_n$. The evaluations of $\hat{\mu}(k)$ and $\hat{\sigma}^2(k)$ are

$$\hat{\mu}(k) = 2.09797\,k + 4.58653\,k^2 + 0(k^3)$$

and $\hat{\sigma}^2(k) = 2.09757 + 19.91015\,k + 248.60225\,k^2 + 0(k^3)$.

Similarly to what happened before we can compute $\hat{E}_n(k<0)$, $\hat{E}_n(k=0)$ and $\hat{E}_n(k>0)$ and show that they converge to zero.

7. FINAL REMARKS

This paper is the first one of a series of statistical choice for extreme models. We hope that the statistic $V_n(z_i)$, and also $\hat{V}_n(x_i)$, are useful outside the region $-1<k<1/4$; in fact, if $V_n(z_i)$ leads to choice of a Fréchet model, say, if $0<k<1/4$ is natural to expect that the same will happen when $k\geq 1/4$ because because the distinction (or distance) between Fréchet and Gumbel model is larger. Appendix 3 shows that asymptotic results seem to be useful for large n $(n\geq 400)$.

ACKNOWLEDGEMENTS

We want to thank Dr. Arne Fransen for correcting the initial values of $\sigma^2(0)$ (and $\hat{\sigma}^2(0)$), for giving the expressions of $\mu(k)$, $\sigma^2(k)$, $\hat{\mu}(k)$ and $\hat{\sigma}^2(k)$ to $O(k^3)$, Prof. M. Ivette Gomes for the computations of Appendix 3, to both, for checking some of the expressions and to Prof. L. Herbach for reading and commenting the paper.

APPENDIX 1: ON MEAN VALUES AND VARIANCES

We know that $v(z)$ has mean value with respect to the distribution function $G(z)$ iff (a) $v(z)\ (1-G(z))\to 0$ if $z\to+\infty$ and (b) $v(z)\ G(z)\to 0$ if $z\to-\infty$; also $v(z)$ has variance (and mean value) iff (a') $v^2(z)\ (1-G(z))\to 0$ if $z\to+\infty$, and (b') $v^2(z)\ G(z)\to 0$ if $z\to\infty$. As we are interested in the existence of both, mean value $\mu(k)$ and variance $\sigma^2(k)$ of $v(z) = z^2/2-z-z^2/2.e^{-z}$ with respect to $G(z|k)$ we will use (a') and (b') to show that we must have $-1<k<1/4$. The existence of $\mu(k)$ only, by the same technique, leads to the weaker condition $-1<k<1/2$. Let us, first, note that $v(z) = z^2/2-z^2/2.e^{-z}-z$ is such that $v(z)/(z^2/2)\to 1$ when $z\to+\infty$ and $v(z)/(-z^2/2.e^{-z})\to 1$ when $z\to-\infty$.

Condition (a') is, thus, equivalent to $z^4\ (1-G(z|k))|0$ when $z\to+\infty$. If $k<0$, as $G(-1/k|k)=1$, the condition is verified; for $k>0$, as $G(z|k)\to 1$ when $z\to+\infty$, we have $G(z|k)=1-(1+kz)^{-1/k} + O((1+kz)^{-2/k})$ so that $z^4(1-G(z|k))\to 0$ is equivalent to $z^4(1+kz)^{-1/k}\to 0$ or, yet, $z^{4-1/k}\to 0$ which is equivalent to $k<1/4$.

Consider now condition (b') equivalent to $z^4\ e^{-2z}G(z|k)\to 0$ when $z\to-\infty$. For $k>0$ we have $G(z|k)=0$ if $z\leq -1/k$ and the condition is verified. If $k<0$ the condition is, thus, $z^4\ e^{-2z-(1+kz)^{-1/k}}\to 0$ if $z\to-\infty$. Taking then $y = (1+kz)^{-1/k}$ or $z = \frac{y^{-k}-1}{k}$ we obtain equivalently $y^{-4k}\ e^{-2/ky^{-k}-y}\to 0$ when $y\to\infty$.

For $-1<k<0$ we obtain $y^{-4k} e^{-2/k} y^{-k} -y \to 0$ and for $k<-1$ we get $y^{-4k} e^{-2/k-y^{-k}-y} \to +\infty$, consequently $\sigma^2(k)$ and $\mu(k))$ exists for $-1<k<1/4$.

APPENDIX 2: SOME LIMITING RESULTS

The basic results contained in this appendix are based on Cramer (1946), Feller (1950), Fraser (1957), and Mann and Wald (1942).

2.1: It is well known the Khintchine convergence of types theorem. If $\{X_n\}$ is a random sequence with distribution functions $\{F_n(x)\}$ and if $F_n(\lambda_n+\delta_n x) \overset{W}{\to} L(x)$, $\delta_n>0$, $L(x)$ proper (i.e., $Prob_L(-\infty) = Prob_L(+\infty)=0$) and non-degenerate (i.e., $Prob_L(a)<1, \forall a$) then $F_n(\bar{\lambda}_n+\bar{\delta}_n x) \overset{W}{\to} \bar{L}(x)$ iff $(\bar{\lambda}_n-\lambda_n)/\delta_n \to \alpha$, $\bar{\delta}_n/\delta_n \to \beta$ and $\bar{L}(x) = L(\alpha+\beta x)$.

Consider, then, a sequence $\{\lambda_n\}$ and let $(d_n-\lambda_n)/\delta_n \to d$ with $|d|<+\infty$. Then $F_n(d_n)\to L(d)$, if $\underline{d}$ is a continuity point of L. In fact, if $(d_n-\lambda_n)/\delta_n \to d$, then for $n>n(\varepsilon)$ we have $\lambda_n+\delta_n(d-\varepsilon)<d_n<\lambda_n+\delta_n(d+\varepsilon)$ so that $F_n(\lambda_n+\delta_n(d-\varepsilon))\leqslant F_n(d_n)<F_n(d_n+\delta_n(d+\varepsilon))$. Letting $n\to\infty$ we get:

$L(d-\varepsilon)\leqslant \underline{\lim}\, F_n(d_n)\leqslant \overline{\lim}\, F_n(d_n)\leqslant L(d+\varepsilon)$ so that by continuity, $F_n(d_n)\to L(d)$. Suppose now that $(d_n-\lambda_n)/\delta_n \to +\infty$. Then for $n>n(M)$, M a continuity point of L, we have $\lambda_n+\delta_n M\leqslant d_n$ so that $F_n(\lambda_n+\delta_n M)\leqslant F_n(d_n)$ and, with $n\to\infty$, $L(M)<\underline{\lim}\, F_n(d_n)$ which implies $F_n(d_n)\to 1 = L(+\infty)$. Analogously we have, if $(d_n-\lambda_n)/\delta_n \to -\infty$, $F_n(d_n)\to 0 = L(-\infty)$.

2.2: In the sequel, denote by $o_p(1)$ a sequence of random variables $\{X_n\}$ such that $X_n \overset{P}{\to} 0$ when $n\to\infty$. Analogously if $\phi(n)X_n \overset{P}{\to} 0$ we will note $\{X_n\}$ by $o_p(\phi(n)^{-1})$. The o_p-notation, similar to the o-notation of Real Analysis, has all classical properties; see Mann and Wald (1942). Then it is well known, Cramer (1942), that if $Y_n-X_n = o_p(1)$ and $Prob\ \{X_n\leqslant x\} \overset{W}{\to} L(x)$ then $Prob\ \{Y_n\leqslant x\}\overset{W}{\to} L(x)$ and if $Y_n/X_n=1+o_p(1)$ and $Prob\ \{X_n\leqslant x\}$

$\overset{w}{\to} L(x)$ then Prob $\{Y_n \leq x\} \overset{w}{\to} L(x)$ if X_n is not $o_p(1)$, i.e., if $L(x)$ is proper and non-degenerate.

APPENDIX 3: APPLICATIONS

A simple simulation for various values of n suggests that only for samples or order 400 or larger the asymptotic result can be used; for smaller n, V_n tends to have negative values. This study will be made in a latter paper.

The analysis of some concrete data seem to confirm this conjecture. Except for the last one, the others seem to fit Gumbel distribution. In fact, if we take data from Gumbel and Goldstein (1964) and compute the $\hat{v}_i = v(\hat{z}_i)$ we get: oldest ages of death in Sweden, men: $n = 54$, $\bar{\hat{v}} = -.2527$, $s^2(\hat{v}) = .4634$; oldest ages of death in Sweden, women: $n = 54$, $\bar{\hat{v}} = -.3340$, $s^2(\hat{v}) = .8540$; floods of Oculgee river, at Macon: $n = 40$, $\bar{\hat{v}} = -.0321$, $s^2(\hat{v}) = .3252$; floods of Ocmulgee river at Hawkinsville: $n = 40$, $\bar{\hat{v}} = -.0306$, $s^2(\hat{v}) = .5039$ which in some way confirms the sketched simulation.

Pluviometric data found in Sneyers (1977) for twenty-four hours leads to $n = 35$, $\bar{\hat{v}} = .1835$, $s^2(\hat{v}) = .8311$.

The data in van Montfort (1970) lead to $n = 47$, $\bar{\hat{v}} = .8037$, $s^2(\hat{v}) = 15.5827$ for the raw data which shows that Gumbel distribution is not a good fit; but for the logarithms of the raw data we have $n = 47$, $\bar{\hat{v}} = .2289$ and $s^2(\hat{v}) = 1.9282$, corresponding to Weibull distribution; the fit to reverse logarithm (Fréchet distribution) is also a bad fit. The forthcoming paper referred to will analyze these data in more detail.

REFERENCES

Cox, D. R. (1961). Tests of separate families of hypothesis. *Proceedings of the 4th Berkeley Symposium, Vol. 1*, 105-123.

Cox, D. R. (1962). Further results on tests of separate families of hypothesis. *Journal of the Royal Statistical Society, Series B*, 24, 406-424.

Cramer, H. (1946). *Mathematical Methods of Statistics*. Princeton University Press.

Feller, W. (1950). *Introduction to Probability Theory, Vol. I*. Wiley, New York.

Fraser, D. A. S. (1957). *Nonparametric Methods in Statistics*. Wiley, New York.

Gumbel, E. J. (1958). *Statistics of Extremes*. Columbia University Press.

Gumbel, E. J. and Goldstein, N. (1964). Analysis of empirical bivariate extremal distributions. *Journal of the American Statistical Association*, 794-816.

Jenkinson, A. F. (1955). The frequency distribution of annual maximum (or minimum) values of meterological elements. *Quarterly Journal of the Royal Meterological Society*, 81, 158-171.

Mann, A. M. and Wald, A. (1942). On stochastic limits of order relationships. *Annals of Mathematical Statistics*, 14, 217-226.

Sneyers, R. (1977). L'intensite maximale des precipitations en Belgique. *Institut Royal Meteorologique de Belgique, Series B*, 86.

Tiago de Oliveira, J. (1968). Quasi-linear prediction. *Annals of Mathematical Statistics*, 87, 1684-1687.

Tiago de Oliveira, J. (1980). The δ-method for the obtention of asymptotic distributions, applications. To be published.

van Montfort, M. N. J. (1970). On testing that the distribution of extremes is of type I when type II is the alternative. *Journal of Hydrology*, 11, 421-427.

[*Received May* 1980. *Revised November* 1980]

AN I-DIMENSIONAL LIMITING DISTRIBUTION FUNCTION OF LARGEST VALUES AND ITS RELEVANCE TO THE STATISTICAL THEORY OF EXTREMES

M. IVETTE GOMES

Faculty of Science of Lisbon
Center of Statistics and Applications
Lisbon, Portugal

SUMMARY. In order to use more efficiently the information generally recorded about extrema, we take into account the asymptotic distribution function of the i largest order statistics, suitably normalized, of a sample of size n from a certain distribution, i a fixed integer, and then, as in Gumbel's approach to the statistical treatment of extremes, we consider a multivariate sample $(\underset{\sim}{X}_1,\ldots,\underset{\sim}{X}_N)$, where $\underset{\sim}{X}_r = (X_{1,r},\ldots,X_{i_r,r})'$ $1 \leq r \leq N$, are independent i_r-dimensional extremal vectors, with suitable unknown parameters to be estimated from the sample. Estimation of the parameters is developed for the three models that arise and comparison is made for different values of i_r, $1 \leq r \leq N$. The new sample of the order statistics of largest values $X_{1,r}$, $1 \leq r \leq N$, and their concomitants is also studied and the results applied to linear estimation of the parameters.

KEY WORDS. extreme value theory, order statistics, concomitants, multivariate extremal processes, estimation in extremal models.

1. GENERAL INTRODUCTION AND PRELIMINARIES

Let $\{X_n\}_{n\geq 1}$ be a sequence of independent, identically distributed (i.i.d.) random variables (r.v.'s) with common distribution function (d.f.) $F(x)$ and for every j, $1 \leq j \leq n$, let $X_{j:n}$ denote the j*th* descending order statistic.

C. Taillie et al. (eds.), Statistical Distributions in Scientific Work, Vol. 6, 389–410.

If there exist real constants $\{a_n\}_{n\geq 1}$ $(a_n>0)$ and $\{b_n\}_{n\geq 1}$ and a non-degenerate d.f. $G(x)$, such that

$$\lim_{n\to\infty} P(X_{1:n} \leq a_n x+b_n) = \lim_{n\to\infty} F^n(a_n x+b_n) = G(x) \qquad (1)$$

for all x in the set of continuity points of $G(x)$, we say that $F(x)$ belongs to the domain of attraction of $G(x)$. As is well-known (Gnedenko, 1943; Galambos, 1978, chapter 2), the only d.f.'s $G(x)$ with non-empty domain of attraction are of one of the following three types:

Type I $\Lambda(x) = \exp(-\exp(-x))$, $-\infty < x < +\infty$ (Gumbel d.f.)

Type II $\Phi_\alpha(x) = \exp(-x^{-\alpha})$, $x > 0$, $\alpha > 0$ (Fréchet d.f.) (2)

Type III $\Psi_\alpha(x) = \exp(-(-x)^\alpha)$, $x < 0$, $\alpha > 0$ (Weibull d.f.).

This asymptotic probabilistic result was used by Gumbel in several papers which culminated in his 1958 book, to give approximations in the statistical analysis of extremes, that is, once we have the validity of (1) Gumbel suggests the approximation of $F^n(x)$, $F(x)$ usually unknonw, by $G((x-b_n)/a_n)$ for large values of n. Inference problems about the maximum values may thus be transferred to inference problems on such asymptotic d.f.'s of largest values.

Although this statistical procedure has proved to be fruitful in the most diversified situations, several criticisms have been made on Gumbel's technique, and one of them is the fact that we are wasting information when using only observed maxima and not further order statistics (o.s.), if available, because the latter would certainly contain useful information about the tail of the d.f. underlying the data. Ramachandran (1974), when applying extreme value theory to large fire losses, points out such a problem and chooses a special model expressed in terms of a Type I d.f. of largest values and in connection with fire insurance. Weissman (1978) also develops a similar model for statistical analysis of extremes. Related results on inference for extremes are those of Pickands (1975). Using the technique developed in this paper it is also possible to sketch an approach different from Gumbel's, similar in spirit to Pickands', and with simpler properties (Gomes, 1978).

With the above mentioned fact in mind, we discuss in section 2 the joint limiting d.f. H_i of the i largest o.s., suitably normalized, of a sample of size n, i a fixed positive integer. A multivariate sample $\underset{\sim}{X}_r = (X_{1,r},\ldots,X_{i,r})'$, $1 \leq r \leq N$, of i.i.d. random vectors with d.f. H_i is there considered and distributional properties of the o.s. of the largest values $X_{1,j}$, $1 \leq j \leq N$, together with their concomitants is therein derived. Finally, we consider statistical estimation methods which take into account not only the maximum but the top few o.s. available for analysis, more precisely, we assume to have a multivariate sample $\underset{\sim}{X}_r = (X_{1,r},\ldots,X_{i_r,r})'$, $1 \leq r \leq N$, of independent random vectors with d.f. H_{i_r}, with suitable unknown location and scale parameters to be estimated from the sample. In section 3 we deal with maximum likelihood estimation of the unknown parameters. However, since in such a model we are outside the field of sufficient statistics and consequently the optimum properties of maximum likelihood estimators are merely asymptotic ones, we shall develop in section 4 estimators of the unknown parameters, based on the o.s. of the largest values $X_{1,j}$, $1 \leq j \leq N$, and their concomitants, when G is of Type I.

2. THE i-DIMENSIONAL EXTREMAL DISTRIBUTION AND CONCOMITANTS IN A RELATED SAMPLE

The following result, concerning the joint limiting d.f. of the i largest o.s. of a sample of size n, of i.i.d. r.v.'s, with d.f. F, follows (see Dwass, 1966; Weissman, 1975).

Theorem 1. Let $\{X_n\}_{n\geq 1}$ be a sequence of i.i.d. r.v.'s with common d.f. F and let $X_{j:n}$ be the j*th* largest among $(X_1,\ldots,X_n)$. If there exist sequences of real constants $\{a_n\}_{n\geq 1}$ $(a_n>0)$ and $\{b_n\}_{n\geq 1}$ and a non-degenerate d.f. $G(x)$, such that (1) holds for all x in the set of continuity points of G, we have for $\lim_{n\to\infty} P(\bigcap_{j=1}^{i} X_{j:n} \leq a_n x_j + b_n)$, the joint d.f. $H_i(x_1,\ldots,x_i)$ given by

$$G(\min_{1\leq j\leq i} x_j) \sum_{\substack{r_j \leq r_{j+1} \leq j \\ 1\leq j\leq i-1}} \cdots \sum \prod_{j=1}^{i-1} \{\log(G(\min_{1\leq k\leq j} x_k)/G(\min_{1\leq k\leq j+1} x_k))\}^{r_{j+1}-r_j} / (r_{j+1}-r_j)! \tag{3}$$

with $r_1=0$. The joint d.f. in (3) is to be taken 0 or 1 if one of the x_j is such that $G(x_j)=0$ or if all the x_j's are such that $G(x_j)=1$, $1 \leq j \leq i$, respectively. The joint density function corresponding to such joint d.f. is

$$h_i(x_1,\ldots,x_i) = \prod_{j=1}^{i-1} (g(x_j)/G(x_j))g(x_i) \quad \text{if } x_1>\ldots>x_i \qquad (4)$$

where $g(x) = G'(x)$.

Using the terminology of Weissman (1978), the joint d.f. given by (3), to which corresponds the density function given by (4) will be called an *i-dimensional extremal distribution* and an i-vector with such joint d.f. will be called an *i-dimensional extremal vector*.

The results of theorem 1 may be easily extended to stationary sequences $\{X_n\}_{n\geq 1}$ in which dependence is not too strong (Gomes, 1978, 1979).

We shall now suppose that we have a multivariate sample $\underset{\sim}{X}_r = (X_{1,r},\ldots,X_{i_r,r})'$, $1 \leq r \leq N$, of independent, i_r-dimensional extremal vectors, where $G \equiv \Lambda$.

In the univariate case the notion of o.s. plays a very important role in statistical methods and is clear and unambiguous. For multivariate samples no reasonable basis exists for a full ordering of the data, but different generalizations of the concept of order can be made in two or more dimensions (see Barnett, 1976, for a critical review). We shall consider here the ordering of the largest values $X_{1,j}$, $1 \leq j \leq N$. We then do not modify the random vectors $\underset{\sim}{X}_r$, $1 \leq r \leq N$, we merely order those vectors according to the ordering of the largest values, in a way similar to the one used by Watterson (1959) for multivariate normal samples. We then get the ordered sample $\underset{\sim}{Y}_j^{(N)} = (Y_{1,j}^{(N)},\ldots,Y_{i_{m(j)},j})'$, $1 \leq j \leq N$, where $Y_{1,1}^{(N)}>\ldots>Y_{1,N}^{(N)}$ and for every j, $1 \leq j \leq N$, $\underset{\sim}{Y}_j^{(N)} = \underset{\sim}{X}_{m(j)}$, $1 \leq m(j) \leq N$ (the m(j)'s all different).

David (1973) and David and Galambos (1974) consider the case $i_r=2$, $1 \leq r \leq N$, in a bivariate normal situation and call the

$Y^{(N)}_{2,j}$, $1 \leq j \leq N$, the concomitants of the o.s. . We shall use this same terminology for the $Y^{(N)}_{s,j}$, $2 \leq s \leq i_{m(j)}$, $1 \leq j \leq N$. More recent results on concomitants have been obtained by Yang (1977).

As is well known $Y^{(N)}_{1,j}$, the jth largest of the i.i.d. r.v.'s $X_{1,j}$, $1 \leq j \leq N$, with d.f. $G(x)$, has the density function $f^{(N)}_{1,j}(x) = j\binom{N}{j}G(x)^{N-j}(1-G(x))^{j-1}g(x)$.

Using, from now on, the notation $\theta_1(\alpha)=\int_{-\infty}^{+\infty}x(G(x))^{\alpha-1}g(x)dx = (\gamma+\log\alpha)/\alpha$, γ Euler's constant, $\alpha > 0$, we have $E(Y^{(N)}_{1,j}) = j\binom{N}{j}\sum_{k=0}^{j-1}\binom{j-1}{k}(-1)^{j-1-k}\theta_1(N-k)$, $1 \leq j \leq N$. When j is large this sum contains terms large in magnitude and alternating in sign. This causes rounding errors in computations and low accuracy of the results. Such difficulties may be overcome by the use of the well-known recurrence relation

$$E(Y^{(N)}_{1,j+1}) = (N\, E(Y^{(N-1)}_{1,j}) - (N-j)\, E(Y^{(N)}_{1,j}))/j,\quad 1 \leq j \leq N-1;\ N \geq 2, \tag{5}$$

that is, once we know $E(Y^{(k)}_{1,1}) = k\theta_1(k)$, $1 \leq k \leq N$, we may obtain $E(Y^{(N)}_{1,j})$, $2 \leq j \leq N$.

We are now interested in the concomitants. Since the rank R of the group containing $X_{s,r}$, $s \geq 2$, depends only on $X_{1,r}$ and is independent of any of the values that the other r.v.'s in the sample may assume, we have

$$f_{Y^{(N)}_{s,j}|Y^{(N)}_{1,j}}(y|x) = f_{X_{s,r}|X_{1,r}}(y|x) = f_{s|1}(y|y)$$

$$= (g(y)/G(x))((\log(G(x)/G(y))^{s-2}/(s-2)!),$$

if $x > y$. From this, and after some manipulation, we may obtain the density function of $Y^{(N)}_{s,j}$ (Gomes, 1978). However, since the formulas are cumbersome, in what follows and for simplicity we shall take $i_r=2$ for $1 \leq r \leq N$.

The density function of $(Y_{1,j}^{(N)},Y_{2,j}^{(N)})$ is $f_{1,j;2,j}^{(N)}(x,y)$ $= f_{1,j}^{(N)}(x)\, f_{2|1}(y|x)$, $x > y$, and $f_{2,j}^{(N)}(y) = \int_y^{+\infty} f_{1,j;2,j}^{(N)}(x,y)dx$ is thus the density function of $Y_{2,j}^{(N)}$. Consequently, a recurrence relation similar to (5) holds for $E(Y_{1,j}^{(N)}Y_{2,j}^{(N)})$ and for $E(Y_{2,j}^{(N)})$. The initial conditions, for this last mean value, are:

$$E(Y_{2,1}^{(N)}) = \begin{cases} \gamma-1 & \text{if } N=1 \\ N(\gamma-\theta_1(N))/(N-1) & \text{if } N>1,\ \gamma \text{ Euler's constant.} \end{cases}$$

Tables of the mean values of the concomitants are then easily obtained.

The same recurrence relation (5) holds obviously for the mean value of any power of the o.s. or concomitants. The initial values for the power two, are:

$$E((Y_{1,1}^{(N)})^2) = N\theta_2(N),\ N \geq 1,$$

$$E((Y_{2,1}^{(N)})^2) = \begin{cases} \pi^2/6+\gamma^2-2\gamma & \text{if } N=1 \\ N(\pi^2/6+\gamma^2-\theta_2(N))/(N-1) & \text{if } N>1, \end{cases}$$

where, for $\alpha > 0$, and from now on, $\theta_2(\alpha) = \int_{-\infty}^{+\infty} x^2 (G(x))^{\alpha-1} g(x)dx$ $= (\pi^2/6+(\gamma+\log\alpha)^2)/\alpha$. On the other hand, for the covariance of $(Y_{1,j}^{(N)},Y_{1,k}^{(N)})$, $1 \leq j < k \leq N$, the use of the well known recurrence relation,

$$E(Y_{1,j+1}^{(N)}Y_{1,k+1}^{(N)}) = (NE(Y_{1,j}^{(N-1)}Y_{1,k}^{(N-1)}) - (k-j)E(Y_{1,j}^{(N)}Y_{1,k+1}^{(N)})$$
$$- (N-k)E(Y_{1,j}^{(N)}Y_{1,k}^{(N)}))/j \qquad (6)$$

enables us to get $E(Y_{1,j}^{(N)} Y_{1,k}^{(N)})$ for every $1 \leq j < k \leq N$, from the knowledge of $E(Y_{1,1}^{(m)} Y_{1,k}^{(m)})$, $2 \leq k \leq m \leq N$.

To obtain such mean values, we need the evaluation of the integral $\iint_{x>y} xyg(x)g(y)(G(x))^{\beta-1}(G(y))^{\alpha-1}dxdy =$

$= \iint_{u>v>0} \log(u)\log(v)\exp(-\alpha u-\beta v)dudv$, which we shall call $A(\alpha,\beta)$. This function was evaluated in Lieblein (1953) and can be expressed as $A(\alpha,\beta) = ((\beta-\alpha)\theta_2(\alpha+\beta) + (\alpha\theta_1(\alpha))^2 - 2L(1+\beta/\alpha) + \pi^2/6)/(2\alpha\beta)$, where $L(1+x) = \int_1^{1+x} (\log(t)/(t-1))dt$ is Spence's integral. Note that $A(\alpha,\beta) + A(\beta,\alpha) = \theta_1(\alpha)\theta_1(\beta)$ and so we merely need to compute $A(\alpha,\beta)$ for $\alpha > \beta$, $A(\alpha,\alpha)$ being given by $(\gamma+\log\alpha)^2/(2\alpha^2)$. To evaluate Spence's integral $L(1+x)$ for $x = \beta/\alpha < 1$, we have used the fact that for $x < 1$, $L(1+x) = \sum_{k\geq 1} (-1)^{k+1} x^k/k^2$. We then get

$$E(Y_{1,1}^{(N)} Y_{1,k}^{(N)}) = k(k-1) \binom{N}{k} \sum_{m=0}^{k-2} \binom{k-2}{m} (-1)^{k-2-m} A(N-1-m,m+1), \qquad 2 \leq k \leq N,$$

and similar computations lead us to

$$E(Y_{1,1}^{(N)} Y_{2,1}^{(N)}) = \begin{cases} \pi^2/6+\gamma^2-\gamma-1 & \text{if } N=1 \\ N\,A(1,N-1) & \text{if } N>1. \end{cases}$$

In order to evaluate the covariances that we still need, we shall first evaluate the integral

$$\iint_{x>y} xy(-\log(G(x))(G(x))^{\beta-1}(G(y))^{\alpha-1}g(x)g(y)dxdy =$$

$$\iint_{u>v>0} v\log(u)\log(v)\exp(-\alpha u-\beta v)dudv,$$

which we shall denote by $B(\alpha,\beta)$. We immediately get $B(\alpha,\alpha) = (\gamma^2-\pi^2/6+4\log 2-2\gamma\log 2-\log^2 2+\log^2\alpha-2\log 2\log\alpha+2\gamma\log\alpha)/(4\alpha^3)$, and $B(\alpha,\beta)$ may be expressed in terms of $A(\alpha,\beta)$ by the use of the formula, $B(\alpha,\beta) = (\beta A(\alpha,\beta) - \theta_1(\alpha) + \theta_1(\alpha+\beta) - \beta(\theta_2(\alpha+\beta) - 2\theta_1(\alpha+\beta))/(\alpha+\beta))/\beta^2$.

Again, from the fact that the ranking depends only on the largest value $X_{1,r}$ and that the $\underset{\sim}{X}_j$, $1 \le j \le N$, are independent, it follows that the density functions of $(Y^{(N)}_{1,j},Y^{(N)}_{2,k})$, of $(Y^{(N)}_{1,k},Y^{(N)}_{2,j})$ and of $(Y^{(N)}_{2,j},Y^{(N)}_{2,k})$, $1 \le j < k \le N$, are respectively,

$$f^{(N)}_{1,j;2,k}(x,y) = \int_y^x f^{(N)}_{1,j;1,k}(x,t)\, f_{2|1}(y|t)dt \quad \text{if} \quad x > y,$$

$$f^{(N)}_{1,k;2,j}(x,y) = \int_{\max(x,y)}^{+\infty} f^{(N)}_{1,j;1,k}(t,x)\, f_{2|1}(y|t)dt, \text{ and finally,}$$

$$f^{(N)}_{2,j;2,k}(x,y) = \iint_{\substack{u>v>y\\ u>\max(x,y)}} f^{(N)}_{1,j;1,k}(u,v)\, f_{2|1}(x|u)\, f_{2|1}(y|v)dudv.$$

It thus follows that the recurrence relation (6) remains valid with $E(Y^{(N)}_{1,j_1}\, Y^{(N)}_{1,j_2})$ replaced by, either $E(Y^{(N)}_{1,j_1}\, Y^{(N)}_{2,j_2})$ or $E(Y^{(N)}_{2,j_1}\, Y^{(N)}_{2,j_2})$ for arbitrary m,j_1,j_2. We need now the initial conditions for these mean values.

Denoting by $R_m(\alpha;p,r;q,s) = \binom{\alpha}{m}(-1)^{\alpha-m}/((p+rm)(q-sm))$, $C_m(\alpha) = A(\alpha-m,m+1) + A(1,\alpha) - \gamma\theta_1(\alpha-m)$, $D(\alpha) = B(1,\alpha) + B(\alpha,1)$, we get, after some manipulation, the following expression for $E(Y^{(N)}_{1,1}Y^{(N)}_{2,k})$, $1 < k \le N$, $N(N-1)A(1,N-1)/(N-k)$ -

$$k(k-1)\binom{N}{k}\sum_{m=0}^{k-2} R_m(k-2;1,0;N-2,1)A(N-1-m;m+1) \quad \text{if} \quad k < N, \quad \text{and}$$

$$N(N-1)(-(\gamma+\psi(N-1))A(1,N-1) - \sum_{m=0}^{N-3} R_m(N-2;1,0;N-2,1)A(N-1-m,m+1) +$$

$(\gamma-1)\theta_1(N-1) - D(N-1))$ if $k=N$, where the sum is to be taken 0 if $N=2$. Also for $1 < k \leq N$, we have $E(Y_{1,k}^{(N)}Y_{2,1}^{(N)}) =$

$$k(k-1)\binom{N}{k}((-1)^{k-2}D(N-1) - \sum_{m=1}^{k-2} R_m(k-2;0,1;1,0)C_m(N-1),$$ where

the sum is to be taken equal to 0 if $k=2$. Finally, we get for $E(Y_{2,1}^{(N)}Y_{2,k}^{(N)})$, $N(N-1)(\gamma^2-2A(1,N-1)/((N-2)(N-k)) +$

$$k(k-1)\binom{N}{k}\{\sum_{m=1}^{k-2} R_m(k-2;0,1;N-2,1)C_m(N-1) - (-1)^{k-2}D(N-1)/$$

$$(N-2)\}\text{ if } 1 < k < N,\quad k(k-1)\binom{N}{k}\{\sum_{m=1}^{N-3} R_m(N-2;0,1;N-2,1)C_m(N-1)$$

$$+ (2A(1,N-1)-\gamma^2)/(N-2)^2 + (2(\gamma+\psi(N-1))A(1,N-1) + (1-\gamma)\theta_1(N-1)$$
$$- \gamma(1-\gamma) + (1-(-1)^{N-2})D(N-1))/(N-2)\} - N(N-1)\gamma^2(\gamma+\psi(N-1))/(N-2)$$

if $k = N > 2$, and $(1-\gamma^2)/2$ if $k=N=2$, where the sums are to be taken equal to zero whenever the upper limit is smaller than the lower limit.

The covariance matrix of the o.s. and concomitants, used later on in section 4, may thus be obtained. For such computations, Fortran IV double precision was incorporated throughout the main program and associated subroutines. The constants γ and π were read with the maximum possible machine precision and two different algorithms were used. A first one involving direct formulas for the mean values, variances and covariances, and a second one involving the recurrence relations pointed out, together with the derived initial conditions.

The second algorithm is much more precise and efficient than the first one. Several checks were made. Among them we mention

(a) $$\sum_{j=1}^{N}\sum_{k=1}^{N} \mathrm{Cov}(Y_{1,j}^{(N)},Y_{1,k}^{(N)}) = N\psi'(1),$$

(b) $$\sum_{j=1}^{N}\sum_{k=1}^{N} \mathrm{Cov}(Y_{2,j}^{(N)},Y_{2,k}^{(N)}) = N\psi'(2),$$ and

(c) $$\sum_{j=1}^{N}\sum_{k=1}^{N} \mathrm{Cov}(Y_{1,j}^{(N)},Y_{2,k}^{(N)})\ N\psi'(2).$$

However, even using the second algorithm and a CDC 7600, the two sides of (a) agree up to 27 decimal figures for $N \leq 6$, 26 decimal figures for $6 < N \leq 12$, 25 decimal figures for $12 < N < 15$ and 24 decimal figures for $15 \leq N \leq 20$. The two sides of either (b) or (c) agree up to 13 decimal figures for $N < 12$ and up to 12 decimal figures for $12 \leq N \leq 20$ and the rounding errors do not permit us to go beyond sample size $N=18$. For $N \geq 18$ the covariance matrix of the o.s. of the largest values and their concomitants becomes ill-conditioned.

3. MAXIMUM LIKELIHOOD ESTIMATION IN AN i-DIMENSIONAL EXTREMAL MODEL

We now assume to have the followint model: a multivariate sample $\underset{\sim}{X}_j = (X_{1,j},\ldots,X_{i,j})'$, $1 \leq j \leq N$, of i.i.d. random vectors with joint density function given by (4), $G(x)$ and $g(x)$ replaced by $G((x-\lambda)/\delta)$ and $g((x-\lambda)/\delta)/\delta$ respectively, λ and δ unknown location and scale parameters to be estimated from the sample. We call this model M1, M2, or M3 according as G is of Type I, II, or III respectively. We shall here deal with the maximum likelihood estimation of the parameters for the three models. For models M1 and M2 and $i=1$ these results agree with those obtained by Tiago de Oliveira (1972).

Remark. More generally we could have considered that the random vectors $\underset{\sim}{X}_j$, $1 \leq j \leq N$, were i_j-variate. This would not interfere with the results presented in the following, apart from some minor alterations. If we had considered that not only the i_j, $1 \leq j \leq N$, were different but also that for $j=1,\ldots,N$, $(X_{1,j},\ldots,X_{i_j,j})$ were the i_j largest o.s. of a sample of size n_j, the estimation procedures dealt with in this section would be different because the parameters would depend on j (they would be functionally related) (Ramachandran, 1974).

3.1 Model M1. $\Lambda(x;\lambda,\delta) = \exp(-\exp(-(x-\lambda)/\delta))$, $-\infty < x < +\infty$. The likelihood of the sample is then

$$L_i(\lambda,\delta;\underset{\sim}{x}_1,\ldots,\underset{\sim}{x}_N) = \exp(-\sum_{j=1}^{N}\sum_{k=1}^{i}(x_{k,j}-\lambda)/\delta$$

$$-\sum_{j=1}^{N}\exp(-(x_{i,j}-\lambda)/\delta))/\delta^{Ni}$$

and consequently the maximum likelihood estimates of λ and δ are given by

$$\tilde{\lambda}_i = \tilde{\delta}_i \log(Ni/(\sum_{j=1}^{N} \exp(-x_{i,j}/\tilde{\delta}_i)))$$

$$\sum_{j=1}^{N} x_{i,j} \exp(-x_{i,j}/\tilde{\delta}_i) = (\sum_{j=1}^{N}\sum_{k=1}^{i} x_{k,j}/(Ni)-\tilde{\delta}_i) \sum_{j=1}^{N} \exp(-x_{i,j}/\tilde{\delta}_i).$$

It is well known that the asymptotic efficiency of maximum likelihood estimation is one and that $(\tilde{\lambda}_i,\tilde{\delta}_i)$ is asymptotically binormal with mean value (λ,δ) and with inverse of covariance matrix $R_i = (r^{(i)}_{k,j})$, where $r^{(i)}_{k,j} = -E(\partial^2 \log L_i(\lambda,\delta;\underset{\sim}{X}_1,\dots,\underset{\sim}{X}_N)/\partial\theta_k\partial\theta_j))$, $1 \leq k, j \leq 2$, $\theta_1 = \lambda$, $\theta_2 = \delta$.

We then get asymptotically, $\sigma^2_{1,i} = \mathrm{Var}_\infty(\tilde{\lambda}_i) = \delta^2(\psi'(i+1)+\psi^2(i+1)+1)/(Ni(1+\psi'(i+1)))$ and $\sigma^2_{2,i} = \mathrm{Var}_\infty(\tilde{\delta}_i) = \delta^2/(Ni(1+\psi'(i+1)))$. Also asymptotically, the <u>correlation coefficient</u> between $\tilde{\lambda}_i$ and $\tilde{\delta}_i$ is $\rho_i = \psi(i+1)/\sqrt{1+\psi'(i+1)+\psi^2(i+1)}$ and the amount of information in the sample is $\Delta_i = |R_i| = (Ni)^2(1+\psi'(i+1))/\delta^4$. In Table 1 we present for $i=1(1)5$ asymptotic characteristics of $(\tilde{\lambda}_i, \tilde{\delta}_i)$. The relative gain in information $((\Delta_2-\Delta_1)/\Delta_2) \times 100\%$, when we change from $i=1$ to $i=2$ is around 70% and when we change from j to $2j$, $j \geq 2$ such relative gain increases slightly as j increases. Comparing $\mathrm{Var}(\tilde{\lambda}_1)$ and $\mathrm{Var}(\tilde{\delta}_1)$ based on N, with $\mathrm{Var}(\tilde{\lambda}_2)$ and $\mathrm{Var}(\tilde{\delta}_2)$ respectively, based on 30 pairs, we get $\mathrm{Var}(\tilde{\lambda}_1) \leq \mathrm{Var}(\tilde{\lambda}_2)$ if and only if $N \geq 40$ and $\mathrm{Var}(\tilde{\delta}_1) \leq \mathrm{Var}(\tilde{\delta}_2)$ if and only if $N \geq 50$.

We have considered also the maximum likelihood estimation of $\chi_\alpha = \lambda+\alpha\delta$, α real and in Figure 1 we present for $i=2,4,6$, $\varepsilon_i(\alpha) = \mathrm{Var}(\tilde{\lambda}_i+\alpha\tilde{\delta}_i)/\mathrm{Var}(\tilde{\lambda}_1+\alpha\tilde{\delta}_1)$.

Table 1. Asymptotic characteristics of maximum likelihood estimators of the unknown parameters λ *and* δ *in model M1.*

i	$N\,Var(\tilde{\lambda}_i)/\delta^2$	$N\,Var(\tilde{\delta}_i)/\delta^2$	ρ_i	$\delta^4\Delta_i/N^2$	Δ_1/Δ_i
1	1.1087	0.6079	0.3131	1.6499	1.0000
2	0.8052	0.3584	0.6157	5.5797	0.2948
3	0.7430	0.2596	0.7425	11.5544	0.1424
4	0.7143	0.2047	0.8062	19.5412	0.0842
5	0.6928	0.1693	0.8434	29.5331	0.0557

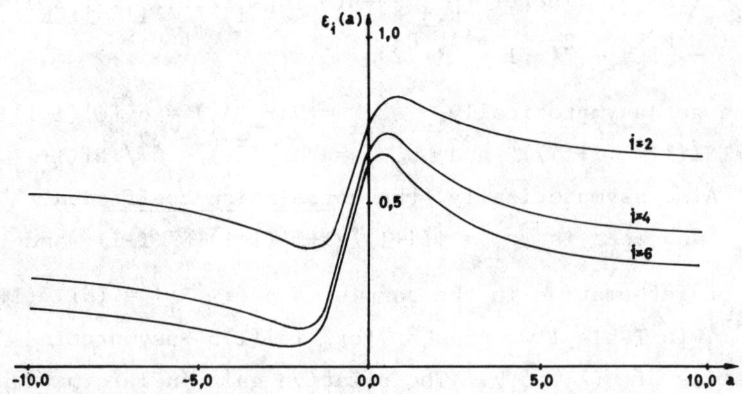

FIG. 1: Graphical representation of $\varepsilon_i(\alpha) = Var(\tilde{\lambda}_i+\alpha\tilde{\delta}_i)/Var(\tilde{\lambda}_1+\alpha\tilde{\delta}_1)$.

The bias and covariance matrix of the maximum likelihood estimators $(\tilde{\lambda}_i,\tilde{\delta}_i)$ for small samples were investigated numerically, for sample sizes $N=5(5)30$ and for $i=1,2,3$. The following steps are an outline of the computational procedure used for this Monte Carlo study and for each value of the pair (i,N).

1. Set $n_c=1$.

2. Set $L=1$.

3. Generate Ni pseudo random numbers U_j, $m_1 \le j \le m_1 + Ni-1$, $m_1 = 500Ni(n_c-1) + Ni(L-1) + 1$, uniformly distributed over the interval (0,1). The algorithm used generates two multiplicative congruential sequences and forms its sum modulus 2^{46}. Satisfactory results were obtained in tests of frequency, runs up, runs down, runs above the mean and runs below the mean.

4. Working then with the set of the Ni random numbers generated in 3, and for $(\lambda,\delta) = (0,1)$ produce $\underset{\sim}{x}_j = (x_{1,j},\ldots,x_{i,j})$, $1 \le j \le N$, i.e., N possible realizations of a i-dimensional random vector with the d.f. required. It is enough to put

$$x_{k,j} = -\log(-\sum_{m=0}^{k-1} \log(U_{i(j-1)+m_1+m})),\quad 1 \le k \le i;\ 1 \le j \le N.$$

5. Using the sample $(\underset{\sim}{x}_1,\ldots,\underset{\sim}{x}_N)_{500(n_c-1)+L}$ previously generated, compute the maximum likelihood estimates $(\tilde{\lambda}_i,\tilde{\delta}_i)_{500(n_c-1)+L}$ of (λ,δ).

6. If L=500 go to step 7; otherwise put L=L+1 and to to step 3.

7. Based on the sample $(\tilde{\lambda}_i,\tilde{\delta}_i)_r$, $1 \le r \le 500n_c$, of the maximum likelihood estimates, compute the desired sample characteristics.

8. Compute the standard deviations of the characteristics evaluated in 7 and stop the process if all standard errors are smaller than 0.0025.

9. If n_c=20 stop the process; otherwise continue.

10. Put $n_c=n_c+1$, go to 2 and repeat the procedure.

The results obtained are summarized in Table 2. The number of replicates used depends on the value of (i,N) and is written on the table. The estimated standard errors up to the scale factor δ are placed between brackets, following the corresponding sample characteristics. The maximum likelihood estimator of the location parameter λ is positively biased for i=1 and negatively biased for i>1 whereas the maximum

TABLE 2: Means, mean square errors and covariances of maximum likelihood estimations of the unknown parameters λ and δ in model M1.

Sample Size N	Dim. of X_j	Number of Replicates	$E(\tilde{\lambda}_i-\lambda)/\delta$	$E(\tilde{\delta}_i)/\delta$	$Cov(\tilde{\lambda}_i,\tilde{\delta}_i)/\delta^2$	M.s.e.$(\tilde{\lambda}_i)/\delta^2$	M.s.e.$(\tilde{\delta}_i)/\delta^2$	Determinant of m.s.e. Matrix
	1	10000	.0912(.0050)	.8423(.0034)	.0391).0018)	.2495(.0041)	.1409(.0018)	.0346
5	2	10000	-.0099(.0040)	.9126(.0027)	.0621(.0013)	.1621(.0024)	.0785(.0010)	.0088
	3	10000	-.0348(.0039)	.9381(.0023)	.0637(.0012)	.1498(.0022)	.0556(.0008)	.0040
	1	10000	.0433(.0034)	.9253(.0025)	.0248(.0009)	.1199(.0018)	.0664(.0009)	.0075
10	2	10000	-.0064(.0028)	.9572(.0019)	.0322(.0007)	.0801(.0012)	.0377(.0005)	.0020
	3	10000	-.0196(.0027)	.9689(.0016)	.0321(.0006)	.0742(.0011)	.0267(.0004)	.0009
	1	10000	.0248(.0027)	.9530(.0020)	.0161(.0006)	.0749(.0011)	.0432(.0006)	.0030
15	2	9000	-.0065(.0024)	.9711(.0016)	.0219(.0005)	.0538(.0008)	.0245(.0004)	.0008
	3	8500	-.0139(.0024)	.9792(.0014)	.0216(.0004)	.0499(.0008)	.0175(.0003)	.0004
	1	9500	.0202(.0024)	.9623(.0018)	.0124(.0005)	.0570(.0008)	.0324(.0005)	.0017
20	2	7000	-.0031(.0024)	.9792(.0016)	.0164(.0004)	.0404(.0007)	.0185(.0003)	.0005
	3	6500	-.0110(.0024)	.9841(.0014)	.0163(.0003)	.0377(.0007)	.0131(.0002)	.0002
	1	7500	.0189(.0025)	.9724(.0018)	.0108(.0004)	.0459(.0008)	.0251(.0004)	.0010
25	2	5500	-.0040(.0024)	.9828(.0016)	.0128(.0003)	.0319(.0008)	.0146(.0003)	.0003
	3	5000	-.0089(.0024)	.9876(.0014)	.0127(.0003)	.0293(.0008)	.0102(.0002)	.0001
	1	6000	.0111(.0025)	.9742(.0018)	.0083(.0004)	.0372(.0007)	.0211(.0004)	.0007
30	2	4500	-.0014(.0024)	.9879(.0016)	.0112(.0003)	.0267(.0006)	.0120(.0002)	.0002
	3	4500	-.0079(.0024)	.9893(.0014)	.0110(.0003)	.0253(.0005)	.0086(.0002)	.0001

likelihood estimator of the scale parameter δ is always negatively biased. For $i=1$ the results agree with those of Harter and Moore (1968).

3.2 M2 Model. $\Phi_\alpha(x;\lambda,\delta) = \exp(-((x-\lambda)/\delta)^{-\alpha})$, $x \geq \lambda$. If we assume that λ is known, the transformation $\underset{\sim}{Y}_j = \log(\underset{\sim}{X}_j - \underset{\sim}{\lambda})$, $1 \leq j \leq N$, where $\underset{\sim}{\lambda}$ is the column vector with all components equal to the known λ, leads to model M1. If all the three parameters are unknown, the likelihood function is,

$$L_i(\lambda,\delta,\alpha;\underset{\sim}{x}_1,\ldots,\underset{\sim}{x}_N) = \alpha^{Ni}\delta^{Ni\alpha} \prod_{j=1}^{N} \prod_{k=1}^{i} (x_{k,j}-\lambda)^{-\alpha-1} \exp(-((x_{i,j}$$

$$-\lambda)/\delta)^{-\alpha}) \quad \text{if} \quad x_{1,j} > \ldots > x_{i,j}; \lambda \leq \min_{1 \leq j \leq N} x_{i,j},$$

and consequently the maximum likelihood estimators $(\tilde{\lambda}_i, \tilde{\alpha}_i)$ of (λ,α) are obtained as solutions of the non-linear system of equations

$$(\alpha+1) \sum_{j=1}^{N} \sum_{k=1}^{i} (X_{k,j}-\lambda)^{-1}/(Ni) - \alpha\left(\sum_{j=1}^{N} (X_{i,j}-\lambda)^{-\alpha-1}\right)/$$

$$\left(\sum_{j=1}^{N} (X_{i,j}-\lambda)^{-\alpha}\right) = 0$$

$$1/\alpha + \left(\sum_{j=1}^{N} (X_{i,j}-\lambda)^{-\alpha} \log(X_{i,j}-\lambda)\right)/\left(\sum_{j=1}^{N} (X_{i,j}-\lambda)^{-\alpha}\right)$$

$$- \sum_{j=1}^{N} \sum_{k=1}^{i} \log(X_{k,j}-\lambda)/(Ni) = 0.$$

The maximum likelihood estimator $\tilde{\delta}_i$ of the scale parameter δ is given by

$$\tilde{\delta}_i = (Ni/(\sum_{j=1}^{N}(X_{i,j}-\tilde{\lambda}_i))^{-\tilde{\alpha}_i}))^{1/\tilde{\alpha}_i}.$$

The asymptotic covariance matrix of $(\tilde{\lambda}_i,\tilde{\delta}_i,\tilde{\alpha}_i)$ is $\Sigma_i = (\sigma_{k,j}^{(i)})$, k,j=1,2,3, given by

$$\sigma_{1,1}^{(i)} = (\delta i)^2(1+\psi'(i+1))/(N\ d_i)$$

$$\sigma_{2,2}^{(i)} = \delta^2 i((\alpha+1)^2(\alpha i+2)(1+\psi'(i+1)+\psi^2(i+1))\Gamma(i+2/\alpha)/((\alpha+2)\Gamma(i))$$
$$- (\alpha+(i\alpha+1)(1/(\alpha+1)+\psi(i+1/\alpha)))(\Gamma(i+1/\alpha)/\Gamma(i))^2)/(N\alpha^2 d_i)$$

$$\sigma_{3,3}^{(i)} = \alpha^2(i(\alpha+1)^2(i\alpha+2)\Gamma(i+2/\alpha)/((\alpha+2)\Gamma(i))-((i\alpha+1)\Gamma(i+1/\alpha)/$$
$$\Gamma(i))^2)/(N\ d_i)$$

$$\sigma_{1,2}^{(i)} = i\delta^2\Gamma(i+1/\alpha)(i\alpha+1)(\psi(i+1)(\alpha/(i\alpha+1)+1/(\alpha+1)+\psi(i+1/\alpha))-1$$
$$-\psi'(i+1)-\psi^2(i+1))/(N\ d_i\alpha\Gamma(i))$$

$$\sigma_{1,3}^{(i)} = i\alpha\delta(i\alpha+1)\Gamma(i+1/\alpha)(\psi(i+1)-1/(\alpha+1)-\psi(i+1/\alpha)-\alpha/(i\alpha+1))/$$
$$(N\ d_i\Gamma(i))$$

$$\sigma_{2,3}^{(i)} = -\delta(i\psi(i+1)(\alpha+1)^2(i\alpha+2)\Gamma(i+2/\alpha)/((\alpha+2)\Gamma(i))$$
$$-(i\alpha+1)(\Gamma(i+1/\alpha)/\Gamma(i))^2(\alpha+(i\alpha+1)(1/(\alpha+1)+\psi(i+1/\alpha))))/(N\ d_i)$$

where $d_i = (i(\alpha+1))^2(i\alpha+2)\Gamma(i+2/\alpha)(1+\psi'(i+1))/((\alpha+2)\Gamma(i))$ -

$$a_i(\Gamma(i+1/\alpha)/\Gamma(i))^2, \quad a_i = i(i\alpha+1)^2(1+\psi'(i+1)+(\psi(i+1)$$

$$-\alpha/(i\alpha+1)-1/(\alpha+1)-\psi(i+1/\alpha))^2).$$

3.3 M3 Model. $\Psi_\alpha(x;\lambda,\delta) = \exp(-(-(x-\lambda)/\delta)^\alpha)$, $x \leq \lambda$. If λ is known, the transformation $\underset{\sim}{Y}_j = -\log(-(\underset{\sim}{X}_j-\lambda))$, $1 \leq j \leq N$, leads again to model M1. If the shape parameter α is not known, we have difficulties for $\alpha \leq 2$, owing to the non-regularity of the estimation. More specifically the asymptotic covariance matrix of the maximum likelihood estimators does not exist. We shall here assume that it is known *a priori* that $\alpha > 2$.

The maximum likelihood estimators $(\tilde{\lambda}_i,\tilde{\alpha}_i)$ of (λ,α) are solutions of

$$(\alpha-1)\sum_{j=1}^{N}\sum_{k=1}^{i}(-(X_{k,j}-\lambda))^{-1}/(Ni)-\alpha\sum_{j=1}^{N}(-(X_{i,j}-\lambda))^{\alpha-1}/$$

$$\sum_{j=1}^{N}(-(X_{i,j}-\lambda))^{\alpha} = 0$$

$$1/\alpha+\sum_{j=1}^{N}\sum_{k=1}^{i}\log(-(X_{k,j}-\lambda))/(Ni)-\sum_{j=1}^{N}(-(X_{i,j}-\lambda))^{\alpha}\log(-(X_{i,j}-\lambda))/$$

$$\sum_{j=1}^{N}(-(X_{i,j}-\lambda))^{\alpha} = 0,$$

and the maximum likelihood estimator $\tilde{\delta}_i$ of δ is given by

$$\tilde{\delta}_i = ((\sum_{j=1}^{N}(-(X_{i,j}-\tilde{\lambda}_i))^{\tilde{\alpha}_i})/(Ni))^{1/\tilde{\alpha}_i}.$$

The information matrix $R_i = \Sigma_i^{-1} = (r_{k,j}^{(i)})$, $k,j=1,2,3$, is given by

$r_{1,1}^{(i)} = N(\alpha-1)^2\Gamma(i-2/\alpha)(\alpha i-2)/(\delta^2(\alpha-2)\Gamma(i))$, $r_{1,2}^{(i)} =$

$-N\alpha(\alpha i-1)\Gamma(i-1/\alpha)/(\delta^2\Gamma(i))$,

$r_{1,3}^{(i)} = N\Gamma(i-1/\alpha)((\alpha i-1)\psi(i-1/\alpha)/\alpha-(\alpha i-1)/(\alpha(\alpha-1))+1)/$

$(\delta\Gamma(i))$, $r_{2,2}^{(i)} = Ni\alpha^2/\delta^2$,

$r_{2,3}^{(i)} = -Ni\psi(i+1)/\delta$, $r_{3,3}^{(i)} = Ni(1+\psi'(i+1)+\psi^2(i+1))/\alpha^2$,

from which the covariance matrix Σ_i of $(\tilde{\lambda}_i, \tilde{\delta}_i, \tilde{\alpha}_i)$ follows by inversion.

4. BEST LINEAR UNBIASED ESTIMATORS OF THE PARAMETERS BASED ON ORDER STATISTICS OF LARGEST VALUES AND CONCOMITANTS-MODEL M1

Concomitants of o.s. as defined in section 2 may obviously have a direct application to problems of estimation in the population considered in section 3 when the sample is ordered with respect to the largest variate. We shall consider Lloyd's matrix approach to linear estimation by o.s., taking into account not only the o.s. of the largest values $X_{1,j}$, $1 \leq j \leq N$, in model M1, but also the corresponding concomitants.

We thus consider $\underset{\sim}{Z}_i = (Y_{1,1}^{(N)},\ldots,Y_{i,1}^{(N)},\ldots,Y_{1,N}^{(N)},\ldots,Y_{i,N}^{(N)})'$, the random vector of dimension Ni of the o.s. and concomitants, corresponding to the multivariate sample $(\underset{\sim}{X}_1,\ldots,\underset{\sim}{X}_N)$ in model M1. For $i=2$, we evaluated in section 2 $\underset{\sim}{\mu}_i$, the column vector of the mean value of $(\underset{\sim}{Z}_i-\underset{\sim}{\lambda})/\delta$, and Σ_i, the covariance matrix of $(\underset{\sim}{Z}_i-\underset{\sim}{\lambda})/\delta$, where $\underset{\sim}{\lambda}$ is the column vector of dimension Ni with all its components equal to λ. The best linear unbiased estimator (B.L.U.E.) $\theta_i^* = (\lambda_i^*,\delta_i^*)'$ of the unknown parameter $\underset{\sim}{\theta} = (\lambda,\delta)'$, based on $\underset{\sim}{Z}_i$, is given by $\theta_i^* = (P_i' \Sigma_i^{-1} P_i)^{-1} P_i' \Sigma_i^{-1} \underset{\sim}{Z}_i$, where $P_i = (\underset{\sim}{1}_{Ni},\underset{\sim}{\mu}_i)$, $\underset{\sim}{1}_{Ni}$ being the column vector of Ni unity elements.

For $i=2$, the weights of the B.L.U.E. of the parameters cannot be obtained beyond $N=18$ due to the impossibility of computing Σ_i^{-1} for such sample sizes, as we stated in section 2. In Table 3 we present the covariance matrix of the B.L.U.E. of λ and δ based on the o.s. alone $(i=1)$ and on o.s. and concomitants $(i=2)$. The comparison of the results of Table 2 and Table 3 for sample sizes $N=5$, 10 and 15 enables us to see that there is not great discrepancy between maximum likelihood estimation and best linear unbiased estimation. For $i=1$ the B.L.U.E. of λ is slightly better than the maximum likelihood estimator of λ. In the other situations for $i=1$ and in all cases for $i=2$ the opposite is true.

For values of N larger than 18, estimates may be obtained in the usual way first discussed by Lieblein and Zelen (1956), that is, by randomly dividing the initial sample into k independent subsamples with sample sizes smaller than 18. A sub-estimate is calculated for each subsample and the results averaged to produce an overall sample estimate. Here, we consider the total sample divided in subsamples of equal size $n_0=15$. We therefore obtain an approximation $\chi^{*'}_{i,\alpha}$ of the B.L.U.E. of $\chi_\alpha = \lambda+\alpha\delta$, which has an asymptotic variance,

$$\mathrm{Var}(\chi^{*'}_{i,\alpha}) = \begin{cases} 15(0.074809+2\alpha 0.015557+\alpha^2 0.045343)/N & \text{if } i=1 \\ 15(0.054137+2\alpha 0.023321+\alpha^2 0.030534)/N & \text{if } i=2 \end{cases}$$

and so, for $i=1$, the asymptotic efficiency of this approximation to the B.L.U.E. of χ_α takes the value 98.8% at $\alpha=0$ and tends to 89.47% as $\alpha \to \pm\infty$. For $i=2$, the asymptotic efficiency of $\chi^{*'}_{2,\alpha}$ takes the value 99.23% at $\alpha=0$ and tends to 78.35% as $\alpha \to \pm\infty$.

As we see from Figure 2 $\mathrm{eff}_a(\chi^{*'}_{2,\alpha}) \geq \mathrm{eff}_a(\chi^{*'}_{1,\alpha})$ for values of α such that $-0.897239 \leq \alpha \leq 0.127712$ to which correspond values of $p = \exp(-\exp(-\alpha))$ such that $0.086051 \leq p \leq 0.414738$.

ACKNOWLEDGEMENTS

I am deeply grateful to Dr. C. W. Anderson and to Professor Tiago de Oliveira for many stimulating discussions and invaluable suggestions, and to Professor J. Galambos for useful criticisms on an earlier draft of this paper. This research was partially supported by the Calouste Gulbenkian Foundation.

TABLE 3: Covariance matrix of best linear unbiased estimators based on order statistics and concomitants (δ=1).

$\sigma_{1,1}^{(1)}$ - Variance of B.L.U.E. of λ based on o.s.

$\sigma_{2,2}^{(1)}$ - Variance of B.L.U.E. of δ based on o.s.

$\sigma_{1,2}^{(1)}$ - Covariance of B.L.U.E.'s of λ and δ based on o.s.

$\sigma_{1,1}^{(2)}$ - Variance of B.L.U.E. of λ based on o.s. and conc.

$\sigma_{2,2}^{(2)}$ - Variance of B.L.U.E. of δ based on o.s. and conc.

$\sigma_{1,2}^{(2)}$ - Covariance of B.L.U.E's of λ and δ based on o.s. and conc.

$\Delta_j = \sigma_{1,1}^{(j)}\,\sigma_{2,2}^{(j)} - \sigma_{1,2}^{(j)\,2}$, j=1,2

N	$\sigma_{1,1}^{(1)}$	$\sigma_{2,2}^{(1)}$	$\sigma_{1,2}^{(1)}$	Δ_1	$\sigma_{1,1}^{(2)}$	$\sigma_{2,2}^{(2)}$	$\sigma_{1,2}^{(2)}$	Δ_2
2	.659547	.711857	.064322	.466366	.049703	.347204	.193322	.104877
3	.402864	.344712	.024772	.138258	.272463	.200496	.124593	.039104
4	.293459	.225283	.034690	.064908	.204032	.139107	.091741	.019966
5	.231395	.166647	.033991	.037406	.163049	.105938	.072534	.012012
6	.191174	.131960	.031373	.024243	.135764	.085320	.059949	.007990
7	.162928	.109096	.028603	.016957	.116296	.071317	.051070	.005686
8	.141983	.092916	.026083	.012512	.101707	.061209	.044474	.004247
9	.125833	.080876	.023881	.009606	.093676	.053579	.039382	.003291
10	.112973	.071573	.021976	.007603	.081302	.047622	.035332	.002623
11	.102509	.064174	.020327	.006165	.073888	.042845	.032036	.002423
12	.093821	.058150	.018894	.005099	.067713	.038930	.029300	.001778
13	.086493	.053152	.017639	.004286	.062490	.035666	.026994	.001500
14	.080228	.048941	.016535	.003653	.058014	.032903	.025024	.001283
15	.074809	.045343	.015557	.003150	.054137	.030534	.023321	.001109

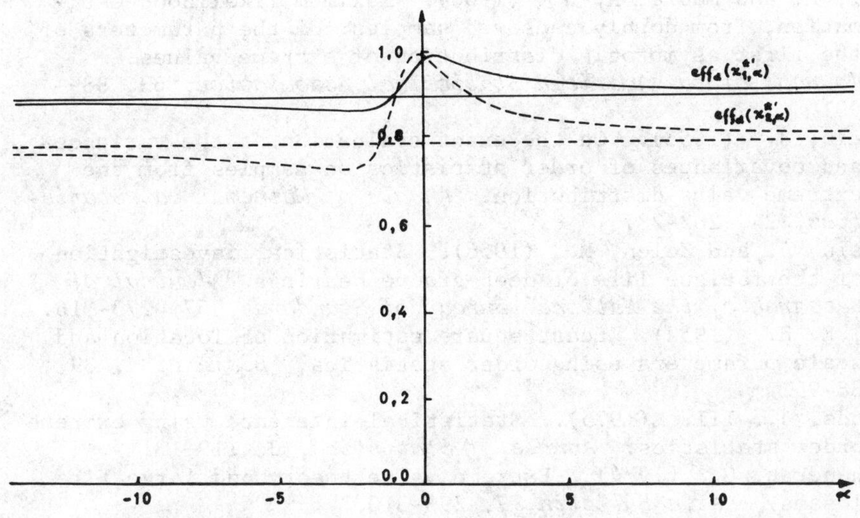

FIG. 2: Asymptotic efficiencies of $\chi^{*'}_{1,\alpha}$ *and* $\chi^{*'}_{2,\alpha}$.

REFERENCES

Barnett, V. (1976). The ordering of multivariate data. *Journal of the Royal Statistical Society, A*, 139, 318-354.

David, H. A. (1970). *Order Statistics*. Wiley, New York.

David, H. A. (1973). Concomitants of order statistics. *Bulletin of the International Statistical Institute*, 45, 295-300.

David, H. A. and Galambos, J. (1974). The asymptotic theory of concomitants of order statistics. *Journal of Applied Probability*, 11, 762-770.

Dwass, M. (1966). Extremal processes, II. *Illinois Journal of Mathematics*, 10, 381-391.

Galambos, J. (1978). *The Asymptotic Theory of Extreme Order Statistics*. Wiley, New York.

Gnedenko, B. V. (1943). Sur la distribution limite du terme maximum d'une série aléatoire. *Annals of Mathematics*, 44, 423-453.

Gomes, M. I. (1978). *Some probabilistic and statistical problems in extreme value theory*. Ph.D. Thesis, University of Sheffield.

Gomes, M. I. (1979). Extremal i-variate laws in stationary sequences. *Revista de la Universidad de Santander*, 2, 1017-1019.

Gumbel, E. J. (1958). *Statistics of Extremes*. Columbia University Press.

Harter, L. and Moore, A. H. (1968). Maximum likelihood estimation, from doubly censored samples, of the parameters of the first asymptotic distribution of extreme values. *Journal of the American Statistical Association*, 63, 889-901.

Lieblein, J. (1953). On the exact evaluation of the variances and covariances of order statistics on samples from the extreme value distribution. *Annals of Mathematical Statistics*, 24, 282-287.

Lieblein, J. and Zelen, M. (1956). Statistical investigation of the fatigue life of deep-groove bearings. *Journal of Research of the National Bureau of Standards*, 57, 273-316.

Lloyd, E. H. (1952). Least square estimation of location and scale parameters using order statistics. *Biometrika*, 39, 88-95.

Pickands, J., III. (1975). Statistical inference using extreme order statistics. *Annals of Statistics*, 3, 119-131.

Ramachandran, G. (1974). Extreme value theory and large fire losses. *Astin Bulletin*, 7, 293-310.

Tiago de Oliveira, J. (1972). Statistics for Gumbel and Fréchet distributions. In *Structural Safety and Reliability*, A. Freudenthal, ed. Pergamon Press.

Watterson, G. A. (1959). Linear estimation in censored samples from multivariate normal populations. *Annals of Mathematical Statistics*, 30, 814-824.

Weissman, I. (1975). Multivariate extremal processes generated by independent non-identically distributed random variables. *Journal of Applied Probability*, 12, 477-487.

Weissman, I. (1978). Estimation of parameters and large quantiles based on the k largest observations. *Journal of the American Statistical Association*, 73, 812-815.

Yang, S. S. (1977). General distribution theory of concomitants of order statistics. *Annals of Statistics*, 5, 996-1002.

[*Received May* 1980. *Revised October* 1980]

WAITING TIMES AND RETURN PERIODS TO EXCEED THE MAXIMUM OF A PREVIOUS SAMPLE

R. S. WENOCUR

Department of Mathematical Sciences
Drexel University
Philadelphia, Pennsylvania 19104 USA

SUMMARY. In this paper, we extend the notion of return period, or expected waiting time, introduced by Gumbel (1958). Gumbel considered return periods to exceed the maximum, Z_N, of a previous sample of size N for the case of $Z_N = x_N$, fixed, and for the case of N large, this analysis being based upon the three classical asymptotic distributions for the maximum. Here, we analyze the case of limited or no available information on the actual value attained by Z_N, where N may be small. Such an analysis applies to situations in which the records on which we must base predictions are incomplete. When no information is available, the return period is infinite. For the case of limited information, we examine conditional return periods and determine conditions under which the return period is finite. Another problem of interest is the magnitude of the future trial at which the previous maximum is first exceeded; therefore. we determine the distribution of this randomly-indexed random variable. Finally, we relate our analysis to the concept of record times.

KEY WORDS. sample maximum, waiting times, return period, record times.

1. INTRODUCTION

Gumbel (1958) introduced the notion of a *return period* in the sense of expected waiting time. He considered return periods to exceed the maximum, Z_N, of a previous sample of size N

C. Taillie et al. (eds.), Statistical Distributions in Scientific Work, Vol. 6, 411–418.

from a population with distribution function F, for the case of $Z_N = x_N$, fixed, $F(x_N)$ known, and for asymptotic cases, N being large, F in the domain of attraction of one of the three classical limit laws for the maximum. As an application, he discussed the prediction of floods.

In this paper, we analyze waiting times and return periods to exceed the maximum, Z_N, of a previous sample of size N, where N may be small, and where we have limited or no information on the actual value attained by Z_N. This method is applicable when we have incomplete records on which we must base our predictions. For further details, see Wenocur (1979).

2. THE CASE OF NO INFORMATION

Suppose we take a random sample $X_1, X_2, \cdots, X_N$ from a population with continuous distribution function F, which may be unknown. Let $Z_N = \max\{X_1, \cdots, X_N\}$. Define W_N to be the number of future independent trials $X_1', X_2', \cdots$, taken from the same distribution as the X_i, needed to exceed Z_N for the first time, where we count the trial at which the exceedance occurs. We readily determine the distribution of the random variable W_N as

$$P(W_N = k) = N/[(N+k)(N+k-1)]. \qquad (1)$$

This can be evaluated directly by setting

$$P(W_N = k) = \int_{-\infty}^{\infty} P(W_N = k \mid Z_N = x)\, dP(Z_N < x),$$

employing the distribution function of Z_N and the fact that W_N, conditional on Z_N, is a geometric random variable with parameter $1-F(Z_N)$. An alternate approach utilizes Gumbel and von Schelling's (1950) distribution of the number of exceedances, which is a specific univariate Pólya distribution, as noted by Sarkadi (1957), again by Morgenstern (1972), and discussed by Wenocur (1979).

From (1) it follows that

$$E(W_N) = +\infty. \qquad (2)$$

However, as shown by Wenocur (1979, 1980),

$$\sum_{k=1}^{\infty} P(W_N = k) = 1.$$

Stated informally: given no information on the actual value attained by Z_N, we expect to wait "forever" to exceed the maximum, while it is almost sure that the maximum will eventually be exceeded after a finite number of future trials.

3. THE CASE OF LIMITED INFORMATION: CONDITIONAL RETURN PERIODS TO EXCEED THE MAXIMUM

Restating (2), if we have no available information on the actual maximum value that was attained among the past N trials, we obtain infinite expected value for W_N. If complete information is available, that is, $Z_N = x_N$ (x_N fixed, $F(x_N) < 1$), $F(x_N)$ known, we get finite expectation

$$E(W_N \mid Z_N = x_N) = 1/[1-F(x_N)].$$

Suppose, instead, we have limited information on what the previous maximum value was; more precisely, suppose it is given that $Z_N \in A$, $A \in \mathcal{B} \subset \mathcal{P}(\mathbb{R})$, where $\mathcal{B}$ denotes the σ-algebra of Borel sets of $\mathbb{R}$, $\mathcal{P}(\mathbb{R})$ the power set of $\mathbb{R}$. We seek conditions on A that will assure a finite return period.

Let P_F be the Borel measure of $\mathbb{R}$ induced by the distribution function F. Then, we say that $A \in \mathcal{B} \subset \mathcal{P}(\mathbb{R})$ is *essentially contained* in $B \in \mathcal{B}$ with respect to P_F if $P_F(A \cap B^c) = 0$, and we employ the notation $A \subseteq_{ess} B(P_F)$. With this notation, we state the following.

Theorem. $E(W_N \mid Z_N \in A)$ is

(a) infinite whenever $A \supseteq_{ess} [z, \infty)(P_F)$, where z satisfies $0 \leqslant F(z) < 1$,

(b) finite whenever $A \subseteq_{ess} (-\infty, z)(P_F)$, with $F(z) < 1$, where

$Z_N = \max\{X_1, X_2, \cdots, X_N\}$, F is the common, continuous distribution function of the X_i, and $P_F(A) > 0$.

Proof. Let $A = [z, \infty)$; i.e., $\{Z_N \in A\} = \{Z_N \geq z\}$. Then

$$P(W_N = k | Z_N \in A) = (\int_{x \in A} P(W_N = k | Z_N = x) dP(Z_N < x))/P(Z_N \in A), \tag{3}$$

which, by direct calculation, becomes

$$P(W_N = k | Z_N \in A) = (N/(1-F(z)^N))(\frac{1-F(z)^{k+N-1}}{k+N-1} - \frac{1-F(z)^{k+N}}{k+N}). \tag{4}$$

For z such that $F(z) = 0$, (4) reduces to the case of no information. From (4), we have

$$[(1-F(z)^N)/N]E(W_N | Z_N \geq z) = \sum_{k=0}^{\infty} (1-F(z)^{N+k})/(N+k). \tag{5}$$

Harmonic divergence implies that the sum (5) is infinite. If $A \underset{ess}{\supset} [z, \infty)(P_F)$, then

$$\int_{x \in A} P(W_N = k | Z_N = x) dP(Z_N < x) \geq \int_z^{\infty} P(W_N = k | Z_N = x) dP(Z_N < x);$$

therefore,

$$P(Z_N \in A) P(W_N = k | Z_N \in A) \geq P(Z_N \geq z) P(W_N = k | Z_N \geq z).$$

Consequently,

$$P(Z_N \in A) E(W_N | Z_N \in A) \geq P(Z_N \geq z) E(W_N | Z_N \geq z) = +\infty,$$

and (a) is proved.

For the second part of the proof, let us begin by setting $A = (-\infty, z)$. By (3), integrating as in the first part of the proof, we obtain

$$P(W_N = k | Z_N \in A) = (N/F(z)^N)(\frac{F(z)^{k+N-1}}{k+N-1} - \frac{F(z)^{k+N}}{k+N}).$$

Then,

$$E(W_N | Z_N \in A) = (N/F(z)^N) \sum_{k=0}^{\infty} F(z)^{k+N}/(k+N) < \infty.$$

Now, let $A \underset{ess}{\subset} (-\infty, z)(P_F)$. By an argument analogous to that used in proving part (a),

$$P(Z_N \in A)E(W_N|Z_N \in A) \leq P(Z_N < z)E(W_N|Z_N < z),$$

which, as shown above, is finite. Therefore, (b) is established.

We evaluate $E(W_1|Z_1 < z) = -\ln(1-F(z))/F(z)$, and, for $N \geq 2$,

$$E(W_N|Z_N < z) = N[-\ln(1-F(z)) - S_N(z)]/F(z)^N, \tag{6}$$

where $S_N(z) = \sum_{k=1}^{N-1} F(z)^k/k$. Since, by convention, $S_1 = 0$, (6) holds for $N \geq 1$. To establish (6), we employ the following facts:

$$\int_0^{F(z)} \left(\sum_{k=0}^{\infty} x^{N-1+k} \right) dx = \sum_{k=0}^{\infty} F(z)^{k+N}/(k+N),$$

and
$$-\ln[1-F(z)] = \int_0^{F(z)} [1/(1-x)]dx.$$

For a given distribution function F, let $m_F = \sup\{x: F(x) = 0\}$; $M_F = \inf\{x: F(x) = 1\}$. Using this notation, we have

Proposition. Given N, $E(W_N|Z_N < z) \to 1$ as $z \to m_F+$; $E(W_N|Z_N < z) \to +\infty$ as $z \to M_F-$, but remains finite for any $z < M_F$.

Proof. Writing

$$E(W_N|Z_N < z) = 1 + N \sum_{k=1}^{\infty} F(z)^k/(N+k),$$

we observe that the proposition follows immediately.

For a discussion of the above proposition and a possible interpretation thereof, see Wenocur (1979).

4. THE DISTRIBUTION OF X'_{W_N}

Another consideration of interest is the distribution of X'_{W_N}, the future trial at which Z_N is first exceeded. We shall evaluate

$$H(x) = P(X'_{W_N} < x) = \sum_{k=1}^{\infty} P(X'_{W_N} < x|W_N = k)P(W_N = k).$$

Before proceeding, we observe that

$$P(X'_{W_N} < x,\ W_N = k)$$

$$= \int_{-\infty}^{x} P(X'_{W_N} < x,\ W_N = k \mid Z_N = z)\,dP(Z_N < z)$$

$$= \int_{-\infty}^{x} F(z)^{k-1}[F(x)-F(z)]N\,F(z)^{N-1}dF(z)$$

$$= N\,F(x)^{N+k}/[(N+k-1)(N+k)],$$

so that

$$P(X'_{W_N} < x \mid W_N = k) = P(X'_{W_N} < x,\ W_N = k)/P(W_N = k) = F(x)^{N+k}.$$

Therefore,

$$H(x) = N \sum_{k=1}^{\infty} F(x)^{N+k}/[(N+k-1)(N+k)]$$

$$= N \int_{F(t)=0}^{F(t)=F(x)} \int_{0}^{F(t)} \Big(\sum_{k=1}^{\infty} u^{N+k-2}du\Big)dF(t)$$

$$= \int_{F(t)=0}^{F(t)=F(x)} E(W_N \mid Z_N < t)F(t)^N dF(t).$$

Consequently,

$$dH(x) = E(W_N \mid Z_N < x)F(x)^N dF(x),$$

and

$$H(x) = N\left(F(x) + [1-F(x)][\ln(1-F(x))] - \sum_{k=2}^{N} \frac{F(x)^k}{(k)(k-1)}\right). \quad (7)$$

It follows that

$$E(X'_{W_N} - Z_N) = \int_{-\infty}^{\infty} x\,F(x)^{N-1}\,[F(x)E(W_N \mid Z_N < x) - N]dF(x).$$

5. RELATIONSHIP TO RECORD TIMES

Exceedances of maxima can be related to record times. If $Y_1, Y_2, \cdots$ are independent, identically distributed random variables with continuous distribution function F, let $L(1) = 1$, and for $n \geqslant 2$, define $L(n) = \min\{j: j > L(n-1)$ and $Y_j > Y_{L(n-1)}\}$. We call $L(n)$ the n*th* record time. The available literature on the theory of record times is vast; the problem seems to have been introduced by Chandler (1952), utilized for statistical analysis by Foster and Stuart (1954), and rediscovered by Rényi (1962). An expository treatment of an entertaining nature is provided by Glick (1978), although his list of references is by no means exhaustive. For a more complete list of references, see Galambos (1978).

For $N = 1$, $W_1 + 1$ is equal to the second record time of the sequence $X_1, X_1', X_2', \cdots$. For $N = 2$, the relationship between W_2 and record times is not so direct as for $N = 1$; $W_2 + 2$ might represent either the second or third record time of the sequence $X_1, X_2, X_1', X_2', \cdots$, depending on whether $X_1 = \max\{X_1, X_2\}$, or $X_2 = \max\{X_1, X_2\}$. Indeed,

$$\begin{aligned} P(W_2 = k) &= P(W_2 = k \mid X_2 \leqslant X_1)P(X_2 \leqslant X_1) \\ &\quad + P(W_2 = k \mid X_2 > X_1)P(X_2 > X_1) \\ &= \tfrac{1}{2}P(L(2) = k+2 \mid L(2) > 2) \\ &\quad + \tfrac{1}{2}P(L(3) = k+2 \mid L(2) = 2). \end{aligned}$$

For $N > 2$, the relationship between W_N and record times becomes increasingly complicated. Although related, the theory of record times and the theory of waiting times developed here are not equivalent. In fact, our waiting times can be viewed as randomly indexed record times, and properties of record times cannot be directly extended to these waiting times without further analysis.

For the case of $N = 1$, (7) agrees with the distribution of the second record value, obtained by Chandler (1952). For $N > 1$, such a comparison of distributions is no longer so direct.

6. CONCLUSION

The solutions to many practical problems are based upon a consideration of the extremes of a previous sample from a continuous distribution. Predictions of floods and other natural phenomena, effects of drugs, and failure of machinery or crops are just a few examples of such a situation. When our sample is small and information is limited -- provided perhaps by old medical records, perhaps by the incomplete observations of farmers or shop foremen -- we can still utilize what information we have to make predictions.

Acknowledgements. Many thanks to both Professor Janos Galambos and Professor Samuel Kotz for their advice and encouragement during the preparation of this paper, and to the referees for several valuable suggestions.

REFERENCES

Chandler, K. N. (1952). The distribution and frequency of record values. *Journal of the Royal Statistical Society, Series B*, 14, 220-228.

Foster, F. G. and Stuart, A. (1954). Distribution-free tests in time series based on the breaking of records. *Journal of the Royal Statistical Society, Series B*, 16, 1-22.

Galambos, J. (1978). *The Asymptotic Theory of Extreme Order Statistics*. Wiley, New York.

Glick, N. (1978). Breaking records and breaking boards. *American Mathematical Monthly*, 85, 2-26.

Gumbel, E. J. (1958). *Statistics of Extremes*. Columbia University Press, New York.

Gumbel, E. J. and von Schelling, H. (1950). The distribution of the number of exceedances. *Annals of Mathematical Statistics*, 21, 247-262.

Morgenstern, D. (1972). Überschreitungswahrscheinlichkeiten, das Polyasche Urnenmodel und ein Wartezeitproblem bei Urnenziehungen. *Math-Phys. Semest.*, Göttingen, 19, 213-215.

Rényi, A. (1962). Théorie des éléments saillants d'une suite d'observations, summary in English; *Colloquium on Combinatorial Methods in Probability Theory*. Matematisk Institut, Aarhus Universitet, Denmark. 104-117.

Sarkadi, K. (1957). On the distribution of the number of exceedances. *Annals of Mathematical Statistics*, 28, 1021-1023.

Wenocur, R. S. (1979). *Waiting times and return periods related to order statistics*. Ph.D. thesis, Temple University, Philadelphia.

Wenocur, R. S. (1980). Rediscovery and alternate proof of Gauss's identity. *Annals of Discrete Mathematics*, 9, 79-82.

[*Received May* 1980. *Revised October* 1980]

WAITING TIMES AND RETURN PERIODS RELATED TO ORDER STATISTICS: AN APPLICATION OF URN MODELS

R. S. WENOCUR

Department of Mathematical Sciences
Drexel University
Philadelphia, Pennsylvania 19104 USA

SUMMARY. Both Sarkadi (1957) and Morgenstern (1972) discuss the relationship between exceedances of order statistics from a continuous parent and a specific univariate Pólya distribution. In this paper, the applicability of a multivariate Pólya scheme to problems involving order statistics is demonstrated and other appropriate urn models are examined. Then, these models are utilized in an analysis of waiting times and return periods related to order statistics. Properties of the multivariate inverse Pólya distribution are employed to derive results which are valid for small, as well as large, sample size; in addition, asymptotic and almost sure laws are determined. Decompositions of waiting times into sums of exchangeable random variables are examined, and the case of observations made at the points of a Poisson process is considered. Emphasis is on exact and limiting distributions, both univariate and multivariate.

KEY WORDS. order statistics, urn models, exceedances, waiting times, return periods, exchangeable random variables, multivariate Pólya distribution, asymptotic distribution, Poisson process, multivariate inverse Pólya distribution.

1. INTRODUCTION

An urn contains b black and r red balls; a ball is drawn at random, then replaced along with an additional ball of the same color. A new random drawing is made from the urn, which now contains r+b+1 balls, and the procedure is repeated. This model is a special case of the general univariate Pólya urn scheme

C. Taillie et al. (eds.), Statistical Distributions in Scientific Work, Vol. 6, 419–433.

(cf. Feller, 1957, p. 109; Johnson and Kotz, 1977, p. 177). In Gumbel and von Schelling (1950), and again, in Gumbel (1958, pp. 58-59), the distribution of the number of exceedances of an order statistic of rank m in t future trials is derived. Both Sarkadi (1957) and Morgenstern (1972) discuss the relationship between this distribution and the specific univariate Pólya urn model described above, their observations being based upon a comparison of Gumbel and von Schelling's (1950) distribution of the number of exceedances and the functional form of a Pólya distribution. We shall extend this analysis by demonstrating the applicability of a multivariate Pólya urn scheme to problems involving order statistics, and by examining alternative urn models that prove to be appropriate. As an application of this approach, we shall analyze waiting times and return periods related to order statistics.

2. MODEL I

If $X_1, X_2, \cdots, X_N$ is a random sample of fixed size from a continuous distribution, the associated order statistics $Y_{N,1} \leqslant Y_{N,2} \leqslant \cdots \leqslant Y_{N,N}$ determine $N+1$ random intervals:

$$U_1 = (-\infty, Y_{N,1}), \quad U_2 = [Y_{N,1}, Y_{N,2}), \quad \ldots, \quad U_{N+1} = [Y_{N,N}, +\infty).$$

We choose to view $U_1, U_2, \cdots, U_{N+1}$ as $N+1$ "urns" into which future items might fall. Let us consider the next k independent observations from the same distribution as the X_i: $X'_1, X'_2, \cdots, X'_k$. For convenience, we shall use the vector notation $\langle k_1, k_2, \cdots, k_{N+1}\rangle$ to indicate the event that k_1 of the X'_i fall into U_1, k_2 of the X'_i fall into $U_2, \cdots, k_{N+1}$ of the X'_i fall into U_{N+1}. For this model, assume that each of possible arrangements $\langle k_1, k_2, \cdots, k_{N+1}\rangle$ has equal probability, for $\sum_{i=1}^{N+1} k_i = k$, fixed. We remark that this is the model leading to Bose-Einstein statistics, (cf. Feller, 1957, pp. 20-22).

But, for any k, $P(\langle k_1, k_2, \cdots, k_{N+1}\rangle)$ depends on $k_1, k_2, \cdots, k_{N+1}$ through the sum $k_1 + k_2 + \cdots + k_{N+1} = k$ only; more precisely, $P(\langle k_1, k_2, \cdots, k_{N+1}\rangle) = 1/\binom{N+k}{N}$, which is the probability of one possible arrangement of ranks for the trials $X_1, X_2, \cdots, X_N$ among a total of N+k observations. As a consequence, the Bose-Einstein urn model of depositing k items into N+1 urns, where all arrangements $\langle k_1, k_2, \cdots, k_{N+1} \rangle$, such that $k_1 + k_2 + \cdots + k_{N+1} = k$, are equally likely, is applicable to problems involving future trials falling below, above, or between the order statistics of a previous random sample.

3. MODEL II

An alternative way to view our first model is by means of the following urn scheme. We again begin with N+1 urns $U_1, \cdots, U_{N+1}$, determined by the order statistics of our previous random sample, where $P(X_1' \in U_i) = 1/(N+1)$. To calculate the probability of the compound event

$$\{X_1' \in U_{i(1)}, X_2' \in U_{i(2)}, \cdots, X_k' \in U_{i(k)}\},$$

where $i(k) \in \{1,2,3,\cdots,N+1\}$, we employ the following algorithm;

$$P(X_1' \in U_{i(1)}) = 1/(N+1);$$

$$P(X_{j+1}' \in U_{i(j+1)} | X_1' \in U_{i(1)}, \cdots, X_j' \in U_{i(j)})$$

$$= (m+1)/(N+1+j), \quad j \geq 1,$$

where m is equal to the number of distinct $i(t)$ such that $i(t) = i(j+1)$, $t = 1,2,\cdots,j$.

We can view this as an urn scheme that varies in the following way: at the $(j+1)st$ stage, there are $(N+1+j)$ urns determined by the order statistics of the random sample $X_1, X_2, \cdots, X_N, X_1', \cdots, X_j'$; hence, the number of urns that correspond to $U_{i(j+1)}$ depends on the U_t's into which $X_1', \cdots, X_j'$ happen to fall.

To calculate the probability $P(X_1' \in V_{i(1)}, \cdots X_k' \in V_{i(k)})$, where the V_n are unions of U_t's, we employ an analogous algorithm:

$$P(X_1' \in V_{i(1)}) = r/(N+1),$$

where r is equal to the number of distinct U_t's such that $V_{i(1)} = \cup U_t$, and for $j \geqslant 1$,

$$P(X'_{j+1} \in V_{i(j+1)} | X'_1 \in V_{i(1)}, \cdots, X'_j \in V_{i(j)}) = m/(N+1+j),$$

where m is determined by the number of distinct U_t's such that $V_{i(j+1)} = \cup U_t$, and by the number of X'_n that fall into the U_t's of this union; $n = 1,2,\cdots,j+1$.

The point of view inherent in this urn scheme emphasizes the dependence of the events $\{X'_i \in U_t\}$, despite the independence of the random variables X'_i, $i = 1,2,\cdots$. Furthermore, it shows that although we are analyzing future events in terms of the order statistics of a previous random sample of fixed size, as soon as we have taken the next j of our future observations, we actually base our calculations on the order statistics of the entire sample $X_1, X_2, \cdots, X_N, X'_1, \cdots, X'_j$.

To verify that this scheme, which employs an increasing number of urns, is applicable to analyses involving falling below, above, or between the order statistics of a previous sample, we must establish that the described algorithm is valid. To do this, it is sufficient to prove that

$$P(X'_{j+1} \in U_{k(j+1)} | X'_1 \in U_{k(1)}, \cdots, X'_j \in U_{k(j)}) = (m+1)/(N+1+j),$$

where m is the number of the X'_i that fall into $U_{k(j+1)}$, $i = 1,2,\cdots,j$. We employ Model I to calculate

$$P(X'_{j+1} \in U_{k(j+1)} | X'_1 \in U_{k(1)}, \cdots, X'_j \in U_{k(j)}) =$$

$$= P(X'_1 \in U_{k(1)}, \cdots, X'_{j+1} \in U_{k(j+1)}) / P(X'_1 \in U_{k(1)}, \cdots, X'_j \in U_{k(j)})$$

$$= \frac{P(<m_1, \cdots, m+1, \cdots, m_{N+1}>)/[(j+1)!/(m_1! \cdots (m+1)! \cdots m_{N+1}!)]}{P(<m_1, \cdots, m, \cdots, m_{N+1}>)/[j!/(m_1! \cdots m! \cdots m_{N+1}!)]},$$

where m_i is the number of X'_n, $n \leqslant j$, that fall into U_i, $\sum_{i=1}^{N+1} m_i = j$, and $m_{k(j+1)} = m$. The divisors

$$\frac{(j+1)!}{m_1! \cdots (m+1)! \cdots m_{N+1}!} \quad \text{and} \quad \frac{j!}{m_1! \cdots m! \cdots m_{N+1}!}$$

are introduced into the numerator and denominator, respectively, to account for a specific way to achieve the given arrangement of m_i's. Substituting and simplifying, we obtain

$$P(X'_{j+1} \in U_{k(j+1)} | X'_1 \in U_{k(1)}, \cdots, X'_j \in U'_{k(j)}) = (m+1)/(N+1+j),$$

as we needed to show.

4. MODEL III

This final urn model represents an alternative, but essentially equivalent, way to view Model II. Suppose, for example, we wish to analyze exceedances of $Y_{N,j}$. We then consider a model consisting of one urn that contains $N+1-j$ balls of color C_0, and one ball each of colors $C_1, C_2, \cdots, C_j$. After each drawing of a ball, the ball drawn is returned along with another one of the same color. For a future observation, X'_i, the event $\{X'_i \in \bigcup_{k=j+1}^{N+1} U_k\}$ (where, as before, the U_k are the "urns" determined by $Y_{N,1} \leqslant \cdots \leqslant Y_{N,N}$) will correspond to the *ith* future drawing of a ball of color C_0 from the urn, and the event $\{X'_i \in U_k\}$, $k = 1,2,\cdots,j$, will correspond to the *ith* future drawing of a ball of color C_k. This model leads to the multivariate Pólya distribution (Janardan and Patil, 1970; Dyczka, 1973), which is examined at length in Johnson and Kotz (1977, pp. 194-200), and is obviously equivalent to Model II, when this model is employed in considering the case of exceeding $Y_{N,j}$.

It is apparent that the balls of color C_0 could represent any number of the urns U_i of Model II, not only the last $N+1-j$. For example, there could be three balls of color C_0 in the urn, taken to represent U_2, U_5, and U_7, and one each of colors $C_1, C_3, C_4, C_6, C_8, \cdots, C_{N+1}$, corresponding to the remaining urns $U_1, U_3, U_4, U_6, U_8, \cdots, U_{N+1}$, respectively. Using this scheme, we could analyze the case of future observations falling within or without $U_2 \cup U_5 \cup U_7$.

An advantage of the second model over the present one is that it serves to emphasize quite clearly that once k future trials are made, the order statistics of the entire sample

$X_1, X_2, \cdots, X_N, X'_1, X'_2, \cdots, X'_k$ determine the number of urns under consideration, and, in addition, the placement of the urns, as random intervals on the real line, with respect to one another; these phenomena become obscured in the final model. Moreover, Model II treats each urn as separate, while in Model III, the balls that represent the urns in question become indistinguishable. The primary advantage of the present model is that we could apply many known results of the multivariate Pólya and multivariate inverse Pólya distributions to the analysis of order statistics. In particular, the results of Janardan and Patil (1971, 1974) could be utilized.

5. AN APPLICATION: WAITING TIMES AND RETURN PERIODS RELATED TO ORDER STATISTICS

Let the *waiting time* $W(N,j,m)$ be the number of future independent trials, from the same distribution as the X_i, needed to exceed $Y_{N,j}$ exactly m times. Then

$$P(W(N,j,m) = m+n)$$

$$= \binom{m+n-1}{n} [(N-j+m)!(j+n-1)!N!]/[(N-j)!(j-1)!(N+m+n)!].$$

However, there is nothing peculiar about exceedances; the applicability of our urn models allows us to generalize $W(N,j,m)$ in the following fashion: define $W(N,j,m;S)$ to be the number of future independent trials until exactly m items fall outside S, where S is defined by $S = U_{k(1)} \cup U_{k(2)} \cup \cdots \cup U_{k(j)}$, where j is equal to the number of distinct urns in the union, S, the U_i being as defined earlier. Then $W(N,j,m) = W(N,j,m;T)$, where $T = U_1 \cup U_2 \cup \cdots \cup U_j$ and $P(W(N,j,m;S) = k) = P(W(N,j,m) = k)$, this probability depending on S through j only; in other words, for a fixed number of previous trials, the distribution of $W(N,j,m;S)$ depends solely upon the number of distinct random intervals ("urns") in S. For an analysis of the relationship between the distribution of $W(N,j,1)$ and the hypergeometric series $F(a,c;b;1) = {}_2F_1(a,c;b;1)$, see Wenocur (1980).

Let us decompose $W(N,j,m;S)$ in the following way: Let $T(N,j,m;S,M)$ denote the number of the next $W(N,j,m;S)$ independent trials whose values fall into $U_{k(M)}$, $M = 1,2,\cdots,j$. Then

$$W(N,j,m;S) = \sum_{M=1}^{j} T(N,j,m;S,M) + m.$$

As a consequence of the applicability of our urn models, in particular, the multivariate Pólya scheme, we obtain

Theorem 1. The random variable $W(N,j,m;S) - m$ can be decomposed as the sum of the j exchangeable random variables $T(N,j,m;S,k)$, $k = 1,2,\cdots,j$. The distribution of any s of the $T(N,j,m;S,k)$'s is given by

$$P(T(N,j,m;S,i_1) = a_1,\cdots,T(N,j,m;S,i_s) = a_s)$$
$$= [(N-j+s)!(N-j+m)!(t_{j-s}+m-1)!]/[(m-1)!(N-j)!(N-j+s+m+t_{j-s})!], \tag{1}$$

where $t_{j-s} = a_1 + a_2 + \cdots + a_s$, and $t_j = 0$. We remark that the probability (1) depends on the values $a_1, a_2,\cdots,a_s$ through their sum, only.

We evaluate

$$E[T(N,N,m;S,i)] = +\infty,$$
$$E[T(N,j,m;S,i)] = m/(N-j),\ j \neq N,$$
$$Var[T(N,N-1,m;S,i)] = +\infty,$$
$$Var[T(N,j,m;S,i)] = [(N-j+m)m(N-j+1)]/[(N-j)^2(N-j-1)],$$
$$j < N-1,$$

and, for $i \neq n$, $j < N-1$,

$$Cov[T(N,j,m;S,i),T(N,j,m;S,n)] = [(N-j+m)m]/[(N-j)^2(N-j-1)],$$

$$Corr[T(N,j,m;S,i),T(N,j,m;S,n)] = 1/(N-j+1).$$

Therefore, the $T(N,j,m;S,i)$'s are positively correlated.

We note that for j fixed as $N \to \infty$, the number of random variables in the decomposition of $W(N,j,m;S)$ remains constant, while they become asymptotically uncorrelated; at the other extreme, for $N-j$ fixed as $N \to \infty$, the number of random variables in question increases, while the correlation is constant and positive. For $j/N \to p$, fixed, $0 < p < 1$, as $N \to \infty$, the number of random variables increases, and they become asymptotically uncorrelated.

The *return period* $E[W(N,j,m;S)]$ associated with $W(N,j,m;S)$ is therefore

$$E[W(N,j,m;S)] = mN/(N-j),\ j \neq N, \tag{2}$$

$$E[W(N,N,m;S)] = +\infty,$$

with associated variance

$$\mathrm{Var}(W(N,j,m;S)) = (mjN(N-j+m))/((N-j)^2(N-j-1)),\ j < N-1,$$

$$\mathrm{Var}(W(N,N-1,m;S)) = +\infty.$$

Utilizing the above results, we obtain

$$W(N,j,m;S) \xrightarrow{\text{pr.}} m \tag{3}$$

for j and m fixed as $N \to \infty$, and

$$W(N,j,m;S)/m \xrightarrow{\text{pr.}} N/(N-j),$$

for N,j fixed as $m \to \infty$. Of course, (3) implies

$$\liminf_N W(N,j,m;S) = m \quad \text{a.s.}$$

For convenience, let $W(N,j;S) = W(N,j,1;S)$. Then, we have

Theorem 2.

(a) For j fixed, $P[\limsup_N W(N,j;S) \leqslant 2] = 1$.

(b) For $\alpha > 1$, $P[\limsup_N \{W(N,N;S)/[N^2(\log N)^\alpha]\} \leqslant 1] = 1$;

for $\alpha > 0$, $P[\limsup_N \{W(N,N;S)/N^{2+\alpha}\} \leqslant 1] = 1$.

(c) For $j = N-m$, m fixed, $m \geqslant 1$, $\alpha > 1/(m+1)$,

$P[\limsup_N \{W(N,N-m;S)/N^{1+\alpha}\} \leqslant 1] = 1$;

in particular, $P[\limsup_N \{W(N,N-m;S)/N^2\} \leqslant 1] = 1$.

(d) For $j/N \to p$, fixed, $0 < p < 1$, as $N \to \infty$,

(i) $P[\limsup_N \{W(N,j;S)/N\} \leqslant 1] = 1$;

(ii) if $0 < p < 1/e$, then

$P[\limsup_N \{W(N,j;S)/\log N\} \leqslant 1] = 1$.

Proof. If $\sum_N P[W(N,j;S) \geq k(N)] < \infty$, then $\{W(N,j;S) \geq k(N)\}$ can occur for only a finite number of N, with probability 1. Hence,

$$P[\limsup_N \{W(N,j;S)/k(N)\} \leq 1] \geq P[\liminf_N \{W(N,j;S)/k(N) < 1\}] = 1. \tag{4}$$

Employment of the appropriate $k(N)$ in (a) through (d) and direct analysis yield that (4) holds in each case, thus establishing the theorem.

Within the proof of Theorem 2, we have actually established somewhat stronger results; namely that $P[\liminf_N \{W(N,j;S)/k(N) < 1\}] = 1$, with $k(N)$ as indicated in each case of the theorem. Since

$$\begin{aligned} 1 &= P[\liminf_N \{W(N,j;S)/k(N) < 1\}] \\ &= \lim_{N\to\infty} P[\bigcap_{n=N}^{\infty} \{W(N,j;S)/k(N) < 1\}], \end{aligned}$$

our results could be applied to make predictions based upon large previous samples. For N large enough, we can predict, with high probability, that $Y_{N,j}$ will be exceeded before a certain number of future trials have been taken. For example, if j is fixed, there exists M sufficiently large so that for any $N \geq M$, $Y_{N,j}$ will be exceeded within two trials, with probability at least $1-\delta$, as close to 1 as we please. Of course, if j becomes large with N, the number of future trials needed to insure that $Y_{N,j}$ will be exceeded with probability $1-\delta$ increases, as indicated by (b), (c) and (d) of the theorem. As mentioned earlier, there is nothing special about exceedances; these comments apply to the situation of future observations' falling into arbitrary unions of the random intervals, U_i.

Up to this point, we have considered a random sample $X_1, X_2, \cdots, X_N$ of size N taken from a continuous distribution, followed by future independent sampling, denoted by $X_1', X_2', \cdots$, from the same distribution. The waiting times that we have introduced have been discrete random variables; the questions posed have involved the number of future independent observations needed until a given event occurs, without explicit mention of the times at which these future trials might be made. Indeed, it has been assumed, more or less, that future observations would be made at unit time intervals (e.g., annually). Now, instead, let us suppose that the sequence of future independent and identically distributed variates $X_1', X_2', \cdots$ occur at the points of a time-homogeneous Poisson process. In other words, $X_1, X_2, \cdots, X_N$ is a previous sample of size N from a population with distribution function F, with corresponding order statistics $Y_{N,1} \leqslant \cdots \leqslant Y_{N,N}$; our future observations will be made at random times $\tau_1, \tau_2, \cdots$, which follow a time-homogeneous Poisson process $\mathcal{P}$ with parameter λ, where $\mathcal{P}$ and the sequence $\{X_N'\}$ are mutually independent. We introduce the continuous random variable $Z(N,j;T)$ as the waiting time until one future trial falls outside T, with $T = \bigcup_{m=1}^{j} U_{k(m)}$, where the $U_{k(m)}$, $m = 1,2,\cdots,j$ are distinct. We assume that time 0 corresponds to the time at which X_N, the last item of the first sample, was taken. Letting $G(t;N,j)$ be the distribution function of $Z(N,j;T)$, we have

$$1 - G(t;N,j) = P(Z(N,j;T) \geqslant t)$$

$$= \sum_{k=0}^{\infty} P(\theta(t) = k)P(X_i' \in T;\ i = 1,2,\cdots,k)$$

$$= \sum_{k=0}^{\infty} [e^{-\lambda t}(\lambda t)^k/k!][j/(N+1)][(j+1)/(N+2)]\cdots[(j+k-1)/(N+k)], \quad (5)$$

where the empty product is equal to 1 and $\theta(t)$ is equal to the number of observations in the time interval $(0,t)$. Letting $H(t;N,j) = 1 - G(t;N,j)$, it follows from (5) that

$$\frac{d}{dt} H(t;N,j) = \lambda(j/(N+1))H(t;N+1,j+1) - \lambda H(t;N,j)$$

and that

$$H(t;N,1) = [N!e^{-\lambda t}/(\lambda t)^N][e^{\lambda t} - S_N(t)],$$

where

$$S_N(t) = \sum_{k=0}^{N-1} (\lambda t)^k/k!.$$

When $j = N = 1$, our result agrees with that of Gaver (1976), for first-record times. For the case of a time-homogeneous Poisson process, the results developed here serve to extend

Gaver's analysis of the first-record time. $L(1)$, since $W(1,1;U_1) + 1 = L(1)$. Since

$$P(X_i' < Y_{N,j}: i = 1,2,\cdots,k)$$

$$= [N!/\{(j-1)!(N-j)!\}] \int_0^1 F(x)^{k+j-1}[1-F(x)]^{N-j}dF(x),$$

an alternative form for $H(t;N,j)$ is given by

$$H(t;N,j) = e^{-\lambda t} M_{N+1-j;j}(\lambda t),$$

where $M_{a;b}(\cdot)$ is the moment-generating function for the Beta distribution with parameters a and b. We deduce from (5) that if $j/N \to p$, p fixed, $0 < p < 1$, as $N \to \infty$, then

$$H(t;N,j) \to e^{-\lambda(1-p)t} = H_1(t) = 1 - G(t),$$

which yields expectation $E_G = 1/[\lambda(1-p)]$. However, for any j,N; in particular, for N small,

$$E[Z(N,j;T)] = \int_0^\infty P[Z(N,j;T) \geqslant t]dt$$

$$= \int_0^\infty \left(\sum_{k=0}^\infty P[\theta(t)=k]P[W(N,j;T) > k]\right)dt$$

$$= (1/\lambda)E[W(N,j;T)] = (1/\lambda)[N/(N-j)]. \quad (6)$$

The result (6), obtained by switching the order of summation and integration, is a consequence of a property of the homogeneous Poisson process, namely that

$$\int_0^\infty P[\theta(t) = k]dt = 1/\lambda$$

is independent of k.

Before concluding this section, we consider a direct generalization. Suppose we are interested in the continuous waiting time until exactly m future items fall outside T, where, again, future observations are made at the points of a homogeneous Poisson process. Denoting this random variable by $Z(N,j,m;T)$, we obtain

$$E[Z(N,j,m;T)] = (1/\lambda)[mN/(N-j)],$$

in the same manner as we obtained (6).

6. A COMPARISON BY MEANS OF ASYMPTOTIC DISTRIBUTIONS

In Gumbel (1958, pp. 30-31), a return period is calculated in the following way: since the observed frequency for $Y_{N,j}$ is equal to j/N, we calculate, for $S^* = U_1 \cup U_2 \cup \cdots \cup U_j$,

$$E[W(N,j;S^*)] = \sum_{k=1}^{\infty} k(j/N)^{k-1}[1-(j/N)] = N/(N-j),$$

which agrees with (2) for $m = 1$. However, this method of calculation assumes that

$$P[W(N,j;S^*) = k] = (j/N)^{k-1}[1-(j/N)],$$

which may be viewed as based on an urn model (of the Maxwell-Boltzmann variety) with N urns, and which requires the independence of the events $\{X'_i \in U_j\}$. This allows zero probability of exceeding $Y_{N,N}$ and assumes the independence of events that we have seen to be dependent. An argument for asymptotic correctness of this method can be presented. But for the case of small N, this method fails to be acceptable, while our methods apply. It is of considerable interest that (2) is an exact equation for any N, rather than being valid only in limit.

Examination of asymptotic behavior will serve to clarify the relationship between our method and that of Gumbel. We have calculated that

$$P[W(N,j;S^*) = k] = \frac{j}{N+1} \cdot \frac{j+1}{N+2} \cdots \frac{j+k-2}{N+k-1} \cdot \frac{N+1-j}{N+k}.$$

For k a fixed positive integer, if $j/N \to p$ as $N \to \infty$, where p is fixed, $0 \leqslant p < 1$, and $q = 1-p$,

$$P[W(N,j;S^*) = k] \to p^{k-1}q. \tag{7}$$

Let $\chi(p)$ denote a p-quantile of F, that is $F(\chi(p)) = p$, and let $W_{\chi(p)}$ be the waiting time to exceed the fixed value $\chi(p)$ for the first time. We can then rephrase (7) as

$$W(N,j;S^*) \overset{d.}{\to} W_{\chi(p)}, \tag{8}$$

for $j/N \to p$ as $N \to \infty$, where p is fixed and satisfies $0 \leqslant p < 1$. Under these assumptions, for N large, j/N provides a good estimate for p; therefore,

$$P[W(N,j;S^*) = k] \simeq (j/N)^{k-1}[(N-j)/N],$$

which justifies Gumbel's method of calculation for large N. Of course, (8) and related remarks are applicable to $W(N,j;S)$ for $S \neq S^*$. Moreover, letting m remain constant as $N \to \infty$, where $j/N \to p$, fixed, $0 \leqslant p < 1$, we observe that

$$P[W(N,j,m;S) = m+n] \to \binom{m+n-1}{n} (1-p)^m p^n .$$

In other words, under the conditions stated above, $W(N,j,m;S)$ converges in distribution to a (modified) negative-binomial random variable, which is, of course, a property of the inverse Pólya distribution under consideration.

7. REMARKS

1. Although Morgenstern (1972) discusses a different waiting-time problem, it is not directly related to our analysis; rather, it involves the connection between Gumbel and von Schelling's distribution of the number of exceedances and the negative hypergeometric distribution. He does not consider the waiting times analyzed in this paper.

2. Related results that have been previously obtained can be reformulated and generalized in view of the applicability of urn models. As an example, Gumbel and von Schelling's distribution of the number of exceedances of $Y_{N,j}$ can be reformulated as the distribution of the number of observations among t future trials that fall into T, where $T = \bigcup_i U_i$, and where the U_i are as defined earlier. The distribution is the same in the more general case as for the case of exceedances. For other applications of this urn model approach, see the thesis of Wenocur (1979).

8. CONCLUSION

We want to predict the occurence of an event that is due to a certain quantity's falling below, above, or between previously observed values. We have not made numerical measurements, but have only the information that we have made N independent observations, the population from which we have taken our sample has an associated underlying continuous distribution function which may be unknown, and that j of the N independent trials resulted in critical values. When might we expect to achieve our next critical value, or perhaps, the next m critical values, based on the limited information we have? For N large and the critical values being extremes of the sample, asymptotic results are known; see, for example, Gumbel (1958) or Galambos (1978). But what if N is small and our critical values are not necessarily extreme? These and related questions were the motivation for the present work.

Examples of such a situation are plentiful. Old medical records may give researchers no more information than that a sample of size N was taken and j occurrences of a malady observed, where it is now assumed that the disease is caused by a critical concentration of a factor in the blood. Farmers, householders, or shop foremen may have the information that j crop or machine failures occurred among N independent trials, failure being due to critical temperature, pressure, or amount of rainfall, be it too high, too low, or somewhere in between, but be unable to supply numerical measurements or further information of any sort. The values of the observations and their ordering unknown, the underlying distribution and critical regions not determined, N possibly small, the j critical values not necessarily extreme, we can still utilize what information we have to make predictions. For further details, see Wenocur (1979).

ACKNOWLEDGEMENTS

I wish to thank both Professor Janos Galambos and Professor Samuel Kotz for their advice and encouragement during the preparation of this paper, and the referees for many valuable suggestions.

REFERENCES

Dyczka, W. (1973). On the multidimensional Polya distribution. *Annales Societatis Mathematicae Polonae*, Series I: *Commentationes Mathematicae XVII*, 43-63.

Feller, W. (1957). *An Introduction to Probability Theory and Its Applications, Vol. I* (second edition). Wiley, New York.

Galambos, J. (1978). *The Asymptotic Theory of Extreme Order Statistics*. Wiley, New York.

Gaver, D. P. (1976). Random record models. *Journal of Applied Probability*, 13, 538-547.

Gumbel, E. J. (1958). *Statistics of Extremes*. Columbia University Press, New York

Gumbel, E. J. and von Schelling, H. (1950). The distribution of the number of exceedances. *Annals of Mathematical Statistics*, 21, 247-262.

Janardan, K. G. and Patil, G. P. (1970). On the multivariate Polya distribution: a model of contagion for data with multiple counts. In *Random Counts in Scientific Work, Vol. 3*, G. P. Patil, ed. Pennsylvania State University Press, Pages, 143-161.

Janardan, K. G. and Patil, G. P. (1971). The multivariate inverse Polya distribution: a model of contagion for data with multiple counts of inverse sampling. In *Studi di Probabilit, Statistica e Ricerca Operativa in Onore di Giuiseppe Pompilij, Gubbio; Odensi,* 327-341.

Janardan, K. G. and Patil, G. P. (1974). On multivariate modified Pólya and inverse Pólya distributions and their properties. *Annals of the Institute of Statistical Mathematics,* 26, 271-276.

Johnson, N. L. and Kotz, S. (1977). *Urn Models and Their Application.* Wiley, New York.

Morgenstern, D. (1972). Uberschreitungswahrscheinlichkeiten, das Polyasche Urnenmodel und ein Wartezeitproblem bei Urnenziehungen, *Math-Phys. Semest.,* Göttingen, 19 (2), 213-215.

Sarkadi, K. (1957). On the distribution of the number of exceedances. *Annals of Mathematical Statistics,* 28, 1021-1023.

Wenocur, R. S. (1979). *Waiting times and return periods related to order statistics.* Ph.D. thesis, Temple University, Philadelphia.

Wenocur, R. S. (1980). Rediscovery and alternate proof of Gauss's identity. *Annals of Discrete Mathematics,* 9, 79-82.

[*Received May* 1980. *Revised October* 1980]

AUTHOR INDEX

SUBJECT INDEX